208

CURRENT ORNITHOLOGY

VOLUME 12

Current Ornithology

Editorial Board

A Continuation Order Plan is available for this series. A continuation order will bring delivery of each new volume immediately upon publication. Volumes are billed only upon actual shipment. For further information please contact the publisher.

CURRENT ORNITHOLOGY

VOLUME 12

Edited by
DENNIS M. POWER
The Oakland Museum
Oakland, California

PLENUM PRESS • NEW YORK AND LONDON

The Library of Congress cataloged the first volume of this title as follows:

Current ornithology.—Vol. 1—
New York: Plenum Press, c1983–
v.: ill.; 24 cm.
Annual.
Editor: Richard F. Johnston.
ISSN 0742-390X = Current ornithology.
1. Ornithology—Periodicals. I. Johnston, Richard F.
QL671.C87 598′.05—dc19 84-640616
(8509) AACR 2 MARC-S

Suggested citation: *Current Ornithology*, Vol. 12
(D. M. Power, ed.). Plenum Press, New York

ISBN 0-306-44978-1

A Division of Plenum Publishing Corporation
233 Spring Street, New York, N. Y. 10013

10 9 8 7 6 5 4 3 2 1

Printed in the United States of America

CONTRIBUTORS

STEVEN R. BEISSINGER, School of Forestry and Environmental Studies, Yale University, New Haven, Connecticut 06511

LES D. BELETSKY, Department of Zoology, University of Washington, Seattle, Washington 98195

ANDRE A. DHONDT, Cornell Laboratory of Ornithology, 159 Sapsucker Woods Road, Ithaca, New York 14850

ERICA H. DUNN, Canadian Wildlife Service, National Wildlife Research Centre, 100 Gamelin Boulevard, Hull, Quebec, Canada K1A 0H3

SCOTT FREEMAN, Burke Museum, University of Washington, Seattle, Washington 98195

DAVID F. GORI, The Nature Conservancy, 300 East University Boulevard, Suite 230, Tucson, Arizona 85705

THOMAS C. GRUBB, JR., Behavioral Ecology Group, Department of Zoology, The Ohio State University, Columbus, Ohio 43210

DAVID J. T. HUSSELL, Ontario Ministry of Natural Resources, Maple, Ontario, Canada L6A 1S9, and Environment Canada (Ontario Region), Nepean, Ontario, Canada K1A 0H3

MARCEL M. LAMBRECHTS, CNRS-CEFE, B.P. 5051, 34000 Montpellier Cedex 1, France.

UDO M. SAVALLI, Department of Biology, Allegheny College, Meadville, Pennsylvania 16335

SCOTT H. STOLESON, School of Forestry and Environmental Studies, Yale University, New Haven, Connecticut 06511

JOHN C. WINGFIELD, Department of Zoology, University of Washington, Seattle, Washington 98195

PREFACE

Behavioral ecology, particularly as it pertains to reproductive biology, continues to attract wide interest in North America and elsewhere. Working together at the University of Washington, Les Beletsky, David Gori, Scott Freeman, and John Wingfield review testosterone and polygyny, further elucidating endocrinological correlates of mating strategies, and hormones and reproductive behavior in general. Scott Stoleson and Steven Beissinger study hatching asynchrony and onset of incubation. Their effort was, in part, to assess the value of "the brood reduction hypothesis" in explaining the "paradox of hatching asynchrony." As a result, they have better organized our thinking about the many hypotheses associated with these phenomena and pointed the way toward better research.

Recording and accounting for changes in bird populations is another field that is of great importance, as ecologists and naturalists continue to be concerned about regional declines. Erica Dunn and David Hussell, both with the Canadian Wildlife Service, explore using migration counts to monitor landbird population change. Along with other protocols such as the Christmas Bird Count and Breeding Bird Survey, migration counts provide important data.

Thomas Grubb of The Ohio State University reviews support for the validity and sensitivity of growth bands in feathers as an indication of nutritional condition, and the use of feather banding in studying growth (ptilochronology). Marcel Lambrechts of the French Center for Scientific Research and Andre Dhondt of the University of Antwerp, Belgium, review studies on the capabilities of birds to discriminate individuals on the basis of song type and vocal cues. Udo Savalli of the University of California outlines every known hypothesis on the evolution of bird coloration and plumage elaboration, emphasizing signifi-

CHAPTER 2

USING MIGRATION COUNTS TO MONITOR LANDBIRD POPULATIONS: REVIEW AND EVALUATION OF CURRENT STATUS

ERICA H. DUNN AND DAVID J. T. HUSSELL

CHAPTER 3

PTILOCHRONOLOGY: A REVIEW AND PROSPECTUS

THOMAS C. GRUBB, JR.

CHAPTER 4

INDIVIDUAL VOICE DISCRIMINATION IN BIRDS

MARCEL M. LAMBRECHTS AND ANDRE A. DHONDT

CHAPTER 5

THE EVOLUTION OF BIRD COLORATION AND PLUMAGE ELABORATION: A REVIEW OF HYPOTHESES

UDO M. SAVALLI

CHAPTER 6

HATCHING ASYNCHRONY AND THE ONSET OF INCUBATION IN BIRDS, REVISITED: WHEN IS THE CRITICAL PERIOD?

SCOTT H. STOLESON AND STEVEN R. BEISSINGER

CURRENT ORNITHOLOGY

VOLUME 12

CHAPTER 1

TESTOSTERONE AND POLYGYNY IN BIRDS

LES D. BELETSKY, DAVID F. GORI, SCOTT FREEMAN, and JOHN C. WINGFIELD

1. INTRODUCTION

The evolution, maintenance, and functioning of avian mating systems and breeding strategies are of great interest to ethologists and behavioral ecologists. Indeed, critical inquiries into the origins of mating systems provided some of the "cornerstones" of the field of behavioral ecology (e.g., Crook, 1964; Verner, 1964; Orians, 1969, 1972). Classically, mating systems were defined with respect to the number of mates obtained by males and females per breeding period (Oring, 1982): monogamous individuals had one mate, polygamous individuals more than one, and promiscuous individuals mated indiscriminately or nearly so. Recently, recognition has grown that various strategies and selective pressures result in the monogamy/polygamy/promiscuity division and that breeding strategies are not discrete entities but form a continuum of relationships between the sexes. Accordingly, mating systems are now classified by the abilities of individuals to monopolize

---

LES D. BELETSKY, DAVID F. GORI, SCOTT FREEMAN, and JOHN C. WINGFIELD • Department of Zoology, University of Washington, Seattle, Washington 98195. *Current affiliation for D.F.G.:* The Nature Consevancy, Tucson, Arizona 85705. *Current affiliation for S.F.:* Burke Museum, University of Washington, Seattle, Washington 98195.

Current Ornithology, Volume 12, edited by Dennis M. Power. Plenum Press, New York, 1995.

resources and mates (Emlen and Oring, 1977; Oring, 1982; Gowaty, 1992). For example, monogamy is seen as the consequence of the inability of members of either sex to monopolize more than a single mate. Males may be polygynous because they can garner more than one mate through the acquisition and defense of sufficient breeding resources (thus the designation "resource–defense polygyny").

Although many investigations of avian mating systems dwell on ultimate causation of breeding strategies and behavior, two recent, innovative avenues of research, which focused initially on proximate behavior, have yielded surprising and significant insights. First, molecular investigations into within-clutch parentage, including those performed with protein electrophoresis and DNA fingerprinting, have demonstrated a high frequency of extra-pair copulations (EPCs) and fertilizations in both monogamous and polygynous systems (reviewed by Westneat *et al.*, 1990). Thus, it can no longer be assumed that pair-bonded adults are the genetic parents of the offspring they attend, a finding that blurs the distinctions among traditionally defined mating systems.

Second, work in the area of "field endocrinology" has identified different temporal patterns of circulating levels of sex steroid hormones in species employing various mating systems. Perhaps most surprising is the discovery that the "boundary" between monogamy and polygyny is more fluid than supposed. Individual males that usually breed monogamously, when given artificially high levels of sex steroids, can be induced to breed polygynously (Wingfield, 1984a). Based on our current knowledge of the relationships among avian physiology, breeding behavior, and reproductive success, this finding is especially intriguing: If one sex can monopolize more members of the other merely by synthesizing and secreting extra molecules of the sex steroid testosterone (T), and if more mates lead, on average, to superior lifetime reproductive success and, hence, higher fitness, then why are more species not regularly polygamous? For example, among the 291 species of North American passerines listed by Verner and Willson (1969), only 14 were considered regularly polygynous.

Here we review current knowledge of the relationships between testosterone and polygyny in birds with the view of elucidating both endocrinological correlates of mating strategies and the potential influences and constraints of hormones on the development and function of reproductive behavior. Our review is limited to those species in which males regularly have more than one mate per breeding season, and thus, includes simultaneously- and sequentially polygynous breeders and promiscuous/lekking species. We also include information on a special case-brood parasitism. The breeding endocrinology of polyandrous species, with behaviors uniquely associated with sex role reversal, has been reviewed in detail elsewhere (Oring and Fivizzani, 1991); thus, we mention it only briefly.

Most early work on the endocrinology of avian breeding, the truly pioneering studies which opened the field for subsequent investigation, was performed in the laboratory with highly domesticated or monogamous species. At first, our understanding of avian reproductive endocrinology was constrained by these biases. There have now been sufficient numbers of studies of wild or almost-wild populations of birds to eliminate the biases introduced by that sample and to review and synthesize what is known specifically of the relationships in males between T and polygyny. The one remaining bias is that most of the species studied to date breed only in the north temperate zone. Tropical species may well be different in several aspects of their endocrinology. Although we concentrate on studies of natural populations, we include some studies of captive animals, especially when field studies are lacking or intraspecific comparisons with free-ranging populations can be made.

It is well established that testosterone, in addition to affecting spermatogenesis, sexual behavior, and stimulating the development of some secondary sex characters (Witschi, 1961; Wingfield and Farner, 1980), is involved fundamentally in the expression of aggressive behavior and social dominance in reproductive contexts (Harding, 1981; Balthazart, 1983; Wingfield *et al.*, 1987; Wingfield, 1990). The exact nature of the relationship between testosterone (T) and reproductive aggression, and the precise action of T on the expression of aggressive behavior, are not yet fully understood, and are areas of current research (Dittami and Reyer, 1984; Wingfield *et al.*, 1987; Harding *et al.*, 1988; Wingfield, 1994). What is known, however, is that circulating levels of T in free-ranging breeding birds are relatively high only during periods when male–male aggressive interactions are frequent and/or sustained, especially times of territory establishment and mate-guarding. This pattern, which is found among polygamous as well as monogamous species, together with some experimental evidence, supports a theory on control of T secretion known as the "Challenge Hypothesis" (Wingfield *et al.*, 1987; Wingfield, 1988): T secretion is strongly stimulated when conspecifics challenge each other for territory ownership, for position within dominance hierarchies, or for access to mates. Therefore, the high peaks of circulating T levels above physiological baseline (the low levels stimulated by vernal increases in daylength and associated with testicular recrudescence and increased spermatogenic activity) during natural breeding activity may not be due to endogenously controlled hormone releases that initiate and regulate behavior, but to responses to environmental (social) stimuli.

The main predictions of the Challenge Hypothesis (CH) are that plasma T levels and aggression should be positively correlated during periods of social instability (Wingfield, 1990), and that patterns of plasma levels of T among species or among individuals within species should be highly variable. In fact, individual T profiles should depend on peculiarities of breeding efforts (e.g., the number of clutches per breeding season, and the extent of mate-guarding and territorial defense) and the ambient social environment (e.g., breeding density, floater–male density).

Our plan here is first to describe various types of T profiles of males associated with species having different polygynous mating patterns. We then discuss several variations in these hormone profiles, such as those associated with differences among individuals of various ages and territory-ownership classes. These variations are especially pertinent because the hormonal patterns involved should provide strong tests of the CH. Next we discuss some recently proposed theoretical considerations that deal with the hormonal regulation of breeding behavior, especially as it affects expression of parental care. To this problem we bring to bear new material on experimental androgen implants to wild birds. We then examine what is known about the relationships between T and the social interactions within and between the sexes. Last, we explore possible costs for polygynous males of maintaining high circulating T levels, including new field experiments. Throughout, we draw extensively on our own work with two species of New World blackbirds (Emberizidae: Icterinae).

2. TYPES OF POLYGYNY AND ASSOCIATED TESTOSTERONE PROFILES

Examination of the breeding season T profiles of various species under our purview reveals not one typical pattern, but rather a general theme and variations. The variations, however, correlate closely with differences in the duration and chronology of bouts of male–male aggression (and, thus, generally provide convincing comparative evidence in support of the CH). *Polygynous breeders, when compared as a group to monogamous ones, have extended periods of male–male aggressive and territorial behavior and decreased parental care contribu-*

tions. These factors may be related and probably represent an evolutionary trade-off (see below). Polygynous species vary in the number of mates males obtain and the temporal sequence of matings or clutch initiations by females.

2.1. Simultaneous Polygyny

Red-winged Blackbirds (*Agelaius phoeniceus*) and Yellow-headed Blackbirds (*Xanthocephalus xanthocephalus*) are strongly polygynous marsh breeders. In the northwestern U.S., Red-winged Blackbird harems average five females per male territory; Yellow-headed Blackbird harems average three females, most of whose breeding overlaps temporally. As a result of the long breeding season of some Red-winged Blackbird populations, up to ten or more females may build nests in a male's territory, but usually not simultaneously (Orians and Beletsky, 1989). (Some populations of Red-winged Blackbirds breed in upland areas, where polygyny is often more weakly expressed.) In contrast to monogamous males such as Song Sparrows (*Melospiza melodia*) (Fig. 1a), with very brief peaks of T (a few days in duration) associated with territory establishment and mate-guarding, breeding territorial male Red-winged and Yellow-headed Blackbirds have prolonged periods of high T levels. There is a high plateau of up to five weeks in the case of the nonmigratory Red-winged Blackbirds of eastern Washington State (Fig. 1b; Beletsky *et al.*, 1989), and an even longer period in a population of Yellow-headed Blackbirds in the same area (Fig. 1b; Beletsky *et al.*, 1990a). These long durations of elevated T coincide with the periods when most nests are initiated on the male territories and, in the migratory Yellow-headed Blackbirds, with the extended period during which territories are established and contested. Time periods surrounding nest initiations are significant because these are the females' fertile periods, when males guard their mates most vigorously from other males.

Relevant hormonal data are available for two other species that fit this mating system classification. Wild Turkeys (*Meleagris galloparvo*), considered polygynous (Lewis, 1973), had high T levels throughout the period when they were sampled during breeding in February through April in Alabama (Lisano and Kennamer, 1977). Ring-necked Pheasants (*Phasianus colchicus*) are polygynous in the wild (Snyder, 1984). When studied in captivity in Japan they had peak T levels during March (Sakai and Ishii, 1986), which coincides with the beginning of their natural breeding season.

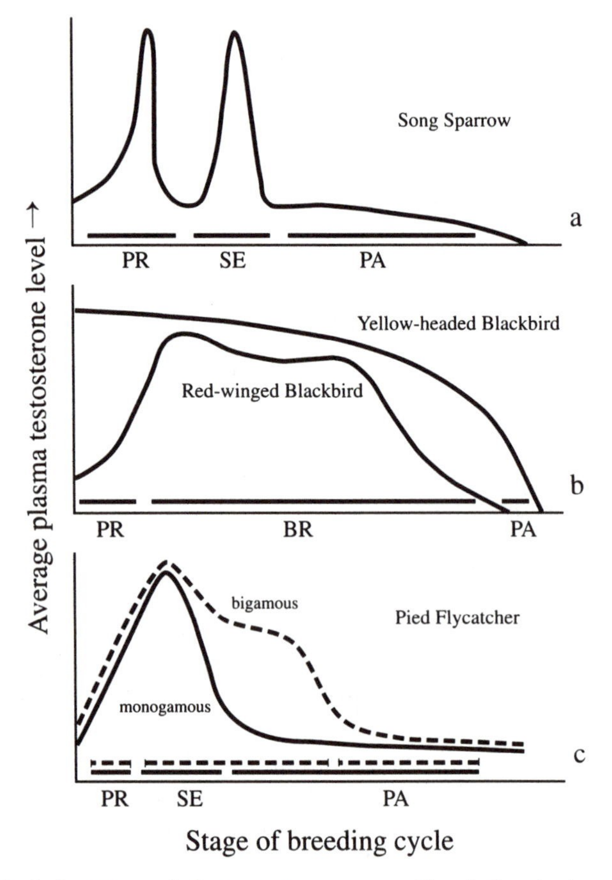

FIGURE 1. Typical patterns of plasma testosterone profiles during the breeding season found during field studies of male (a) Song Sparrows; (b) Red-winged and Yellow-headed Blackbirds; (c) monogamous and bigamous Pied Flycatchers. Stages of breeding cycle: PR, prebreeding; SE, sexual; PA, parental; BR, breeding. The Y axes, denoting increasing testosterone level, are meant to range linearly from zero to approximately six ng of testosterone per ml of plasma.

2.2. Sequential Polygyny

The breeding endocrinology of wild Pied Flycatchers (*Ficedula hypoleuca*) has been studied extensively in Sweden (Silverin, 1980, 1983, 1993; Silverin and Wingfield, 1982). This cavity-nester makes an important contribution to understanding the relationship between T and breeding behavior, and provides an interesting comparison with the previously discussed species, because it has a sequentially polygynous mating system (or "polyterritoriality"; von Haartman, 1969; Alatalo *et al.*, 1981; Silverin, 1983). Males establish a territory in the spring and attract a mate. When egg-laying commences, some males leave to establish second territories to which they attempt to attract second mates. When the second clutch is underway, the bigamous males return to their first mates and help provision their young. Thus, a natural experiment exists in which a single population has both monogamous and sequentially polygynous males. Monogamous males

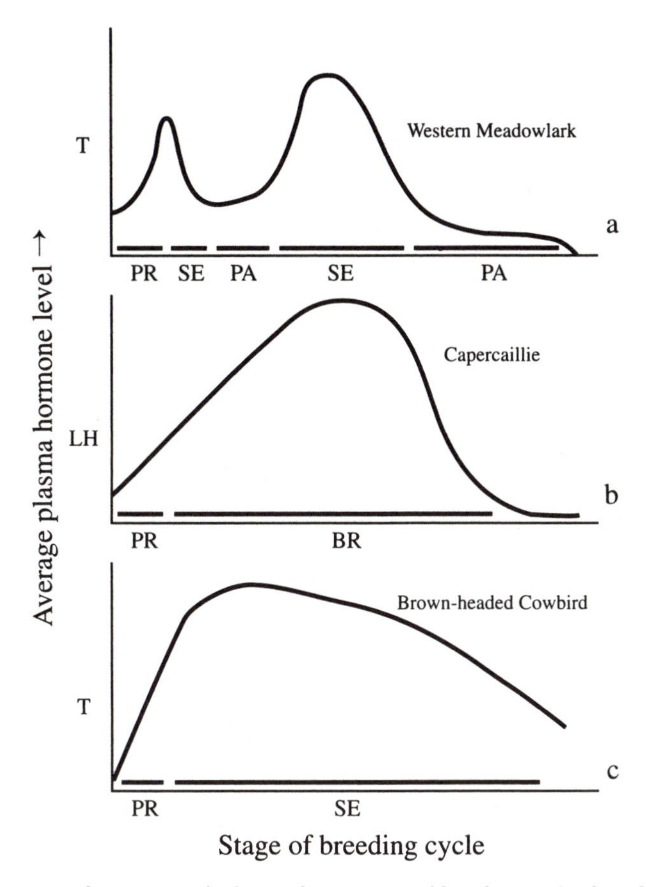

FIGURE 2. Typical patterns of plasma hormone profiles during the breeding season found during field studies of male (a) Western Meadowlarks (T = testosterone); (b) Capercaillie (LH = luteinizing hormone); (c) Brown-headed Cowbirds (T). Stages of breeding cycle: PR, prebreeding; SE, sexual; PA, parental; BR, breeding. The Y axes, denoting increasing hormone levels, are meant to range for T from zero to approximately six ng per ml of plasma, and for LH, to range from zero to approximately eight ng per ml of plasma.

have a T profile generally indistinguishable from that of monogamous breeders in other species, but bigamous males maintain high T levels until second females begin incubating (Fig. 1c), i.e., when territory defense declines and mate-guarding is no longer necessary.

Western Meadowlarks (*Sturnella neglecta*) are at least occasional polygynous breeders (Lanyon, 1957; Verner and Willson, 1969). The breeding season T profile from a Montana population that showed evidence of polygyny had two high peaks, the first coincident with territory establishment, courtship and copulation, and the second with sexual activity associated with the second female (Wingfield and Farner, 1993; Fig. 2a). The low T levels between the two peaks occurred when the first clutch was incubated and the young fed. Male Vesper Sparrows (*Pooecetes gramineus*), thought to be occasionally polygynous, with two broods per male territory common (Best and Rodenhous, 1984), had a breeding season T pattern in Montana almost identical to that found for Western Meadowlarks (Wingfield and Farner, 1993). Males of all sequentially polygynous breeders described here participate to some degree in parental care, be it incubation or feeding young, a facet of their breeding biology relevant to subsequent discussions.

2.3. Promiscuity/Leks

The endocrinology of several promiscuous or lekking species has been studied under free-ranging or semiwild conditions. These systems are characterized by extended male–male aggression and the lack of male parental care. Females and males associate only briefly for copulation. As predicted by the CH, males in these species have high T levels essentially throughout their breeding periods. Captive male Capercaillie (*Tetrao urogallus*), a Eurasian galliform, had very high circulating luteinizing hormone (LH) levels when sampled during the months they would be breeding in the wild (May and June), and had near-baseline levels at all other times of the year (Fig. 2b; Hissa *et al.*, 1983). LH, the pituitary glycoprotein that stimulates testicular androgen synthesis and secretion, is an indicator of circulating androgen levels. Androgens usually, but not always, track LH levels after a brief delay (e.g., Wingfield *et al.*, 1982; Silverin and Wingfield, 1982; Ball and Wingfield, 1987). LH may also directly influence aggressive behavior (e.g., Wingfield *et al.*, 1992). A low percentage of wild Capercaillie (approximately 0.7% of individuals in Finland) exhibit profoundly abnormal, hyperaggressive, behavior that has been associated with abnormally high circulating T levels (Milonoff *et al.*, 1992; see below).

Male Australian Satin Bowerbirds (*Ptilonorhynchus violaceus*) build "bowers"—stick structures—that attract females. A male can mate with many different females in a single breeding season (females probably mate only once). In a recent study, males with bowers had very high plasma T levels throughout the breeding season, significantly higher than adult males without bowers. Also, T levels varied as a function of bower quality: Males with higher quality bowers, as scored by the size and density of sticks used in construction, had higher T levels than males with lower-quality bowers (Borgia and Wingfield, 1991). T levels were also related to reproductive success (see below). Lastly, studies of Mallards (*Anas platyrhynchos*) have shown that, while wild birds are monogamous, captive, domestic individuals are often promiscuous. Associated with this change in mating system is an extension in duration of high plasma androgen in domestic males, sometimes with months-long periods of high circulating T levels (Balthazart and Hendrick, 1976; Donham, 1979).

2.4. Brood Parasitism

Brown-headed Cowbirds (*Molothrus ater*) are North American passerines that are brood parasites, with no parental care. Their mating system apparently varies regionally (Darley, 1982; Dufty, 1982; Teather and Robertson, 1986). In some areas including the West, males do not defend territories as such, but form dominance hierarchies that control access to females and spend the breeding season guarding their mates from other males. There is a tendency in some populations for polygynous or promiscuous matings (Elliott, 1980). In concordance with the CH, male Brown-headed Cowbirds typically show high T levels through much of the breeding season (Fig. 2c; Dufty and Wingfield, 1986).

In summary, polygynous males tend to have high plasma T levels for extended periods, usually much longer than monogamous males. We note that it is only the longer durations of high T levels that differentiate polygynous from monogamous T profiles, not the absolute values of the circulating T concentrations. In fact, the ranges of breeding plasma T levels across species are remarkably similar (Table I), within the same order of magnitude. (It is possible that receptor populations in target tissues may vary more markedly; future research should address this issue.) The long, high T levels of polygynous males are closely associated with prolonged periods of advertisement, mate acquisition, mate-guarding, male-male aggression, and/or territory defense. The

TABLE I
The Ranges of Mean Circulating Testosterone (T) Levels Found for Some Free-Ranging Male Birds during Breeding Seasons

Species	Range of $\bar{X}$ T levels (ng T/ml plasma)	Breeding system	Reference
White-crowned Sparrow[a] (*Zonotrichia leucophrys*)	0.5–4.0	Monogamy	Wingfield and Farner, 1978a
Song Sparrow[a] (*Melospiza melodia*)	0.5–5.8	Monogamy	Wingfield, 1984b
Red-winged Blackbird[a] (*Agelaius phoeniceus*)	1.5–4.0	Simultaneous polygyny	Beletsky *et al.*, 1989
Yellow-headed Blackbird[a] (*X. xanthocephalus*)	0.9–5.7	Simultaneous polygyny	Beletsky *et al.*, 1990a
Wild Turkey[b] (*Meleagris galloparvo*)	1.0–5.5	Simultaneous polygyny	Lisano and Kennamer, 1977
Ring-necked Pheasant[b] (*Phasianus colchicus*)	0.6–2.5	Simultaneous polygyny	Sakai and Ishii, 1986
Pied Flycatcher[a] (*Ficedula hypoleuca*)	0.1–1.6	Sequential polygyny	Silverin and Wingfield, 1982
	3.4[c]		Silverin, 1993
Western Meadowlark[a] (*Sturnella neglecta*)	1.0–4.0	Sequential polygyny	Wingfield and Farner, 1993
Vesper Sparrow[a] (*Pooecetes gramineus*)	0.5–4.6	Sequential polygyny	Wingfield and Farner, 1993
Satin Bowerbird[a] (*Ptilonorhynchus violaceus*)	5.7[c]	Promiscuity	Borgia and Wingfield, 1991
Capercaillie (*Tetrao urogallus*)	0.2–1.0	Promiscuity	Milonoff *et al.*, 1992
Brown-headed Cowbird (*Molothrus ater*)	0.3–1.3	Brood parasite	Dufty and Wingfield, 1986
	2.0[c]		Dufty, 1989
Spotted Sandpiper[a] (*Actitis macularia*)	0.1–6.1	Polyandry	Fivizzani and Oring, 1986
Wilson's Phalarope[a] (*Phalaropus tricolor*)	0.2–3.6	Polyandry	Fivizzani *et al.*, 1986; Oring *et al.*, 1988

[a]Only known breeders sampled.
[b]Captive birds or some captive birds sampled.
[c]Only one mean T value reported for breeding or assumed breeding males.

patterns of prolonged high T levels characteristic of these polygynous species support the CH: T levels are high during periods when male–male aggression is frequent.

3. TESTOSTERONE LEVELS AND BREEDING ECOLOGY, AGE, AND REPRODUCTIVE SUCCESS

So far we have considered androgen levels only in breeders. However, advances have been made in understanding the endocrinology of different classes of individuals, including those not yet of breeding age and those of breeding age that do not yet own territories ("floaters").

3.1. Migratory Versus Nonmigratory Males

Red-winged and Yellow-headed Blackbirds in eastern Washington State show differences in between-year territory stability. Male Red-winged Blackbirds are resident in the area, visiting their territories occasionally even during winter; females are migratory. Year-to-year fidelity to territories is high (Beletsky and Orians, 1987) and relatively few territories are lost or gained during the three-month breeding season. In contrast, the migratory Yellow-headed Blackbirds reestablish territories each spring. Although there is high fidelity to general areas, males frequently change territories within localities (Beletsky and Orians, 1994). Furthermore, during at least one year, nonterritorial male Yellow-headed Blackbirds challenged territory owners successfully for parts of their holdings essentially throughout the breeding season (Beletsky *et al.*, 1990a).

One endocrinologic characteristic of the nonmigratory Washington Red-winged Blackbirds is that male T levels are low at the start of the breeding season, both before females are present and then after they are present but before they are sexually receptive. T levels rise in most males only when nesting commences (Beletsky *et al.*, 1989). The relatively low T levels during the prenesting periods may be because resident territorial males usually do not have to settle boundaries with new neighbors. If this is the case, then the CH predicts early low T levels in a resident population but higher T in migratory populations where territorial boundaries are reestablished annually. However, contra the CH, males in a migratory Red-winged Blackbird population at Yellowwood Lake, Indiana, had a seasonal average T profile very similar to the one exhibited by nonmigratory Washington males: relatively low before females arrived on the breeding grounds and a peak when females became receptive (Johnsen, 1991; Fig. 3a). This between-population comparison of T levels and behavior does not support the CH (unless return rates to territories for the Indiana population are high and boundaries between years are relatively stable). In contrast, T levels of male Yellow-headed Blackbirds are high when they arrive in spring on the breeding

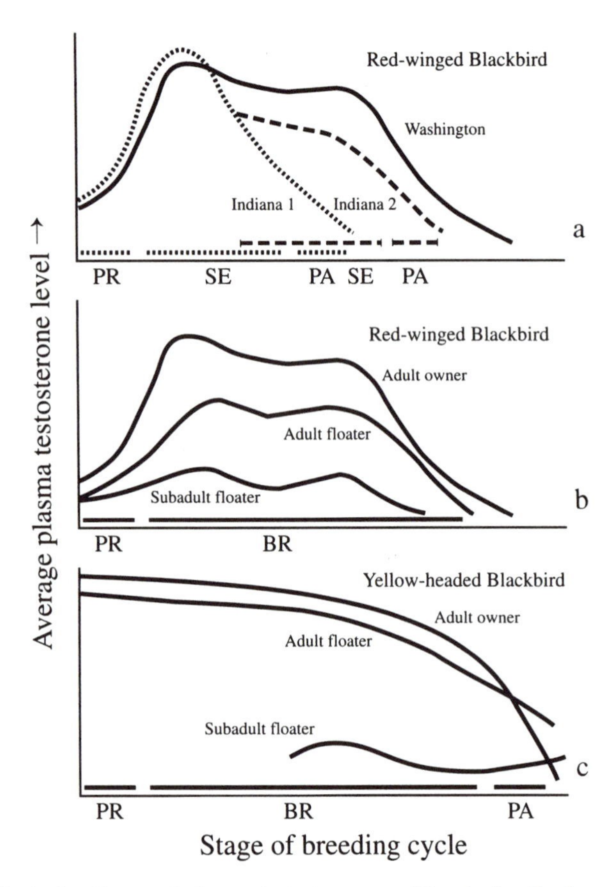

FIGURE 3. Typical patterns of plasma testosterone profiles during the breeding season found during field studies of (a) Washington and Indiana populations of Red-winged Blackbirds (Indiana 1 = single-brooded males, Indiana 2 = multiple-brooded males); (b) adult territory owner, adult floater, and subadult floater Red-winged Blackbirds; (c) adult territory owner, adult floater, and subadult floater Yellow-headed Blackbirds. Stages of breeding cycle: PR, prebreeding; SE, sexual; PA, parental; BR, breeding. The Y axes, denoting increasing testosterone level, are meant to range linearly from zero to approximately six ng of testosterone per ml of plasma.

grounds and levels remain elevated during the long period of territory challenges until eggs hatch and males begin to feed young (up to eight weeks; Beletsky *et al.*, 1990a; Fig. 1b).

3.2. Territory Owners Versus Floaters

Territory owners in many polygynous species aggressively defend their territories from challengers and guard their multiple mates from other males, both neighbors and floaters, over long periods. Thus, the CH predicts sustained (or frequent) high T levels in polygynous territory owners. But what of adult floaters? These individuals are fully capable of expressing territorial and reproductive behavior, but are prevented from doing so by shortages of resources, such as suitable breeding habitat. Floaters may reside in the same local area as owners (or even live on territories, as "underworld" individuals; Smith, 1978; Arcese, 1987), have equivalent experience in the area, and trespass on territories to challenge owners.

Should floaters have high T levels? According to CH, the prediction is that only floaters that regularly challenge owners for sustained periods should show high T levels; this appears to be the case. Among male Red-winged Blackbirds, adult floaters have significantly lower T levels than territory owners; subadult floaters, who only rarely acquire territories, have even lower levels (Fig. 3b). (The fact that territorial adult and nonterritorial adult male Red-winged Blackbirds have different seasonal circulating T levels offers convincing evidence that different T profiles are not simply age-related, but are also influenced by social or reproductive behavior such as territory ownership and defense.) Although adult floaters do challenge territory owners, most challenges are brief. Usually contests are won quickly and decisively by owners. Plasma T levels in birds are known to be responsive to challenges, but contests must be prolonged; T levels usually do not rise for at least 10 minutes after the start of a challenge (Wingfield and Wada, 1989).

Escalated contests for territory ownership, involving long bouts of threat displays and even combat, do occur, but they are rare. The CH predicts high T levels during these contests for all participants. For example, among Yellow-headed Blackbirds, floaters sometimes continue to insert themselves into already-occupied habitat during much of the breeding season. The CH therefore predicts that Yellow-headed Blackbird floaters should often be found to have high breeding season T levels. During one breeding season this was found to be the case: floaters had average T levels that were high and indistinguishable in seasonal pattern from those of territory owners (Beletsky *et al.*, 1990a). According to the CH, this was because Yellow-headed Blackbird floaters were persistent challengers of territory owners.

3.3. Testosterone and Age

In many species, juvenile males are behaviorally and physiologically capable of reproduction, i.e., they produce viable sperm and they can establish territories and attract females, but they do not regularly breed because adult males monopolize females or territories. This is presumably because adults are larger, have different, functionally advantageous, plumage, are more aggressive, and/or have more experience. Male Red-winged and Yellow-headed Blackbirds generally do not obtain breeding territories until they are at least two years old and in adult plumage. For both species, yearlings, or "subadults," in Washington only rarely obtain territories naturally although, at least in Red-winged Blackbirds, if they get a territory, they can attract mates and breed. As predicted by the CH, these subadult floaters have very low circulating T levels during the breeding season (Beletsky *et al.*, 1989, 1990a; Fig. 3b,c). The CH also makes the testable, but as yet untested, prediction that those few subadults that do obtain territories should have higher T levels, perhaps equivalent to those of adult territorial males.

Moore (1991) considers subadult and adult stages of male Red-winged and Yellow-headed Blackbirds to be alternative reproductive tactics and suggests that androgens may play causal roles in developing the two reproductive strategies, or "phenotypes." However, Moore's hypothesis assumes that subadult floaters, usually with femalelike plumage, obtain "sneaky" matings (EPCs) and thus achieve some reproductive success. However, among Red-winged Blackbirds, DNA studies strongly indicate that young are sired only by territory owners, although not necessarily by the owners of the territories on which the nests are located (Gibbs *et al.*, 1990; Westneat, 1992).

Juvenile male Satin Bowerbirds generally do not build bowers or breed. In fact, they do not normally molt to fully adult plumage until seven years of age (Borgia and Wingfield, 1991). Juveniles induced with T implants to have circulating T levels even higher than those of adults, showed increases in reproductive behavior and male–male aggression, but they still could not dominate adult males and thus enter the breeding population (Collis and Borgia, 1992). Juvenile experimentals acted more like adults. They molted to adultlike plumages, built bowers, and were dominant to untreated, control juveniles. However, their adultlike behaviors were somewhat uncoordinated, and they were behaviorally subordinate to breeding adult males. Apparently, coordinated adult behavior that leads to successful breeding requires several years of juvenile experience during which young males interact with and observe adult males, in addition to hormonally-dependent attributes such as heightened aggressiveness and adult plumage.

Peak seasonal T levels in adult males apparently do not vary with advancing age. Red-winged Blackbird males breeding for their first, second or third-plus year (mostly two, three, and four plus years old) have comparable plasma T levels (Beletsky *et al.*, 1992). Likewise, T levels of adult male Satin Bowerbirds do not change with age, although males with bowers, who are usually older, have significantly higher T levels than those without bowers (Borgia and Wingfield, 1991).

3.4. Testosterone Manipulations

In most species, possessing a territory is a prerequisite for breeding and in some populations, a territory virtually guarantees a male access to females. Thus, knowledge of factors that determine which males are successful competitors for territories is crucial for understanding the breeding biology of birds. Influences such as size and fighting ability ("Resource Holding Potential"; Parker, 1974), as well as relative investments in and valuations of different areas ("Value Asymmetries"; Parker, 1974; Maynard Smith and Parker, 1976) are being investigated as general explanations of which animals win territorial contests. However, because T is intimately tied to the type of aggressive behavior displayed in territorial contests, relative T levels, on a proximate basis, may have an important role in territory establishment and maintenance. Thus, if territory ownership is restricted to males capable of, say, physiologically sustaining high T levels, territorial males given implants that maintain plasma T at peak levels should expand their territories at the expense of their unmanipulated neighbors. Also, floaters given high T should obtain territories.

The general pattern after territorial or hierarchically organized male birds are given T implants is for overt aggressive behavior to increase, but for their social status, dominance position, or territory position/size to remain stable. Several field investigations support this trend. Three California Quail (*Callipepla californica*) implanted with exogenous T became more aggressive toward conspecifics, but they did not advance in social position (Emlen and Lorenz, 1942). Similarly, six Sharp-tailed Grouse (*Tympanuchus phasianellus*) showed heightened aggression, but their territory positions and boundaries changed little following implants (Trobec and Oring, 1972). Males in a British population of Red Grouse (*Lagopus lagopus scoticus*) are territorial and each usually has one or two females. Three territorial males implanted with androgen exhibited elevated levels of aggressive and sexual behavior and also expanded their territories. Two nonterritorial males given androgen implants managed to obtain small territories from other owners (Watson and Parr, 1981).

In Red-winged Blackbirds, T-implanted territory owners also behaved more aggressively, with increases in the number of boundary disputes with neighbors and male–male chases (Beletsky *et al.*, 1990b). In one study, a few implanted Red-winged Blackbirds increased their territory sizes (five of eighteen males; Searcy, 1981), but in another, T-implanted males (n = 9) did not enlarge their territories (Beletsky *et al.*, 1990b). Furthermore, floaters with artificially elevated T levels were unable to obtain territories (Beletsky *et al.*, 1990b). The sizes of territories of male Yellow-headed Blackbirds given either T implants (n = 9) or empty, control implants (n = 11) generally did not change (D. Gori, unpublished data; see below). Thus, the predominant pattern for the several species that have been investigated is that, although T levels influence aggressive interactions with other males, elevated plasma T levels alone are insufficient in most cases to overcome previously established social relationships between territory owners or between owners and floaters. In general, these experimental results support the idea that when relative social stability prevails, whether enforced by "social inertia," individual recognition of status, or accepted territory boundaries, increases in aggressiveness associated with higher T levels are usually insufficient to alter dominance relationships (Rohwer and Rohwer, 1978; Ramenofsky, 1984; Wingfield and Ramenofsky, 1985; Hegner and Wingfield, 1987a). If there is a difference related to mating system, it is that strongly polygynous males, with their continually high T levels, may be less "sensitive" to T implants than are monogamous males, and so less likely to be strongly behaviorally affected by the "extra" T. Thus, male Red-winged and Yellow-headed Blackbirds that were given T implants tended not to increase their territory sizes, but monogamous males, such as Song Sparrows, with usually lower T levels, did expand their territories after T treatment (Wingfield, 1984a).

3.5. Testosterone-Induced Changes in Mating Systems

We have noted that differences in mating status within a population are correlated with different T-level profiles (e.g., monogamous versus bigamous Pied Flycatchers), but can altered androgen levels alone cause a shift in mating system? In three species which usually breed monogamously, giving territorial males androgen implants to change their circulating T patterns from monogamous to polygynous types increased the number of females attracted to their territories. White-crowned Sparrows (*Zonotrichia leucophrys*) and Song Sparrows became bigamous by expanding their territories at the expense of their neighbors (Wingfield, 1984a), as did a few Red Grouse (Watson and Parr, 1981). In addition, all territorial male Pied Flycatchers given intramuscular T injections became polyterritorial and exhibited prolonged territorial and courtship behavior (Silverin, 1980); about 30% never returned to their first territories, but continued on their second ones. This result suggests that second territories in this species and, thus, regular polygamous matings, depend on sustained high T levels. The prevalent mating systems in some species are therefore evidently hormonally mediated and subject to change with shifting hormone balances—whether artificially or naturally induced.

3.6. Testosterone and Breeding Density

Two aspects of the known or suspected relationships between T and breeding behavior suggest that male birds in high density breeding situations should generally have higher circulating T levels than conspecifics, even those in the same population, breeding at lower densities. First, when males compete for breeding resources, the frequency of aggressive interactions should be positively correlated with male density. If so, the CH predicts higher T levels in males breeding in densely than in sparsely settled areas. Second, dense breeding situations should lead to greater exposure of males to females, interactions which in at least some species can lead to elevated male T levels (Moore, 1984). Several studies have reported positive correlations between male breeding density and plasma T (European Starling, *Sturnus vulgaris*, Ball and Wingfield, 1987; Yellow-headed Blackbird, Beletsky *et al.*, 1990a; Red-winged Blackbird, Beletsky *et al.*, 1992; Song Sparrow, Wingfield and Hahn, 1994), but no experiments have been performed. Thus it is still unclear whether the relationship is causal or incidental. Polygynous males, with variable harem sizes and sometimes colonial breeding systems, should be especially useful subjects for probing the relationship.

3.7. Testosterone and Reproductive Success

Plasma T levels and reproductive success may be positively correlated in polygynous species because (1) the frequency of mate-guarding, which often includes aggressive interactions with other males, is positively correlated with number of mates (the CH would predict high T levels); (2) the most aggressive males, which obtain the best territories, may be subjected to more frequent trespassing and challenges by floaters (another CH prediction); (3) males with larger harems, which lead to higher reproductive success, interact more frequently with females, leading to higher T levels (Moore, 1984); and/or (4) in promiscuous species, the most aggressive males obtain the most copulations. Support for a relationship between T and breeding success has been found for a population of Red-winged Blackbirds and for Satin Bowerbirds. Among Red-winged Blackbirds, breeding male T levels were significantly correlated with harem size and the number of nests on the males' territories (Beletsky *et al.*, 1989, 1992) and Satin Bowerbird T levels were correlated with the actual number of copulations males obtained (Borgia and Wingfield, 1991). On the other hand, in species with male parental care, experimentally raised or sustained T levels in males during periods when they naturally decline can lower reproductive success (Silverin, 1980; Wingfield, 1984a; Hegner and Wingfield, 1987b; see below).

4. TESTOSTERONE, MALE–MALE AGGRESSION, AND PARENTAL CARE

4.1. Theory and Background

A theoretical framework for the role of T in reproductive behavior has been advanced that relates male–male aggression to mating systems and male parental care (Wingfield *et al.*, 1990). The major argument of the hypothesis is that high circulating T levels, essential for the expression of sustained aggression, are physiologically incompatible with the expression of parental care, particularly feeding young. Thus, according to the postulated associations of behavior and hormone levels, the typical breeding season temporal patterns of T secretion should be predictable from the degree and chronology of male breeding aggression and parental care in a population. Plasma T levels should be high and sustained for males in populations with extensive male–male aggression and limited or no male parental care, and low (or high for only brief periods) where males show little male–male aggression and extensive parental care. In field testing the hypothesis for polygynous species, two crucial predictions are that T levels should fall when and if unmanipulated males commence feeding young, and males with artificially sustained high T levels should show less parental care, and hence, reduced breeding success.

Evidence supporting these predictions has been found in some monogamous species. Among White-crowned Sparrows, male T levels are high during their mate's first laying period, but remain low when their mates lay second clutches, because, according to the theory, males at the time are feeding first-clutch nestlings (Wingfield and Farner, 1978b; Wingfield, 1984a; Wingfield *et al.*, 1990). The same pattern is normally found in Song Sparrow males, but if first nests are depredated or fail for other reasons, then male T levels are high during the females' renesting laying period (Wingfield and Farner, 1979; Wingfield, 1985b). Male House Sparrows (*Passer domesticus*) given T implants fed young less frequently than controls and participated in more male–male aggression. Conversely, parental care was prolonged by administration of an androgen blocker (Hegner and Wingfield, 1987b). A result of the switch from feeding young to increased aggression was relatively low reproductive success for implanted males. Similar effects on reproductive success were obtained when T-implanted Song and White-crowned Sparrows became polygynous (Wingfield, 1984a). These latter results underscore the importance of precise timing between breeding behaviors and endocrinologic changes.

Polygynous birds offer perhaps the best opportunities to test the general applicability of a model of tradeoffs between male aggression and parental care because they exhibit a wide variety of parental care patterns and specializations. Bigamous Pied Flycatchers, which normally return to their first territories to feed young after mating with a second female on their second territory, fed their young at significantly lower rates following injection of exogenous T. T-treated males also produced fewer fledglings than control males (Silverin, 1980). Artificially high T levels in polyandrous male Spotted Sandpipers (*Actitis macularia*) had the same generally negative effect on parental care: incubating males given T implants tended to desert eggs or incubate less often than did controls (Oring *et al.*, 1989). As for Red-winged Blackbirds, males in eastern Washington do not regularly feed young (Beletsky and Orians, 1990), and therefore T profiles from that population cannot be used to test the hypothesis of Wingfield and colleagues. However, the hypothesis can be tested in midwestern and eastern populations, where males regularly feed young.

4.2. Experiment: Testosterone Implants to Yellow-Headed Blackbirds

Male Yellow-headed Blackbirds in eastern Washington normally assist females in feeding young at the primary nest (i.e., belonging to the first female to nest) on their territories (Willson, 1966; Gori, 1988).

This feeding behavior is labile and highly sensitive to environmental conditions, as demonstrated by experiments in which reductions of clutch size result in males switching nests at which they assist females (Patterson *et al.*, 1980; Gori, 1988). The hypothesis that males trade off between two options for increasing their reproductive success by either advertising for additional mates (requiring high T) or participating in parental care activities (feeding young, requiring low T) was tested by implanting territorial males with T at the time they could either continue trying to attract additional mates or switch to feeding young in their first nest.

This study differs from previous ones of the effect of exogenous T on parental care in one important respect: When Pied Flycatcher (Silverin, 1980; Alatalo *et al.*, 1981) and House Sparrow (Hegner and Wingfield, 1987b) young hatched, the availability of unmated females was relatively low so that males had potentially little to gain by continuing high levels of investment in mate attraction and territory defense. However, in this study, more than 20% of the breeding female Yellow-headed Blackbirds were still unmated when young in the primary nests began hatching.

Male Yellow-headed Blackbirds devote significant amounts of time to courtship and territory defense early in the season when females begin settling in marshes to breed. However, song rates decline rapidly when young in the primary nest hatch, and males normally begin to feed nestlings when they are four to five days old. Coincident with their provisioning behavior is a decline in circulating T levels (Beletsky *et al.*, 1990a). Paternal care often begins when significant numbers of unmated females are still setting in marshes to nest, suggesting that males may benefit by devoting time and energy to attracting additional mates rather than providing parental care (Gori, 1984).

The experiment was conducted from 5 May to 15 June 1987 at Beda Lake, Grant County, Washington. During the initial stages of female settlement we mapped territories by observing perch sites and boundary disputes between adjacent males. Marshes were censused for new nests and nests were checked at three-day intervals. All males and most females were color-banded. Females normally begin nest-building two to four days after settling on a territory. A female is said to be "settled" when she spends most of the day on a single territory. Prior to settlement, females visit marshes only in the early morning and late evening. At these times, they move from territory to territory where they are courted by males and interact aggressively with already-settled females.

Several days before the first young hatched in the marsh we implanted two 20 mm lengths of Silastic tubing (inside diameter 1.47 mm, outside diameter 1.96 mm) subcutaneously in each male. The implants were placed adjacent and parallel to the pectoralis muscle, one on each side. One group of males (experimentals, n = 9) received implants packed with crystalline T; the other group (controls, n = 11) received empty implants. For the most part, experimental males had other experimentals as neighbors and control males had other controls as neighbors. Both groups were surrounded by untreated territorial males.

Four to six days after the implants were inserted, we observed each male for a period of 30 minutes, from 05:00 to 07:30, and recorded the time spent in various activities (Table II). At this time, primary females were still incubating eggs. When young in the primary nest were seven days old, we observed the nest for a period of one to two hours between 09:00 to 12:00 or 14:00 to 17:00 and recorded the time budget of the male, his feeding rate to the nest, and, when time permitted, the female's feeding rate. On the evening before young fledged (i.e., 10 to 11 days after hatching), we counted the number of young in the nest and then returned several days later to check if any young had failed to fledge. Secondary and later nests on experimental and control territo-

TABLE II
Comparison of Pre-Hatching Behaviors ($\overline{X}$ +/− SD) Recorded during 30-Minute Observation Periods for Male Yellow-Headed Blackbirds that Received Testosterone Implants and Those Receiving Empty (Control) Implants. Observations Were Made before Young in the Primary Nest Hatched.

Behavior	T-implanted males (n = 9)	Control males (n = 11)	t-test/Mann-Whitney U-test[a]
No. of songs	43.9 ± 17.5	16.5 ± 12.6	t = 4.1***
Songs/min.	1.98 ± 0.51	1.02 ± 0.37	t = 5.1***
Time up (min.)	22.1 ± 5.0	16.2 ± 4.3	t = 2.4*
Time down (min.)	4.6 ± 5.5	9.7 ± 4.2	t = 2.3*
Time off territory (min.)	2.1 ± 5.3	4.0 ± 5.3	t = 0.8, n.s.
No. of boundary disputes	0.55 ± 0.82	0.11 ± 0.33	U = 63, n.s.
Time in boundary disputes (min.)	0.4 ± 0.6	0.1 ± 0.3	U = 62, n.s.
No. of sexual chases	0.2 ± 0.4	0.4 ± 0.5	U = 56.5, n.s.
Time in sexual chases (min.)	0.1 ± 0.1	0.5 ± 1.0	U = 53, n.s.

[a] All probabilities are for 2-tailed tests; * = $P < 0.05$, *** = $P < 0.001$.

ries were also observed for evidence of male feeding for one to two hours when young were seven days old.

We captured several experimental and control males 20 days after they were implanted and took blood samples to ascertain T levels. As expected, average plasma levels of T were significantly greater in experimental males than in controls (0.91 ± 0.20 ng/ml plasma, $n = 6$, and 0.37 ± 0.27 ng/ml, $n = 4$, respectively; Mann-Whitney U-test, $U = 22$, $P < 0.025$). Although T-implanted males had higher plasma T levels, their levels were below the average level of untreated males measured early in the breeding season (Beletsky *et al.*, 1990a), suggesting that the T implants may have been nearly empty by the time blood samples were taken.

4.2.1. Prehatching Investment in Male–Male Competition

Males that received T implants invested more time and energy in mate attraction and territory defense early in the morning, when these activities are normally at their peak, than did control males (Table II). Although the two groups spent an equal amount of time on territory, experimentals spent significantly more time up in the marsh vegetation, from where they advertise, and less time down in the marshes, where they forage, than did control males. Experimental males also had higher song rates than control males. There were no significant differences in the number of boundary disputes and sexual chases or in the time males spent engaging in these activities. We assume that the greater time and energy devoted to singing by experimental males represents greater investment in mate attraction and territory defense because males direct their songs and displays that accompany them at receptive females and other males (Orians and Christman, 1968).

4.2.2. Post-Hatching Investment in Male–Male Competition and Parental Care

After hatching, when young in primary nests were seven days old, T-implanted males continued to invest more heavily in mate attraction and territory defense (Table III). Experimental males spent twice as much time up in the marsh vegetation and half as much time off the territory as did controls. T-implanted males had higher song rates and took part in significantly more sexual chases. Associated with their increased investment in male–male competition, experimentals had a significantly lower probability of feeding nestlings: only four of nine experimental males fed nestlings in the primary nest, whereas all con-

TABLE III
Comparisons of Post-Hatching Behaviors ($\overline{X}$ +/− SD) Recorded during One-Hour Observation Periods for Experimental Male Yellow-Headed Blackbirds that Received Testosterone Implants and Control Males with Empty Implants. Males were Observed when Young in Their Primary Nests Were Seven Days Old.

Behavior	T-implanted males (n = 9)	Control males (n = 11)	t-test/Mann-Whitney U-test[a]
No. of songs	32.6 ± 13.9	10.5 ± 14.7	t = 3.4**
Songs/min.	1.38 ± 0.6	0.42 ± 0.5	t = 4.0***
Time up (min.)	25.6 ± 14.4	12.4 ± 12.5	t = 2.2*
Time down (min.)	15.5 ± 11.7	11.0 ± 8.3	t = 1.0, n.s.
Time off territory (min.)	17.7 ± 12.8	37.1 ± 12.8	t = 3.6**
No. of boundary disputes	0.1 ± 0.3	0.1 ± 0.3	U = 50.5, n.s.
Time in boundary disputes (min.)	0.1 ± 0.3	0.4 ± 1.3	U = 50, n.s.
No. of sexual chases	0.56 ± 0.73	0.0	U = 77*
Time in sexual chases (min.)	0.96 ± 1.6	0.0	U = 79*

[a] All probabilities are for 2-tailed tests; * = P <0.05, ** = P <0.01, *** = P <0.001.

trol males did (Fisher's exact test, P = 0.008). Among males that did provide parental care, experimentals fed their young at a significantly lower rate than did control males, whether the rate was expressed as number of feeding trips per nest per hour (U-test, U = 44, P < 0.005; Fig. 4a) or on a per-nestling basis (U = 44, P < 0.005; Fig. 4b). In addition, three controls fed at secondary nests; no other nests were fed by males of either group.

4.2.3. Mate Attraction and Breeding Success

Although experimental males invested more time and energy singing and displaying on territory, the probability of obtaining an unmated female following implantation was similar for control and experimental males. Five new nests were initiated on experimental territories compared to eight new nests on controls (X^2 = 0.22, 1 df, P > 0.05). These late nests were initiated over a one-week period beginning nine days after implants were inserted. Because nest-building normally begins one to four days after settlement, with one- to two-day intervals

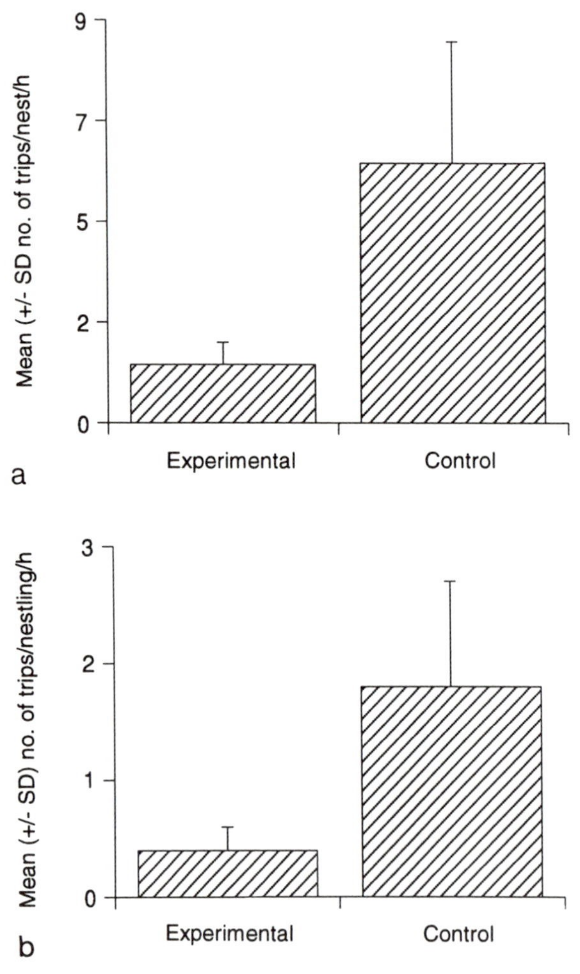

FIGURE 4. Mean (+/− SD) rates that testosterone-implanted experimental and control male Yellow-headed Blackbirds fed their young: (a) number of feeding trips per nest per hour; (b) number of feeding trips per nestling per hour.

common later in the season, these late nests most likely belonged to females that had settled after males were implanted.

The reduced feeding rates of T-implanted males resulted in significantly fewer young fledged from primary nests (Table IV). This difference was not a result of differences between treatment groups in clutch size, hatching success, or female "competence" at feeding young (i.e., the number of young in nests on day five, when parental care by males

TABLE IV
Mean (± SD) Clutch Size, Hatching Success and Fledging Success of Primary Nests on Territories of Experimental Male Yellow-Headed Blackbirds Receiving Testosterone Implants and Control Males with Empty Implants

	T-implanted males (n = 9)	Control males (n = 11)	Mann-Whitney U-test[a]
Clutch size (no. eggs/nest)	3.6 ± 0.5	3.7 ± 0.5	U = 58, n.s.
% hatching	97.2 ± 8.3	100.0 ± 0.0	U = 55, n.s.
No. of young, day 5	3.4 ± 0.5	3.7 ± 0.5	U = 58.5, n.s.
No. of young, day 8	2.9 ± 0.6	3.4 ± 0.5	U = 69, n.s.
No. of young fledged	2.2 ± 0.7	3.4 ± 0.5	U = 88.5*

[a] One-tailed tests; * = $P < 0.001$.

normally begins; Willson, 1966; Gori, 1984; Table IV). The number of young alive on day eight in control nests was greater than in experimental nests, although the difference was not quite significant ($0.05 < P < 0.10$). A difference in this case would not be surprising because it presumably reflects greater offspring mortality in experimental nests between days five and eight as a result of reduced parental care by males. Females did not compensate for the reduced care by experimental males. Primary females fed young at a rate of 7.2 ± 4.4 trips per nest per hour on territories of experimental males compared with 8.6 ± 2.6 trips per nest per hour on territories of control males ($U = 54$, $P > 0.5$).

All experimental and control nests fledged at least one young, suggesting that the reduced fledging success from experimental nests was not due to increased predation. Predation in this area typically results in the removal of all young from nests (Orians, 1980; Patterson *et al.*, 1980). Data on nestling growth rates indicate that young that disappeared from "successful" nests first stopped gaining weight and then lost weight, whereas other young in the nest continued to gain weight, suggesting starvation (D. Gori, unpublished data).

These results are consistent with the hypothesis that the variable pattern of investment shown by male Yellow-headed Blackbirds during the breeding season represents a trade-off between the competing demands of parental care and competition for mates. The decline in investment in mate attraction and territorial defense after young hatch, and the corresponding increase in parental care, appears to benefit males because they can produce more offspring by caring for young at this time than they can by continuing high levels of investment in sexual advertisement and territorial defense. Thus, greater investment in sexual advertisement later in the season did not increase a male's chance of attracting unmated females enough to offset the increase in young fledged due to paternal care. These results with Yellow-headed Blackbird implants also support the hypothesis proposed by Wingfield *et al.* (1990) because, by artificially altering the breeding T profiles of the experimental males, we caused shifts in their balance between time and energy devoted to parental care versus T-mediated behavior such as territorial advertisement. Observational data also support the hypothesis: T levels of male Yellow-headed Blackbirds that were sampled repetitively during a breeding season declined only after they were first observed to feed young (n = 22; Beletsky *et al.*, 1990a).

5. TESTOSTERONE LEVELS AND SOCIAL INTERACTIONS

Although the correlation between elevated plasma T levels and male breeding activity, including territorial aggression and sexual behavior, has been long recognized (see Wingfield *et al.*, 1987; Wingfield and Marler, 1988 for reviews), there have been few studies aimed at separating cause from effect or discriminating among various environmental factors that might mediate changes in sex steroid secretions.

Accumulating evidence indicates that agonistic interactions with other males are largely responsible for high circulating T levels during breeding. But is male–male aggression the only factor, or do other social interactions affect sex steroid secretion? And do species vary in their endocrine responsiveness to social stimuli such as male–male aggression?

Territorial male Red-winged Blackbirds that behaved aggressively toward an intruding male conspecific in a cage had average circulating T levels that were equivalent to or lower than males sampled while foraging (Harding and Follett, 1979), although a few males with the highest androgen levels tended to behave the most aggressively (Harding, 1981). Thus, social interactions with the caged male may have had only marginal effects on hormone levels. Likewise, male challenges in Japanese Quail (*Coturnix coturnix*) resulted in no change in circulating T levels (Ramenofsky, 1984). On the other hand, monogamous, territorial Song Sparrows experienced significant elevations in T levels after about 10 minutes when they were challenged by and attacked a caged "decoy" male conspecific placed in their territory centers (Wingfield, 1985a; Wingfield and Wada, 1989). Furthermore, when male Song Sparrows were removed from territories and replaced by previously nonterritorial males, T levels were elevated not only in the replacement males, but also in unmanipulated adjacent neighbors. This suggests that agonistic interactions involved in settling new boundaries stimulated T secretion (Wingfield, 1985a).

Dufty and Wingfield (1990) have shown that captive male Brown-headed Cowbirds housed together in groups under natural daylengths experience elevated T levels during the breeding season, but isolated males have relatively low levels. This finding strongly suggests that social interactions alone can lead to higher T. Dufty and Wingfield also found that it did not matter if the male group vocalized: intact males with devocalized companions had the same high T levels as males in groups of intact, vocalizing males, suggesting that visual presence of conspecifics was sufficient to stimulate T secretion. Thus, social strife among males or, at least, visual exposure to conspecific males (Dufty 1993), apparently can lead directly to higher concentrations of plasma T.

A critical question that emerges from these studies is whether in natural situations it is *solely* male–male interactions that affect T secretion, or whether intersexual, and perhaps interspecific, interactions also elevate T. Male birds can and do engage in sexual activity and even copulate after castration, i.e., essentially in the absence of circulating T (Moore, 1983; Moore and Kranz, 1983). Thus, males do not require high T for sexual activity, but free-ranging males usually have high T levels during the period when their mates are receptive. Do interactions with females lead directly to these higher male T levels? Or are they triggered by male–male interactions as the males step-up territory defense to guard their mates from conspecific males?

Moore (1984) has shown that captive male White-crowned Sparrows with sexually receptive females have higher T levels than males with females in nonreceptive states. Thus, female sexual displays may stimulate T secretion in males. The benefit of this burst of T may not be due to regulation or enhancement of male sexual activity but to facilitation of mate-guarding. To discriminate between these alternatives, Moore (1984) demonstrated that free-ranging males with sexually receptive females were also more aggressive toward conspecific males, which suggests that the benefit of elevated T levels in this case was to prevent EPCs by mates. In a study of the Brown-headed Cowbird, however, captive males exposed to females underwent more rapid gonadal development than controls, but their plasma T levels were not different. This seeming anomaly may be explained by the fact that these brood parasites express no parental behavior (Dufty and Wingfield, 1986,

1990). Females are sexually receptive for prolonged periods and can lay throughout the breeding season. Thus, there may have been selection for males that "mate-guard" continuously; hence, in some situations, T secretion may be high regardless of whether females are present.

The studies described above indicate that, although males of some species respond to changes in the immediate social environment with rapid changes in circulating T, others do not. Polygynous species, brood parasites, and other species in which males seldom brood or feed nestlings, may be inherently less responsive endocrinologically to social interactions because their breeding systems favor high T levels throughout much of their breeding seasons (Wingfield *et al.*, 1990; Wingfield, 1991). For example, Willow Ptarmigan (*Lagopus lagopus lagopus*) and Japanese Quail, which show little or no male parental care, do not respond to receptive females by elevating T levels (Stokkan and Sharp, 1980; Delville *et al.*, 1984). Endogenous rhythms or photoperiodic cues in these species may play larger roles in regulating T secretion such that social interactions have only minor modifying effects. Thompson and Moore (1992), who term the Wingfield *et al.* (1990) hypothesis the "social insensitivity hypothesis" of endocrinologic responsiveness, acknowledge the hormone/behavior patterns it seeks to explain, but they propose an as yet untested alternative hypothesis. They suggest that in systems in which males participate in parental care, males have "an unknown physiological mechanism that inhibits full expression of aggressive reproductive behavior" which activates when they must tend young. Males that never tend young would lack the mechanism and therefore would have relatively high T levels and express high levels of aggression throughout their breeding seasons (and thus would appear endocrinologically unresponsive to social stimuli such as territorial intrusions).

6. COSTS OF HIGH TESTOSTERONE

6.1. Theory and Background

Evidently, it is beneficial for male birds to have high circulating T levels during portions of their breeding efforts, yet males of many species apparently maintain low plasma T levels unless they are challenged. Perhaps the most striking evidence that there are potential reproductive benefits to maintaining high T is that males of several species that usually breed monogamously have been induced to breed polygynously (simultaneously, or sequentially by lengthening a breeding period) by artificially elevating and sustaining high T (Watson and Parr, 1981; Wingfield, 1984a). That high T levels are usually not maintained for long periods despite these potential benefits suggests that there may be high costs associated with this strategy.

In addition to potentially lower fledging success due to a reduced ability to express parental care, high T may cause inappropriately high aggressiveness (Dufty, 1989; Milonoff *et al.*, 1992), increased metabolism (Högstad, 1987; but Hänssler and Prinzinger [1979] found no effect on metabolism of T in Japanese Quail) and thus increased breeding season weight loss, delayed post-breeding molt (Nolan *et al.*, 1992), and increased visibility to predators, all of which may lead to decreased survival. Marler and Moore (1988) demonstrated that artificially elevated T levels led to decreased survival among lizards because the resulting increase in aggressive activity exposed individuals to more predation. Similarly, Dufty (1989) found that male Brown-headed Cowbirds with artificially elevated T levels suffered higher mortality and more wounding than did control males, suggesting that it is dangerous to be continually aggressive. Juvenile male Satin Bowerbirds usually have lower T levels than breeding adults. When given T implants they molted to more adultlike plumage and were more aggressive than controls, but they could not dominate adults and they received more threats and attacks from adults than did untreated juveniles. Thus, high T levels in juveniles may be costly because, although T-implanted juveniles exhibited adult-level aggression, they did not have the experience to exploit it reproductively or to deal with its social consequences (Borgia and Wingfield, 1991).

In contrast, Song Sparrows whose T levels were artificially raised during one breeding season suffered no increase in over-winter mortality when compared to untreated controls (Wingfield, 1984a; but see Nolan *et al.*, 1992: T-implanted Dark-eyed Juncos, *Junco hyemalis*, exhibited prolonged breeding, delayed molt, and reduced over-winter survival). However, such an effect on mortality could be more pronounced in polygynous breeders, because they sustain high T levels for much longer periods than do (monogamous) Song Sparrows. Furthermore, some polygynous breeders apparently experience single years of very high reproductive success (e.g., Sage Grouse, *Centrocercus urophasianus*) that, if related to very high circulating T, could, through associated metabolic or behavioral effects, reduce survival. In an extreme case of naturally occurring, abnormally high, T levels, male Capercaillie in Scandinavia with T levels up to five times normal attack humans and motor vehicles—inappropriate aggressive behavior that is obviously costly, although individual survival information is unavailable (Milonoff *et al.*, 1992).

6.2. Experiment: Testosterone Implants to Red-winged Blackbirds

We conducted a field study to determine if territorial fighting is actually risky for Red-winged Blackbirds. We artificially raised circulating T levels in territory owners and then assessed whether experimental males lost weight more rapidly, behaved more aggressively, or were wounded more often than controls.

Our study was conducted from 8 April to 31 May 1987 on two breeding marshes located in Grant Co., Washington. Between 8 and 15 April, we trapped all 24 males with territories on the two marshes. Each was (1) given a Silastic tube implant (25 mm long; inside diameter 1.47 mm, outside diameter 1.96 mm) adjacent to the left pectoralis muscle; (2) provided with colored legbands for individual identification; (3) weighed and released. Experimental males ($n = 12$), with contiguous territories that were spatially separated from the control males' territories, were given implants packed with crystalline testosterone; control males ($n = 12$) received empty implants. All males returned to their territories after release.

Four days after implants we began observations of the males on their territories. We observed control and experimental males on alternate days for 30-minute periods during the hours of peak territorial activity (04:15–08:30, 17:30–19:30). Between 16 and 29 April we observed each male during at least three 30-minute periods. We recorded time on or off territory, song rates, song spread display intensity (degree of feather erection accompanying songs, greater erection signalling greater threat [Yasukawa, 1978]), boundary disputes with other males, and other time budget information. On 30 April we recaptured as many of the males as we could (nine experimentals, 10 controls), weighed them, examined them for wounds, and gave them additional implants—a second empty implant to controls and an implant containing testosterone proprionate (more soluable than crystalline T) to the experimentals. We made additional observations subsequent to the second implants. On 15 May we again caught as many of the males as we could, weighed them, and looked for wounds.

Males with T implants spent more time on territory, gave higher intensity song spread displays, and were involved in more frequent and lengthier boundary disputes than control males (Table V). Other behav-

TABLE V
Comparison of Mean (± SD) Rates of Behaviors of Experimental (Testosterone Implants) and Control (Empty Implants) Male Red-Winged Blackbirds during 30-Minute Focal Observations

	Experimentals (n = 45 30-min periods)	Controls (n = 51 30-min periods)	Mann-Whitney U-test[a]
Time on territory ($\bar{X}$ ± SD min/30-min period)	25.6 ± 5.1	22.7 ± 6.7	U = 845.5**
Song rate ($\bar{X}$ no./min on territory)	2.6 ± 1.0	2.4 ± 0.8	U = 1160.0, n.s.
Song spread display intensity ($\bar{X}$)[b]	1.85 ± 0.08	1.04 ± 0.08	U = 1368.0*
Boundary disputes			
Rate (no./min on territory)	0.06 ± 0.08	0.03 ± 0.05	U = 1479.0***
Duration (s)	57 ± 85	38 ± 100	U = 1428.0**

[a]One-tailed tests; * = P <0.05, ** = P <0.025, *** = P <0.005
[b]Song spread display intensity recorded as 1 = low, 2 = medium, 3 = high.

iors recorded, such as numbers of songs, call notes, and female sex chases, were not different between the two groups, but they are less pertinent to the current discussion. The differences in behaviors associated with male–male aggression between the experimentals and controls clearly indicate the effects of the elevated T levels. Although males with T implants lost weight at a slightly greater average rate than did control males (0.14 ± 0.18 grams per day versus 0.12 ± 0.12 grams per day), the difference was not significant (t = 0.29, 1-tailed P = 0.39). More of the T-implanted males that maintained territories throughout the experiment experienced major boundary changes (4 of 11) than did controls which maintained territories (2 of 11), but the difference was not significant (Fisher's exact test, P = 0.32). We observed few fights, but the two long fights we did observe (and one short one) involved experimental males, whereas we observed only one short fight among control males.

One male from each group disappeared. We do not know the reason for the control male's disappearance, but the experimental male disappeared after participating in a long fight with another male. We found a total of five wounds, each on a different experimental male. One was a four mm square scar three mm from an eye; the others consisted of large (e.g., 10 × 13 mm) defeathered patches on the back of the skull and/or neck. Thus, significantly more T-implanted males than controls suffered wounds (5 of 11 versus 0 of 11; Fisher's exact test, two-tailed $P = 0.035$).

We conclude from this investigation (1) that continually high plasma T levels are associated with increased male–male aggressive interactions in territorial males (see also Beletsky *et al.*, 1990b); (2) that the increased aggressiveness and presumed increase in fighting leads to wounding; and (3) that it is therefore risky for these birds to fight (perhaps explaining why so few aggressive interactions between territorial neighbors and between territory owners and floaters actually culminate in physical combat); (4) that continually high T levels in these birds may indeed be costly, in terms of increased wounding and perhaps decreased survival; and (5) the present results, in conjunction with results of studies cited above, provides support for the hypothesis that sustained high plasma T levels are costly to male birds.

7. CONCLUSIONS

Although birds that regularly breed polygynously comprise a minority of species, to date they have contributed disproportionately to our knowledge of the relationships between androgens and avian breeding behavior. This is largely due to the "natural" experiments afforded by the range of mating behaviors found among these animals.

Although a good number of species already have been examined for breeding T patterns, further work is needed because comparative information provides powerful tests of models of hormone and behavior interactions. The importance of the comparative approach in this area cannot be stressed too strongly, because the relevant hypotheses predict breeding T profiles and T responsiveness of tested species. Also, exceptions to even general theories can exist. Tests that compare the relationships between breeding strategies and T physiology among closely related species will be especially informative because confounding effects of genetic differences that can constrain physiology and behavior are reduced. More data on territorial species from south temperate and tropical latitudes would also be helpful, especially since recent studies suggest that the relationships between T and reproductive aggression in tropical breeders may differ from those identified in temperate zone breeders (Levin and Wingfield, 1992; Wingfield and Lewis, 1993). Hormonal comparisons of a single species throughout its range also could provide valuable information. Red-winged Blackbirds, for example, breed from Alaska to Costa Rica, with breeding seasons of various durations, with both resident and migratory populations, and with varying degrees of polygyny. Comparative information of this kind will eventually enable us to determine the generality of the ideas that have been developed about T and breeding behavior, and also permit testing of the general applicability of the "trade-off" hypothesis of T secretion (Wingfield *et al.*, 1990). As for the Challenge Hypothesis, the preponderance of evidence compiled here is supportive.

Because of the intimate association of reproductive behavior and androgen secretion and action, knowledge of the relationships between behavior and hormone levels will aid not only the understanding of avian physiology, but will provide insights into ultimate causation of aspects of territorial and breeding strategies. Because T influences the type of aggressive behavior involved in territorial contests, of special interest will be determining in various species the role of T in all phases of territory establishment and maintenance. If territories are limited in number, do males with higher T levels have a competitive advantage? How does T affect the extremely rapid but crucial behavioral transition from floater to territory owner when floaters find vacant territories? Similarly, given the recent discovery of the high frequency of EPCs in "monogamous" as well as polygynous systems, are males with higher T better "mate-guarders," i.e., do they experience fewer instances of cuckoldry? Or are they better at garnering EPCs on adjacent territories? If regularly monogamous species can be induced to become polygynous by the administration of exogenous T, might some groups of species be permanently relegated to monogamy by endocrinologic constraints as well as ecological ones? Investigations of the relationship between T and polygynous breeding should also prove useful for improving our understanding of polygyny itself—the trade-offs associated with its evolution and maintenance. The manipulation of testosterone levels in wild birds in particular permits new lines of experimental inquiry where previously only observational and correlational approaches existed.

Many of the questions posed above hinge on potential advantages that accrue to males able to achieve, sustain, and exploit high plasma T levels. Future studies that address benefits but also physiological and reproductive costs of high T will be especially valuable. Such studies may eventually yield insights into the question of why so few temperate zone species regularly breed polygynously, and could complement ecological explanations of the phenomenon. Finally, future investigations may be most beneficial when they examine androgen cycles in the field, formulate hypotheses of hormone–behavior relationships, and then test the hypotheses in both natural settings and in the more controlled environmental conditions of the laboratory.

8. SUMMARY

We reviewed known and hypothesized relationships between breeding-season testosterone (T) levels and avian polygyny by describing and comparing seasonal male T profiles in species employing different nonmonogamous breeding strategies, within-species variations in T profiles, and relationships between T levels and breeding ecology. We found that:

1. Compared to monogamous males, polygynously-breeding males generally have much longer periods of high T levels, which coincide with prolonged periods of advertisement, mate acquisition, mate-guarding, and/or territory defense.
2. Mean and peak seasonal male T levels do not vary greatly among species with different mating systems.
3. Male territory owners have higher T levels than nonterritorial adult males (floaters) when floaters do not strongly challenge owners for territories (e.g., Red-winged Blackbird), but that T levels of owners and floaters are equivalent when floaters do challenge owners (e.g., Yellow-headed Blackbirds).
4. Preterritorial juvenile males generally have lower T levels than adults. Peak seasonal T levels do not change as territorial males age.
5. Experimentally increasing circulating T levels in males generally leads to individuals that are more aggressive than are controls, but not to changes in territorial status or social rank. However, experimentally increasing T levels in territorial males that are normally monogamous can shift their mating toward polygyny.
6. Annual reproductive success of males may in some cases be correlated with plasma T levels.
7. Many of these findings provide support for the idea that high breeding season levels of circulating T in birds are stimulated by aggressive social interactions, i.e., the Challenge Hypothesis.
8. Results of a field experiment in which territorial male Yellow-headed Blackbirds were given T implants to increase their cir-

culating T levels supported the hypothesis that high T levels, needed for territorial aggression, are physiologically incompatible with the expression of male parental care. Males with artificially elevated T levels spent more time trying to attract additional mates, had a lower probability of feeding their nestlings, and fledged fewer young than did control males.

9. Males of species which usually breed polygynously may be inherently less sensitive endocrinologically to immediate social interactions, with respect to T secretion, because their breeding systems favor continual high T levels.
10. Results of a field experiment in which male Red-winged Blackbirds were given T implants to increase their circulating T levels supported the hypothesis that there are costs associated with maintaining high plasma T levels. Males given high T levels suffered more wounding than did controls, suggesting that they engaged in more fighting with conspecifics.

We conclude that polygynous birds have contributed much to our current understanding of the relationships between T and breeding behavior, but that more comparative studies are needed to test current hypotheses of how T and avian mating systems influence each other.

ACKNOWLEDGMENTS. We wish to thank Sievert Rohwer, Diane Steeck, and Wendy Schweizer for assistance in the field and Alfred Dufty, Gordon Orians, and Lewis Oring for reviewing this manuscript. Preparation of the manuscript was supported by National Institute of Mental Health grant MH44609.

REFERENCES

Alatalo, R. V., Carlson, A., Lundberg, A., and Ulfstrand, S., 1981, The conflict between male polygamy and female monogamy: The case of the Pied Flycatcher *Ficedula hypoleuca*, *Am. Nat.* **117**:738–753.

Arcese, P., 1987, Age, intrusion pressure, and defence against floaters by territorial male Song Sparrows, *Anim. Behav.* **35**, 773–784.

Ball, G. F., and Wingfield, J. C., 1987, Changes in plasma levels of luteinizing hormone and sex steroid hormones in relation to multiple-broodedness and nest-site density in male starlings, *Physiol. Zool.* **60**:191–199.

Balthazart, J., 1983, Hormonal correlates of behavior, in *Avian Biology*, vol. 7 (D. S. Farner and K. C. Parkes, eds.), Academic Press, New York, pp. 221–365.

Balthazart, J., and Hendrick, J., 1976, Annual variation in reproductive behavior, testosterone, and plasma FSH levels in the Rouen Duck, *Anas platyrhynchos*, *Gen. Comp. Endocrinol.* **28**:171–183.

Beletsky, L. D., and Orians, G. H., 1987, Territoriality among male Red-winged Blackbirds. I. Site fidelity and movement patterns, *Behav. Ecol. Sociobiol.* **20**:21–34.

Beletsky, L. D., and Orians, G. H., 1990, Male parental care in a population of Red-winged Blackbirds, 1983–1988, *Can. J. Zool.* **68**:606–609.

Beletsky, L. D., and Orians, G. H., 1994, Site fidelity and territorial movements in a rapidly declining population of Yellow-headed Blackbirds, *Behav. Ecol. Sociobiol.*, **34**:257–265.

Beletsky, L. D., Orians, G. H., and Wingfield, J. C., 1989, Relationships of steroid hormones and polygyny to territorial status, breeding experience, and reproductive success in male Red-winged Blackbirds, *Auk* **106**:107–117.

Beletsky, L. D., Orians, G. H., and Wingfield, J. C., 1990a, Steroid hormones in relation to territoriality, breeding density, and parental behavior in male Yellow-headed Blackbirds, *Auk* **107**:60–68.

Beletsky, L. D., Orians, G. H., and Wingfield, J. C., 1990b, Effects of exogenous androgen and antiandrogen on territorial and nonterritorial blackbirds (Aves: Icterinae), *Ethology* **85**:58–72.

Beletsky, L. D., Orians, G. H., and Wingfield, J. C., 1992, Year to year patterns of circulating levels of testosterone and corticosterone in relation to breeding density, experience, and reproductive success of the polygynous Red-winged Blackbird, *Horm. Behav.* **26**:420–432.

Best, L. B., and Rodenhouse, N. L., 1984, Territory preference of Vesper Sparrows in cropland, *Wilson Bull.* **96**:72–82.

Borgia, G., and Wingfield, J. C., 1991, Hormonal correlates of bower decoration and sexual display in the Satin Bowerbird (*Ptilonorhynchos violaceus*), *Condor* **93**:935–942.

Collis, K., and Borgia, G., 1992, Immediate and long-term effects on male aggression from testosterone implants in juvenile male Satin Bowerbirds, *Auk* **109**:422–434.

Crook, J. H., 1964, The evolution of social organization and visual communication in the weaverbirds (Ploceinae), *Behav. Suppl.* **10**:1–178.

Darley, J. A., 1982, Territoriality and mating behavior of the male Brown-headed Cowbird, *Condor* **84**:15–21.

Delville, Y., Sulon, J., Hendrick, J.-C., and Balthazart, J., 1984, Effect of the presence of females on the pituitary-testicular activity in male Japanese Quail (*Coturnix coturnix japonica*), *Gen. Comp. Endocrinol.* **55**:295–305.

Dittami, J. P., and Reyer, H.-U., 1984, A factor analysis of seasonal, behavioral, hormonal and body weight changes in adult male Barheaded Geese, *Anser indicus*, *Behaviour* **90**:114–124.

Donham, R. S., 1979, Annual cycle of plasma luteinizing hormone and sex hormones in male and female Mallards (*Anas platyrhynchos*), *Biol. Reprod.* **21**:1273–1285.

Dufty, A. M., 1982, Movements and activities of radio-tracked Brown-headed Cowbirds, *Auk* **99**:316–327.

Dufty, A. M., 1989, Testosterone and survival: A cost of aggressiveness?, *Horm. Behav.* **23**:185–193.

Dufty, A. M., 1993, Testosterone concentrations in males of an atypical species: the brood parasitic Brown-headed Cowbird, in *Avian Endocrinology: Proceedings of the Vth International Symposium on Avian Endocrinology* (P. J. Sharp, ed.), Journal of Endocrinology Ltd, Bristol, pp. 61–72.

Dufty, A. M., and Wingfield, J. C., 1986, Temporal patterns of circulating LH and steroid hormones in a brood parasite, the Brown-headed Cowbird, *Molothrus ater*. 1. Males, *J. Zool. Lond. A* **208**:191–203.

Dufty, A. M., and Wingfield, J. C., 1990, Endocrine response of captive Brown-headed Cowbirds to intrasexual social cues, *Condor* **92**:613–620.

Elliott, P. F., 1980, Evolution of promiscuity in the Brown-headed Cowbird, *Condor* **82**:138–141.
Emlen, J. T., and Lorenz, F. W., 1942, Pairing responses of free-living Valley Quail to sex-hormone pellet implants, *Auk* **59**:369–378.
Emlen, S. T., and Oring, L. W., 1977, Ecology, sexual selection, and the evolution of mating systems, *Science* **197**:215–223.
Fivizzani, A. J., and Oring, L. W., 1986, Plasma steroid hormones in relation to behavioral sex role reversal in the Spotted Sandpiper, *Actitis macularia, Biol. Repro.* **35**:1195–1201.
Fivizzani, A. J., Colwell, M. A., and Oring, L. W., 1986, Plasma steroid hormone levels in free living Wilson's Phalaropes, *Phalaropus tricolor, Gen. Comp. Endocrinol.* **62**:137–144.
Gibbs, H. L., Weatherhead, P. J., Boag, P. T., White, B. N., Tabak, L. M., and Hoysak, D. J., 1990, Realized reproductive success of polygynous Red-winged Blackbirds revealed by DNA markers, *Science* **250**:1394–1397.
Gori, D., 1984, Evolution of parental care and coloniality in Yellow-headed Blackbirds, *Xanthocephalus xanthocephalus*, Ph.D. thesis, University of Arizona, Tucson.
Gori, D. F., 1988, Adjustment of parental investment with mate quality by male Yellow-headed Blackbirds (*Xanthocephalus xanthocephalus*), *Auk* **105**:672–680.
Gowaty, P. A., 1992, The battles of the sexes: Frequency dependent reproductive tactics of females and males and the evolution of mating systems in birds and mammals, Abstract, *Intl. Behav. Ecol. Conf.*
Haartman, L. von, 1969, Nest-sites and evolution of polygamy in European passerine birds, *Ornis. Fenn.* **46**:1–12.
Hännsler, I., and Prinzinger, R., 1979, The influence of the sex-hormone testosterone on body temperature and metabolism of the male Japanese Quail (*Coturnix coturnix japonica*), *Experientia* 35:509–510.
Harding, C. F., 1981, Social modulation of circulating hormone levels in the male, *Amer. Zool.* **21**:223–231.
Harding, C. F., and Follett, B. K., 1979, Hormone changes triggered by aggression in a natural population of blackbirds, *Science* **203**:918–920.
Harding, C. F., Walters, M. J., Collado, D., and Sheridan, K., 1988, Hormonal specificity and activation of social behavior in male Red-winged Blackbirds, *Horm. Behav.* **22**:402–418.
Hegner, R. E., and Wingfield, J. C., 1987a, Social status and circulating levels of hormones in flocks of House Sparrows, *Ethology* **76**:1–14.
Hegner, R. E., and Wingfield, J. C., 1987b, Effects of experimental manipulation of testosterone levels on parental investment and breeding success in male House Sparrows, *Auk* **104**:462–469.
Hissa, R., Saarela, S., Balthazart, J., and Etches, R. J., 1983, Annual variation in the concentrations of circulating hormones in Capercaillie (*Tetrao urogallus*), *Gen. Comp. Endocrinol.* **51**:183–190.
Högstad, O., 1987, It is expensive to be dominant, *Auk* **104**:333–336.
Johnsen, T. S., 1991, Steroid hormones and male reproductive behavior in Red-winged Blackbirds (*Agelaius phoeniceus*): Seasonal variation and behavioral correlates of testosterone, Ph.D. thesis, Indiana University, Bloomington.
Lanyon, W. E., 1957, The comparative biology of the meadowlarks (*Sturnella*) in Wisconsin, *Publ. Nuttall Ornithol. Club*, No. 1.
Levin, R. N., and Wingfield, J. C., 1992, The hormonal control of territorial aggression in tropical birds, *Ornis Scand.* **23**:284–291.
Lewis, J. C., 1973, *The World of the Turkey*, J. B. Lippincott, Philadelphia.

Lisano, M. E., and Kennamer, J. E., 1977, Seasonal variations in plasma testosterone level in male eastern Wild Turkeys, *J. Wildl. Manage.* **41:**184–188.

Marler, C. A., and Moore, M. C., 1988, Evolutionary costs of aggression revealed by testosterone manipulations in free-living male lizards, *Behav. Ecol. Sociobiol.* **23:** 21–26.

Maynard Smith, J., and Parker, G. A., 1976, The logic of asymmetric contests, *Anim. Behav.* **24:**159–175.

Milonoff, M., Hissa, R., and Silverin, B., 1992, The abnormal conduct of Capercailles *Tetrao urogallus, Horm. Behav.* **26:**556–567.

Moore, M. C., 1983, Effect of female sexual displays on the endocrine physiology and behavior of male White-crowned Sparrows, *Zonotrichia leucophrys, J. Zool.* (*London*) **199:**137–148.

Moore, M. C., 1984, Changes in territorial defense produced by changes in circulating levels of testosterone; a possible hormonal basis for mate-guarding behavior in White-crowned Sparrows, *Behaviour* **88:**215–226.

Moore, M. C., 1991, Application of organization-activation theory to alternative male reproductive strategies: A review, *Horm. Behav.* **25:**154–179.

Moore, M. C., and Kranz, R., 1983, Evidence for androgen independence of male mounting behavior in White-crowned Sparrows (*Zonotrichia leucophrys gambelli*), *Horm. Behav.* **17:**414–423.

Nolan, V., Ketterson, E. D., Ziegenfus, C., Cullen, D. P., and Chandler, C. R., 1992, Testosterone and avian life histories: Effects of experimentally elevated testosterone on prebasic molt and survival in male Dark-eyed Juncos, *Condor* **94:**364–370.

Orians, G. H., 1969, On the evolution of mating systems in birds and mammals, *Am. Nat.* **103:**589–603.

Orians, G. H., 1972, The adaptive significance of mating systems in the Icteridae, *Proc. 15th Intl. Ornithol. Congr.*, pp. 389–398.

Orians, G. H., 1980, *Some Adaptations of Marsh-nesting Blackbirds*, Princeton University Press, Princeton, NJ.

Orians, G. H., and Beletsky, L. D., 1989, Red-winged Blackbird, in *Lifetime Reproduction in Birds* (I. Newton, ed.), Academic Press, New York, pp. 183–197.

Orians, G. H., and Christman, G. M., 1968, A comparative study of the behavior of Red-winged, Tricolored, and Yellow-headed Blackbirds, *Univ. Calif. Publ. Zool.* **84:**1–85.

Oring, L. W., 1982, Avian mating systems, in *Avian Biology*, vol. 6 (D. S. Farner, J. R. King, and K. C. Parkes, eds.), Academic Press, New York, pp. 1–92.

Oring, L. W., and Fivizzani, A. J., 1991, Reproductive endocrinology of sex role reversal, *Proc. 20th Intl. Ornithol. Congr.*, Christchurch, pp. 2072–2080.

Oring, L. W., Fivizzani, A. J., Colwell, M. A., and El Halawani, M. E., 1988, Hormonal changes associated with natural and manipulated incubation in the sex-role reversed Wilson's Phalarope, *Gen. Comp. Endocrinol.* **72:**247–256.

Oring, L. W., Fivizzani, A. J., and El Halawani, M. E., 1989, Testosterone-induced inhibition of incubation in the Spotted Sandpiper (*Actitis macularia*), *Horm. Behav.* **23:**412–423.

Patterson, C. B., Erckmann, W. J., and Orians, G. H., 1980, An experimental study of parental investment and polygyny in male blackbirds, *Am. Nat.* **16:**757–769.

Parker, G. A., 1974, Assessment strategy and the evolution of fighting behavior, *J. Theor. Biol.* **47:**223–243.

Ramenofsky, M., 1984, Agonsitic behavior and endogenous plasma hormones in male Japanese Quail, *Anim. Behav.* **32:**698–708.

Rohwer, S., and Rohwer, F. C., 1978, Status signalling in Harris' Sparrows: Experimental deceptions achieved, *Anim. Behav.* **26:**1012–1022.

Sakai, H., and Ishii, S., 1986, Annual cycles of plasma gonadotropins and sex steroids in Japanese Common Pheasants, *Phasianus colchicus versicolor, Gen. Comp. Endocrinol.* **63:**275–283.

Searcy, W. A., 1981, Sexual selection and aggression in male Red-winged Blackbirds, *Anim. Behav.* **29:**958–960.

Silverin, B., 1980, The effects of long-acting testosterone treatment on free-living Pied Flycatchers, *Ficedula hypoleuca*, during the breeding period, *Anim. Behav.* **28:** 906–912.

Silverin, B., 1983, Population endocrinology and gonadal activities of the male Pied Flycatcher (*Ficedula hypoleuca*), in *Avian Endocrinology: Environmental and Ecological Perspectives* (S. Mikami, K. Homma, and M. Wada, eds.), Japan Scientific Societies Press, Tokyo, and Springer-Verlag, Berlin, pp. 289–305.

Silverin, B., 1993, Territorial aggressiveness and its relation to the endocrine system in the Pied Flycatcher. *Gen. Comp. Endocrinol.* **89:**206–213.

Silverin, B., and Wingfield, J. C., 1982, Patterns of breeding behavior and plasma levels of hormones in a free-living population of Pied Flycatchers, *Ficedula hypoleuca, J. Zool. Lond. A* **198:**117–129.

Smith, S. M., 1978, The "underworld" in a territorial sparrow: Adaptive strategy of floaters, *Am. Nat.***112:**571–582.

Snyder, W. D., 1984, Ring-necked Pheasant nesting ecology and wheat farming on the high plains, *J. Wildl. Manage.* **48:**878–888.

Stokkan, K.-A., Sharp, P. J., 1980, Seasonal changes in the concentrations of plasma luteinizing hormone and testosterone in Willow Ptarmigan (*Lagopus lagopus lagopus*) with observations on the effects of short days, *Gen. Comp. Endocrinol.* **40:**109–115.

Teather, K. L., and Robertson, R. J., 1986, Pair bonds and factors influencing the diversity of mating systems in Brown-headed Cowbirds, *Condor*, **88:**63–69.

Thompson, C. W., and Moore, M. C., 1992, Behavioral and hormonal correlates of alternative reproductive strategies in a polygynous lizard: Tests of the relative plasticity and challenge hypotheses, *Horm. Behav.*, **26:**568–585.

Trobec, R. J., and Oring, L. W., 1972, Effects of testosterone propionate implantation on lek behavior of Sharp-tailed Grouse, *Am. Midl. Nat.* **87:**531–536.

Verner, J., 1964, Evolution of polygamy in the Long-billed Marsh Wren, *Evolution* **18:** 252–261.

Verner, J., and Willson, M. F., 1969, Mating systems, sexual dimorphism, and the role of male North American Passerine birds in the nesting cycle, *Ornithol. Monogr.* **9:**1–76.

Watson, A., and Parr, R., 1981, Hormone implants affecting territory size and aggressive and sexual behaviour in Red Grouse, *Orn. Scand.* **12:**55–61.

Westneat, D. F., 1992, Do female Red-winged Blackbirds engage in a mixed mating strategy?, *Ethology* **92:**7–28.

Westneat, D. F., Sherman, P. W., and Morton, M. L., 1990, The ecology and evolution of extra-pair copulations in birds, in *Current Ornithology*, vol. 7 (D. M. Power, ed.), Plenum Press, New York, pp. 331–370.

Willson, M. F., 1966, Breeding biology of the Yellow-headed Blackbird, *Ecol. Monogr.* **36:**51–76.

Wingfield, J. C., 1984a, Androgens and mating systems: Testosterone-induced polygyny in normally monogamous birds, *Auk* **101:**665–671.

Wingfield, J. C., 1984b, Environmental and endocrine control of reproduction in the Song Sparrow, *Melospiza melodia*. I. Temporal organization of the breeding cycle, *Gen. Comp. Endocrinol.* **56:**406–416.

Wingfield, J. C., 1985a, Short-term changes in plasma levels of hormones during estab-

lishment and defense of a breeding territory in male Song Sparrows *Melospiza melodia*, *Horm. Behav.* **19**:174–187.
Wingfield, J. C., 1985b, Influences of weather on reproductive function in male Song Sparrows, *Melospiza melodia*. *J. Zool. Lond.* **205**:525–544.
Wingfield, J. C., 1988, The Challenge Hypothesis: Interrelationships of testosterone and behavior, *Proc. 19th Intl. Ornithol. Congr.*, Ottawa, pp. 1685–1691.
Wingfield, J. C., 1990, Interrelationships of androgens, aggression, and mating systems, in *Endocrinology of Birds: Molecular and Behavioral* (M. Wada, S. Ishii, and C. G. Scanes, eds.), Japan Scientific Societies press, Tokyo, and Springer-Verlag, Berlin, pp. 185–205.
Wingfield, J. C., 1991, Mating systems and hormone-behavior interactions, *Proc. 20th Intl. Ornithol. Congr.*, Christchurch, pp. 2055–2062.
Wingfield, J. C., 1994, Hormone-behavior interactions and mating systems in male and female birds, in *The Difference Between the Sexes* (R. V. Short and E. Balaban, eds.), Cambridge University Press, Cambridge, pp. 303–330.
Wingfield, J. C., and Farner, D. S., 1978a, The endocrinology of a natural breeding population of the White-crowned Sparrow (*Zonotricia leucophrys pugetensis*), *Physiol. Zool.* **51**:188–205.
Wingfield, J. C., and Farner, D. S., 1978b, The annual cycle in plasma irLH and steroid hormones in feral populations of the White-crowned Sparrow, *Zonotrichia leucophrys gambelli*, *Biol. Reprod.* **19**:1046–1056.
Wingfield, J. C., and Farner, D. S., 1979, Endocrine correlates of renesting after loss of clutch or brood in the White-crowned Sparrow, *Zonotrichia leucophrys gambelii*, *Gen. Com. Endocrinol.* **38**:322–331.
Wingfield, J. C., and Farner, D. S., 1980, Control of seasonal reproduction in temperate zone birds, *Prog. Reprod. Biol.* **5**:62–101.
Wingfield, J. C., and Farner, D. S., 1993, Reproductive endocrinology of wild species, in *Avian Biology*, Vol. 9 (D. S. Farner, J. R. King, and K. C. Parkes, eds.), Academic Press, New York, pp. 163–327.
Wingfield, J. C., and Hahn, T. P., 1994, Testosterone and territorial behaviour in sedentary and migratory sparrows, *Anim. Behav.* **47**:77–89.
Wingfield, J. C., and Lewis, D. M., 1993, Hormonal and behavioural responses to simulated territorial intrusion in the cooperatively breeding White-browed Sparrow Weaver, *Plocepasser mahali*, *Anim. Behav.* **45**:1–11.
Wingfield, J. C., and Marler, P. R., 1988, Endocrine basis of communication in reproduction and aggression, in *The Physiology of Reproduction* (E. Knobit and J. Neill, eds.), Raven Press, New York, pp. 1647–1677.
Wingfield, J. C., and Ramenofsky, M., 1985, Testosterone and aggressive behavior during the reproductive cycle of male birds, in *Neurobiology*, (R. Gilles and J. Balthazart, eds.), Springer-Verlag, Berlin, pp. 92–104.
Wingfield, J. C., and Wada, M., 1989, Changes in plasma levels of testosterone during male-male interactions in the Song Sparrow, *Melospiza melodia*: Time course and specificity of response, *J. Comp. Physiol. A* **166**:189–194.
Wingfield, J. C., Smith, J. P., and Farner, D. S., 1982, Endocrine responses of White-crowned Sparrows to environmental stress, *Condor* **84**:399–409.
Wingfield, J. C., Ball, G. F., Dufty, A. M., Hegner, R. E., and Ramenofsky, M., 1987, Testosterone and aggression in birds, *Am. Sci.* **75**:602–608.
Wingfield, J. C., Hegner, R. E., Dufty, A. M., and Ball, G. F., 1990, The "challenge hypothesis": Theoretical implications for patterns of testosterone secretion, mating systems and breeding strategies, *Am. Nat.* **136**:829–846.

Wingfield, J. C., Hegner, R. E., and Lewis, D. M., 1992, Hormonal responses to removal of a breeding male in the cooperatively breeding White-browed Sparrow Weaver, *Plocepasser mahali*, *Horm. Behav.* **26**:145–155.

Witschi, E., 1961, Sex and secondary sexual characteristics, in *Biology and Comparative Physiology of Birds* (A. J. Marshall, ed.), Academic Press, New York, pp. 115–168.

Yasukawa, K., 1978, Aggressive tendencies and levels of graded displays: Factor analysis of response to song playback in the Red-winged Blackbird (*Agelaius phoeniceus*), *Behav. Biol.* **23**:446–459.

CHAPTER 2

USING MIGRATION COUNTS TO MONITOR LANDBIRD POPULATIONS: REVIEW AND EVALUATION OF CURRENT STATUS

ERICA H. DUNN and DAVID J.T. HUSSELL

1. INTRODUCTION

Increased knowledge of North American landbird populations and concern for their status (e.g., Robbins *et al.*, 1989; Askins *et al.*, 1990; Askins, 1993) has led to greater interest in methods of monitoring population change in nongame birds. Without monitoring we are unable to document long-term change, to determine whether short-term fluctuation is within a normal range, or to evaluate the effectiveness of management.

Counts of breeding birds are generally accepted as the most useful abundance monitoring method, because they are directly tied to a particular breeding population. The Breeding Bird Survey (BBS) is the main such program in North America (Erskine, 1978; Robbins *et al.*, 1986). Breeding bird counts have limitations, however, and there is

---

ERICA H. DUNN • Canadian Wildlife Service, National Wildlife Research Centre, Hull, Quebec, Canada K1A 0H3. DAVID J.T. HUSSELL • Ontario Ministry of Natural Resources, Maple, Ontario, Canada L6A 1S9, and Environment Canada (Ontario Region), Nepean, Ontario, Canada, K1A 0H3

Current Ornithology, Volume 12, edited by Dennis M. Power. Plenum Press, New York, 1995.

lively discussion about the most appropriate protocols for counting and analysis (e.g., Sauer and Droege, 1990). Moreover, BBS coverage is sparse in many areas.

It is important to have additional types of monitoring surveys to obtain independent data to compare with BBS and to fill gaps in BBS species and population coverage. Counts of birds during migration offer one such independent monitoring method. Migration counts in North America could be especially valuable in tracking population trends in species whose breeding density is very low, that do not breed primarily in habitats sampled by a roadside survey such as BBS, or whose summer ranges are in remote areas such as the boreal forest zone. Many of the species missed by BBS also winter south of the U.S., so are unavailable for winter monitoring in the U.S. by programs such as Christmas Bird Counts. For these species, migration counting is currently the only available monitoring option.

Although they are potentially valuable for population monitoring, migration counts have been relatively little used for this purpose (Hussell, 1981). High variability in some migration counts, associated mainly with weather effects (Richardson, 1978), leads to concern that annual indices and trends derived from such data may lack precision and some authors have concluded that other methods of tracking population change are more efficient (e.g., Svensson, 1978). In part to address such concerns, the multi-agency Neotropical Migratory Bird Conservation Program ("Partners in Flight") recommended that a workshop be held to evaluate the potential of migration counts as tools in population monitoring (Butcher *et al.*, 1993). That workshop, held in 1993, resulted in recommendations to expand our ability to monitor populations during migration (Blancher *et al.*, 1993).

The aims of this paper are to summarize what we currently know about the use of migration counts to monitor populations of landbirds, to evaluate that knowledge, and to clarify what we still need to learn. Data collection and analysis methods are an important component of the review, because the strengths and weaknesses of different procedures have a bearing on the probable ability of a given migration count to track populations. The review includes European literature, but concentrates on the status and future of migration monitoring in North America.

2. WHAT IS A "MIGRATION COUNT?"

For the purposes of this paper, a *migration count* includes any day's tally of landbirds on spring or fall passage; a broad definition that covers nocturnal or diurnal migrants counted at stop-over sites, flying past a fixed point, captured, or seen on radar screens. *Landbird* refers to all species that are mainly terrestrial (including raptors), and thus excludes waterfowl, shorebirds, seabirds, and other waterbirds.

The broadness of our definitions for *migration count* and *landbird* mean that this paper will review data from a variety of species groups, gathered by very different methods. Although we limit the scope of this paper to landbirds, some of the methods and ideas we discuss may be applicable to other groups of birds. This section examines the degree to which all migration counts are similar and can thus be discussed together.

Nocturnal migrants are usually counted at stopover sites following a migratory flight, but may also be detected during migration, for example by night-time trapping (Slack *et al.*, 1987; Duffy and Kerlinger, 1992), with radar (Richardson, 1978), or with other special methods for counting birds at night (e.g., Lowery and Newman, 1955; Graber and Cochrane, 1960; Russell *et al.*, 1991; Svazas, 1991). Nocturnal migrants may also be counted in flight when migration continues into daylight (Bingman, 1980; Hall and Bell, 1981; Wiedner *et al.*, 1992). Diurnal migrants are generally counted only during active flight (Bednarz and Kerlinger, 1989; Eckert, 1990).

Regardless of species, migration mode or method of tally, a daily *migration count* is only a sample of the number of birds actually migrating that day. Moreover, the proportion counted will vary each day with weather and perhaps other factors. Because data gathering and analysis procedures have to be designed accordingly, it is important to be clear on the sources of variation. We therefore define several terms in detail (illustrated schematically in Fig. 1).

The *total population* is all the individuals of a species. In most cases changes in total population cannot be monitored at a single site.

The *monitored population* is the segment of the total population that can potentially be counted at a given monitoring site. This is the population whose changes we hope to monitor. It will usually not be possible to define precise geographic boundaries for monitored populations within a broader total population (as indicated by the broken lines in Fig. 1). Although there will be some core area in which a high proportion of individuals belongs to the monitored population, there will be peripheral areas from which a low proportion of individuals is potentially countable at a given monitoring site.

It is assumed that birds follow traditional migration routes, such that the same monitored population is sampled at a given site each year. Monitoring sites across the migratory path of the total population may sample monitored populations that are distinct or overlapping, depending on the extent to which individuals breeding or wintering in a specific area follow the same or different routes.

The *migrating population* is the daily pulse of the monitored population that can potentially contribute newly-arrived individuals to a count on a given day (Fig. 1). The sum of daily migrating populations over the entire season equals the monitored population for that site.

Although the migrating population passes over, through or near the monitoring site, many individuals in it cannot be counted, because they pass over the site in nocturnal migration without landing, fly too high or too far away to be counted in diurnal migration, or appear outside a standardized counting or trapping period. The number of birds that is potentially countable, we call the *count population* (shaded in Fig. 1).

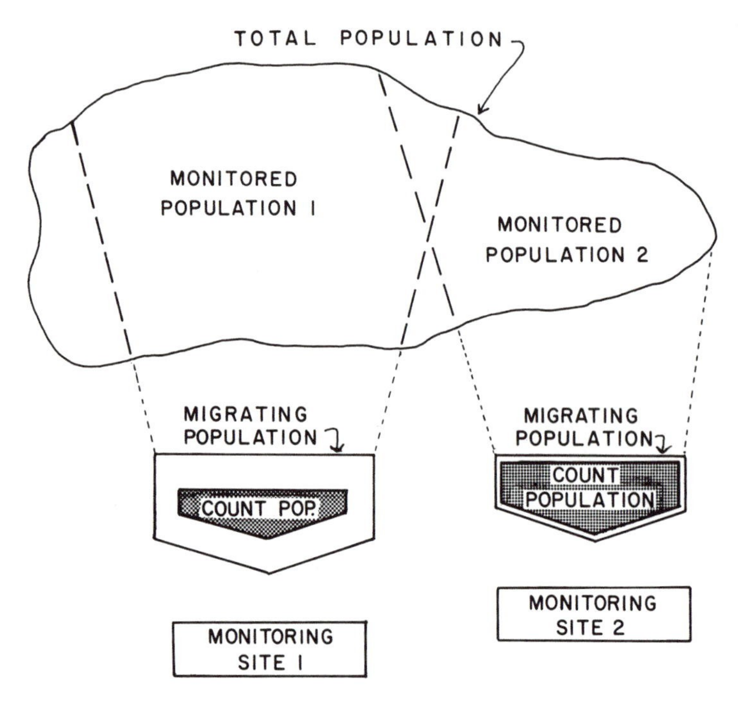

FIGURE 1. Schematic illustration of terms (see text). *Monitored population* is that portion of the *Total population* that may migrate over a given monitoring site. *Migrating population* is the portion of the monitored population moving into or past the monitoring site in 24 hours. *Count population* (shaded) is the portion of the daily migrating population potentially countable at the monitoring site. The *migration count* is that portion of the count population actually detected during the daily count period.

The daily *migration count* is the subset of individuals from the count population that is actually detected and recorded. As defined here, then, each day's count is an independent sample of the monitored population, as assumed by most analysis methods (see Section 4). To the extent that daily counts include nonmigrants or lingering stopovers from earlier influxes, this assumption is violated.

The relationships among these defined groups is not constant. For example, the proportion of the count population included in the migration count will vary from day to day depending on effort, observer skills, weather (e.g., visibility) and seasonal effects (such as degree of leaf-out in spring). Hunger levels of migrants or local food supply might affect foraging behavior and therefore visibility of migrants. The objective of standardizing procedures for counting or trapping is to ensure as high a level as possible of day-to-day and year-to-year consistency in the proportion of the count population that is included in the migration count.

In addition, the *count population* is not a constant proportion of the *migrating population* (e.g., Graber and Cochrane, 1960; Zalakevicius, 1991). An extreme example for nocturnal migrants is the "fall-out," when a large migratory flight ends abruptly at a weather front, causing a much larger proportion of migrants than usual to land at a particular spot (Richardson, 1978; Riddiford, 1985). Weather conditions may cause diurnal migrants such as hawks to change altitude of migration (adding or removing them from the count population), or may concentrate them along the coastlines or mountain ridges where many hawk counts take place (e.g., Murray, 1964; Alerstam, 1978; Kerlinger and Gauthreaux, 1984; Gauthreaux, 1991). The proportion of the migrating population represented in the count population may also vary because of changes in attractiveness of habitat.

Lastly, the *migrating population* is not a constant proportion of the *monitored population*, because migration often proceeds in uneven pulses in response to weather conditions and seasonal phenology (e.g., Richardson, 1978; Titus and Mosher, 1982). A day of heavy migration is often followed by several days of relative inactivity.

A principal problem to be addressed for any type of migration count, then, is whether the "noise" introduced by variation in the proportions of the monitored population available to be counted is an insurmountable obstacle to detecting change in the true size of that population.

When migrating birds are counted from 24-hour films of radar screens, the *migration count* potentially equals the *migrating population*, removing much confounding variation. However, birds may fly at different heights according to weather conditions, and radar picks up targets at each altitude with varying sensitivity. Low altitude migrants are especially hard to detect. Moreover, flocks of birds are frequently recorded as a single radar signal, even though flock size varies (Richardson, 1978). Even so, radar should provide as consistent a method of sampling migration as other methods (trapping and visual counts), provided that the equipment itself remains unchanged in its sensitivity from day to day and year to year.

The most serious problem with radar counts from the perspective of population monitoring is that signals can rarely be assigned to species (Richardson, 1978). Methods that might be used in conjunction with radar to identify the nocturnal migrants on radar screens include looking at birds flying through strong lights or across the face of the moon, or recording flight calls (Lowery and Newman, 1955; Graber and Cochrane, 1960; Svazas, 1991). However, most individuals observed cannot be identified to species (Richardson, 1978). Many calls cannot yet be ascribed to particular species, nor is it known whether all species call at constant rates (Graber and Cochrane, 1960; Hamilton, 1962). An identified sample of birds close to the ground may not represent species composition at higher altitudes. Because radar methodology is not currently useful for monitoring individual species, it will not be discussed further in this review.

In summary, a *migration count* is defined here as a day's tally of migrants at a single site. No matter how the count is made, it will not correspond consistently to the *migrating population* (numbers actually migrating at or past the site that day), and the daily *migrating population* is not a consistent proportion of the *monitored population* (the population whose changes we wish to detect). Our challenge is to determine whether these problems can be overcome by appropriate data collection and analysis. The problems and solutions are similar for all species groups, whether they are nocturnal or diurnal migrants, so hereafter we discuss them together.

3. COLLECTION OF MIGRATION COUNT DATA

This section describes count methods, their advantages and disadvantages in estimating numbers of migrants, and the ways in which they can be standardized. For each count method we note the major sources of data currently available.

3.1. Choice of Site

Many long-term migration counts have been established at sites where migrants are concentrated by topography, in part because volunteers are easily attracted to such sites to count birds year after year. Coastal sites are favored by bird observatories (e.g., Hagan et al., 1992; Hussell *et al.*, 1992; Pyle *et al.*, 1994), but migrants can also be concentrated at inland sites. Mountain ranges form leading lines and areas of updrafts where hawks migrate (e.g., Bednarz *et al.*, 1990), and mountain passes can be good sites for observation of nocturnal migrants (Hall and Bell, 1981; Jenni and Naef-Daenzer, 1986). Concentrations can also occur because of habitat configuration, particularly when the site represents an isolated area of relatively good stopover habitat within an otherwise unfavorable landscape (e.g., urban greenbelt, oases, or forested areas surrounded by agricultural land or marshes; Nisbet *et al.*, 1963; Nisbet and Drury, 1969; Martin, 1980; Laske and Helbig, 1987; Bairlein, 1992). Concentration of migrants is a matter of degree, and can vary from mild (e.g., where suitable stopover habitat is patchily distributed) to extreme (usually due to topographical features).

Where concentration is caused by topography, birds tend to concentrate regardless of the suitability of the stopover habitat. This offers an advantage for long-term monitoring, because minor changes in habitat over time are unlikely to affect the proportion of the *migrating population* represented in the *count population* (Section 2). A further advantage of many (but not all) concentration sites is that they are often at exposed coastal or montane localities lacking dense vegetation. At such sites birds are visible and easily counted during the day, and are likely to continue migration as soon as conditions allow (Kuenzi *et al.*, 1991; Ehnbom *et al.*, 1993; Jean and Razin, 1993). There may be few local breeders to confuse with migrants, and daily counts are minimally contaminated with repeat counts of residents or of the same transient individuals. These characteristics make analysis more straightforward (Section 4). Moreover, migration counts may be more consistent in open habitat where ingress to or egress from the study area can be distinguished from milling about. This may occur not only with ongoing migration, but when birds that are stopping over seek preferred habitat (Bingman, 1980; Herremans, 1989; Moore *et al.*, 1990; Bastian and Berthold, 1991; Wiedner *et al.*, 1992), or move in response to food distribution or competitive interactions (Martin, 1980).

Migration counts at coastal concentration points are subject to particularly large weather-related variations in the proportion of the *mi-*

grating population that is represented in the *count population* (Lack, 1959; DeSante, 1983; Riddiford, 1985). This is partly a matter of catchment area; coastal locations not only get birds "falling out" from migration overhead, but they also collect birds that find themselves over water when ready to land (Wiedner *et al.*, 1992; Ehnbom *et al.*, 1993). Autumn samples of migrants at coastal sites are often biassed towards young of the year (Ralph, 1971; Dunn and Nol, 1980; Ehnbom *et al.*, 1993), and coastal "waves" may actually represent relatively small movements of inexperienced birds (Nisbet and Drury, 1969). If coastal sites are on the edges of migration routes (Ralph, 1981), those sites might sometimes be bypassed by most migrants. On the other hand, coastal and nearby inland sites may show parallel fluctuations in bird numbers despite differences in total volume of birds (Forsythe and James, 1971). Moreover, the exaggerated variation in numbers at coastal sites may not affect ability to detect population change (Section 5).

Migration counts are also conducted at sites where birds are not concentrated, including many mist-netting operations run by individuals or nature centers. Day-to-day variation in bird numbers may be relatively low, and total migratory volume can be 5–10 times lower than at concentration points (Nisbet *et al.*, 1963; Ehnbom *et al.*, 1993). There are some disadvantages of such sites for long-term monitoring. Sample sizes may be too small to analyze or to hold the long-term interest of volunteers. Local breeders, dispersing young-of-the-year and molting birds may be confused with true migrants (Rappole and Ballard, 1987; Young, 1991; Winker *et al.*, 1992; Baker, 1993), which complicates analysis, whereas concentration sites have a higher proportion of true migrants. At suitable stopover sites, habitat probably has a stronger effect on species composition, abundance and stopover time than it does at more exposed concentration points (Hutto, 1985; Weisbrod *et al.*, 1993; Dunn *et al.*, *in press* a). This raises the danger that long-term trends in bird numbers at good stopover sites may reflect change in habitat rather than in size of the monitored population. Nonetheless, at least some sites without particular concentrating effects on migrants may prove suitable for detecting long-term population change (see Section 5).

3.2. Trapping

In this paper *trapping* refers to any method of capture, including mist-netting.

There are advantages to using the number of birds trapped as the sample of the count population: data collection is readily standardized, and consistent results can be obtained from personnel with a wide range of experience beyond basic bird identification and handling skills. Handling birds allows aging and sexing, which is important in certain analyses of population dynamics, and it may be possible to identify the breeding ground through external or genetic characteristics (Bergstrom and Drury, 1956; H. Lisle Gibbs, *pers. comm.*). Some species that are hard to record in other ways can be sampled by trapping (e.g., skulkers, owls). Lastly, recapture data can be used to estimate the number of birds stopping over from previous days. This is an important consideration, as variation in stopover times of migrants (day to day or year to year) could distort migration counts. Moreover, certain analyses assume that each daily count is an independent sample of the *monitored population*, an assumption that is violated if the same individuals are included in successive counts.

On the other hand, total reliance on capture totals has certain disadvantages. Wind or rain can preclude trapping even though large numbers of migrants may be present. Some habitats and/or species are poorly sampled (such as aerial insectivores). Trap location and vegetation height affect numbers and kinds of species caught (Payevsky, 1990), so trap location must be standardized and vegetation controlled. Foraging behavior of birds that affects catchability may vary with weather and seasonal factors such as phenology of vegetation.

Trapping can be standardized in several ways to reduce variability in capture totals caused by uneven effort, and habitat can be managed to keep it the same over time. Standards should address number and type of nets and other traps (including net length and mesh size; see Pardieck and Waide, 1992), hours of operation, frequency of trap checks and type of allowed attractants (such as baits, bird feeders or water drips). Supervision of participants may be an important part of attaining and maintaining standards (training volunteers, preparing an operations manual, checking results, providing feedback to volunteers, etc.).

The aim of trapping for population monitoring purposes is to get consistent samples (not necessarily maximum numbers) of the *count population* (the potentially countable birds in or passing through an area in a day). Standards of coverage should therefore be set at a level achievable with the personnel normally available. Captures resulting from extra, nonstandard efforts should be coded separately in the records from those obtained by standard coverage so they can be excluded from analysis.

There are dozens of individual banders in North America who could potentially provide monitoring data, but relatively few have operated regularly during migration periods for a decade or longer (probably fewer than 25; Dawson, 1990). Moreover, degree of consistency within and between seasons has probably been low in most cases. Where good quality data exist, results from stations run by individuals may have monitoring potential.

North America's bird observatories and cooperatively run banding stations generally have longer runs of complete banding data than do individuals (Hagan *et al.*, 1992; Hussell *et al.*, 1992; Pyle *et al.*, 1994). Banding is done nearly every day and net number is usually fairly stable. However, some stations vary net numbers and locations on a short-term basis according to weather conditions, and on a longer-term basis with habitat change. Daily effort may also vary, and use of supplementary traps and attractants may be wholly unstandardized. Banding totals are also available from certain hawk-watch sites, but the data are probably less standardized on average than are the longer songbird data sets.

The German-Austrian "Mettnau-Reit-Illnitz-Programm" is one of the most highly standardized systems for obtaining effort-independent banding totals. Three widely separated field stations on the same north–south migration pathway have been operated since 1974, in shore-edge and reed-bed habitat that is similar at all sites and is managed to prevent change. Nets of the same type and number are set in similar patterns at each site, about the same distance apart. They are kept open 24 hours a day and are checked hourly from dawn to dusk during the entire migration season (Berthold and Schlenker, 1975; Berthold *et al.*, 1986a, 1991). Volunteers are trained and other details are standardized to the extent possible, to minimize count variation resulting from different methods or degrees of effort.

3.3. Observations

The advantages and disadvantages of visual counting of migrants tend to be opposite to those for banding. Some species are more readily seen than captured, and more species are likely to be recorded overall. If the study area has varied habitat, observations may allow more uniform sampling than does trapping. Counting can take place in weather that precludes banding.

On the negative side, no recapture and few age/sex data are collected, and counts may often include lingering individuals as opposed to new arrivals. If individual birds are milling about, they may be counted more than once (e.g., Holthuizjen and Oosterhuis, 1985). Some species are harder to see than to trap. Variation among observers can cause greater differences in numbers observed than in numbers captured (Källender and Rydén, 1974).

Observational data should be standardized in many of the same ways as trapping, such as predetermining the number and location of watches and the times at which they should be carried out. As with captures, the objective should be to get consistent samples day to day and year to year. Different types of counts may be needed to sample different species (Wiedner *et al.*, 1992), and habitat may affect choice of methods (such as area search, transects or point counts, modified for migration counting; Ralph *et al.*, 1993). Rules can state explicitly what may be counted (sightings versus birds heard, distance of observation). Watches can be continuous, or sample various time periods (e.g., one morning and one afternoon hour-long count, or 15-minute watches every hour). Records should allow calculation of birds per unit time of watching or total birds in the area covered. If one or few observers do most counts, there should be sufficient overlap to allow observer differences to be quantified. If turnover is high, an alternative approach is to rotate observers randomly through the season to avoid a consistent observer bias. Training and/or testing could be required to ensure a minimum level of competence for all participants. Manuals help ensure that standards are understood and adhered to in a uniform way (e.g., Fuller and Titus, 1990; McCracken *et al.*, 1993). As with trapping, supervision of an organized effort that communicates goals, checks results frequently, and gives feedback to participants can be an important part of maintaining standards.

The North American migration data sets that are based on observations vary considerably in standardization. Legions of birders, particularly in spring, go out regularly to observe migrants. Many visit the same sites repeatedly each season, year after year, and keep careful records of numbers and dates. Some of these individual birders' records may be suitable for tracking populations (Bennett, 1952; Martsching, 1986, 1987; Hill and Hagan, 1991).

Considerably more standardized are a few pioneering programs that organize birders in more focused counts of migrants. The Toronto Ornithological Club, for example, has conducted a spring migrant warbler count since 1970. Volunteers select a route in a woodlot or vegetated ravine within urban Toronto that takes one-half to one hour to walk. On as many early mornings as possible from late April through early June, observers count all individuals of 21 warbler and selected other species along the route (George Fairfield, *personal communication*). BBS-style counts during migration (Van Tighem and Burns, 1984) also provide a suitable protocol for monitoring, but are probably too time-intensive for the near-daily coverage necessary for adequate sampling.

There is a long tradition of counting diurnally migrating hawks at concentration points. Early hawk watches varied tremendously in standards and degree of coverage (e.g., Hackman and Henny, 1971; Mueller *et al.*, 1988; Stedman, 1990), but most now follow standards set by the Hawk Migration Association of North America. Approximately 10–20 stations across North America have reasonably standardized long-term data (Fuller and Titus, 1990).

Some sites have substantial diurnal songbird flights (e.g., Scheider and Crumb, 1985; Laske and Helbig, 1987; Eckert, 1990; Wiedner *et al.*, 1992). The Netherlands has a network of sample points where volunteers have observed diurnal migration of passerines (and other birds) since the early 1980's. Simultaneous sunrise-to-sunset watches take place on one day every two weeks during migration seasons. A few sites have early morning counts only, but are covered much more frequently (Ward Hagemeijer, *personal communication*). In North America, however, consistent long-term counts of diurnal songbird migration have been done very rarely (e.g., Eckert, 1990).

Most bird observatories record numbers of birds seen but not banded, especially at sparsely-vegetated sites where migrants are concentrated and easily visible. Operation Baltic, a series of a half-dozen or more stations established in 1960 along the coasts of Poland and adjacent countries, assigns an observer to record sightings for 15 minutes of every hour (Busse, 1979). Other stations assign watchers to continuous observation starting near dawn and lasting up to six hours (e.g., Thunder Cape Bird Observatory, Ontario). Long Point Bird Observatory (Ontario) conducts an hour-long count each morning along a standard route at each of three sites. Nearly every observatory also records observations made less formally over the rest of the day.

Observations discussed to this point are those made at specific locations that are visited repeatedly. In addition, several regional programs have systematically collected observations from general birding activities. The North American Migration Count is a recently launched annual, one-day survey of counties, scheduled for the second Saturday in May (along the lines of Christmas Bird Counts; James Stasz, *personal communication*), and there are several checklist programs that compile observations year round (e.g., Quebec's long-running ÉPOQ program, Cyr and Larivée, 1993; the Wisconsin Checklist Project that began in 1982, Temple and Carey, 1990a). Advantages of using such data sets for monitoring (compared to systematic observations at specific sites) include large numbers of observers and broad geographic coverage. Disadvantages are lower standardization (e.g., as to date, duration, sites and habitats visited, and observer skill). Single-day spring counts probably have low monitoring potential for species on migration, however, because one day's *migrating population* is not a consistent sample of the *monitored population* (see Section 2.).

3.4. Daily Estimated Totals

"Daily estimated totals," or "DET's," are daily estimates of the total numbers of each species present in a clearly defined coverage area. All participants in the day's activities gather at day's end and jointly arrive at DET's, based on banding totals, standardized counts, and other incidental observations (Hussell *et al.*, 1992; McCracken *et al.*, 1993). The system was developed by Long Point Bird Observatory, but Point Reyes Bird Observatory (Farallons Station), California, has long recorded what are in effect DET's (DeSante, 1983; Pyle *et al.*, 1993; Pyle *et al.*, 1994), and several other North American observatories are adopting the system. British bird observatories use a similar procedure, though lacking standardized counts and usually encompassing larger areas.

The main advantage of DET's is that data from a variety of counting procedures probably give the best overall estimate of birds in an area, by drawing on the advantages of each method (Bergstrom and Drury, 1959). As examples, records of recaptured birds can be used to exclude lingering individuals from numbers of birds sighted (e.g., Pyle *et al.*, 1994), and observations can be used to adjust for variation in foraging height of birds that affects capture totals.

DET's are most successful in small, manageably censused areas with relatively open habitat, where personnel living in the area are making more or less continuous observations. They will be most consistent when based on standardized effort (e.g., a set number of hours of netting and standard observation periods). When data collection is not completely standardized, however, DET's can overcome that deficiency at least in part (Dunn et al., *in press* b), because the estimation procedure allows adjustment of numbers for variable effort (e.g., when weather conditions preclude netting).

A limitation of DET's is that they may add little to banding totals in heavily vegetated areas. Moreover, they are estimates, and are therefore somewhat subjective. Even though results are likely to give a better index of numbers present than are trapping or observation alone, individual variation doubtless affects the estimates. Training of personnel, manuals and supervision are especially important in maintaining consistency, and component procedures (banding, observation) should be as standardized as possible.

3.5. Summary of Data Collection Section

Sites for migration stations require careful assessment, as many of their features can affect ability to collect consistent, long-term data sets suitable for tracking population change.

Migration count data can be collected in many ways, including capture or visual counts, or with a combination of several procedures. Different strengths and weaknesses are associated with each method. Count methods and effort will have effects on the proportion of the *count population* that will be included in the *migration count*. Therefore standardization in all of these components is crucial for obtaining consistent counts from day to day and year to year. Regardless of the methods used in counting, the end product is one day's *migration count*. Thus the same types of analyses will be appropriate for all kinds of counts and methods of data collection.

4. ANALYSIS OF MIGRATION COUNT DATA

4.1. Annual Indices

The first step in analysis of migration counts to detect change in bird populations is to calculate an annual index of abundance. This index should reflect the size of the *monitored population* in a given year. However, there are many reasons to believe that the daily *migration count* does not represent a constant proportion of the monitored population (see Section 2.). Methods of calculating annual indices should address this problem by attempting to remove variability that is unrelated to population size.

Many authors calculate annual indices of abundance simply by summing daily numbers across the migration season (Spofford, 1969; Dalberg Peterson, 1976; Nagy, 1977; Hjört and Lindholm, 1978; Svensson, 1978, 1985; Jones, 1986; Stewart, 1987; Titus and Fuller, 1990; Jean and Razin, 1993). Other authors have calculated the sum of daily bird numbers recorded per unit of effort (Hackman and Henny, 1971; Abraszewska-Kowalczyk, 1974; Österlöf and Stolt, 1982; Titus and Mosher, 1982; Hussell, 1985; Bednarz *et al.*, 1990). Effort-corrected indices are usually limited to 90% of the season (or some other cut-off point) to ensure that variation of effort in the "tails" of the season does not have a disproportionate effect on the annual index. Unadjusted sums of birds may be better than effort-corrected indices for certain species, however, such as those that are about equally countable regardless of effort (Butcher and McCulloch, 1990; Titus and Fuller, 1990).

There are several potential problems in summing daily counts to obtain annual indices. Gaps in coverage can have a large effect on the seasonal sum, particularly if days missed are at the peak of a migration season. Missing days can be dealt with to some extent by using the mean number of birds per day as the index (instead of the sum), or by substituting the average number expected on a given date before calculating the sum.

Annual indices will be strongly influenced by unusually high counts that occur occasionally in some years but not others. Daily migration counts taken over a season at major concentration points are typically distributed as in the example in Fig. 2, with a few very high counts skewing the distribution. These big counts are the ones that most likely result from special weather conditions causing fallouts in nocturnal migrants or causing unusual spatial concentration in diurnal migrants (Murray, 1964; Richardson, 1978; DeSante, 1983; Kerlinger and Gauthreaux, 1984; Riddiford, 1985). In these conditions the *count population* represents a higher than normal proportion of the daily *migrating population*. When daily numbers are summed, the annual totals and daily mean can be substantially altered simply by addition or subtraction of a few large counts, even though overall distribution of

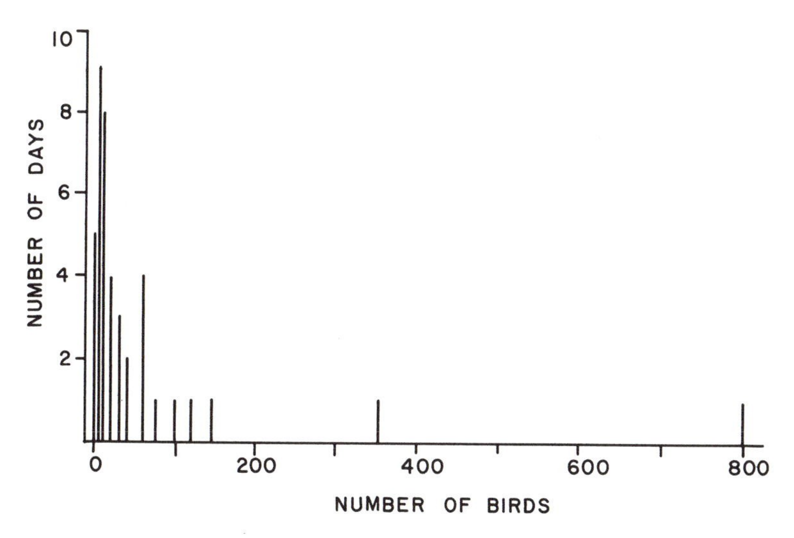

FIGURE 2. An example of skewed distribution in daily migration counts. "Number of birds" are Daily Estimated Totals (DET's, see text) for White-throated Sparrow at one site at Long Point Bird Observatory in spring, 1971 (from first through last date of sighting in season). Vertical axis shows frequency of occurrence of each value of the DET.

daily numbers scarcely changes. In Fig. 2, for example, removal of the single count of 800 reduces the mean from 34 to 15, a decrease of 56%.

Such dramatic, weather-induced day-to-day variation is sometimes considered "inherent" to migration count data, with no remedy for reducing it beyond standardizing the data collection methods (e.g., Titus *et al.*, 1989). However, there are simple and statistically justifiable methods for reducing the problem. Most commonly used is log transformation, which tends to normalize daily migration counts (Blokpoel and Richardson, 1978; Hussell, 1981; Titus and Mosher, 1982; Payevsky, 1990; Hagan *et al.*, 1992; Hussell *et al.*, 1992; Berthold et al., 1993; Pyle *et al.*, 1994). The annual index is calculated as the geometric mean (arithmetic mean of the transformed counts, back-transformed). Missing days can be dealt with as discussed earlier. In practice, a constant (often 1) is first added to the migration count to permit transformation of counts of zero. Thus, the annual index is:

$$\text{Antilog}\left\{\frac{\Sigma \log (n_i + 1)\}}{I}\right\} - 1$$

where n_i is the *migration count* on day i and I is the number of counts (i = 1, 2, . . . I).

Compared to arithmetic means, the geometric mean is relatively insensitive to occasional unusually high counts, while still reflecting small increases or decreases that are consistently present in daily counts across the entire migration season. Log-transformation also converts variability from a multiplicative scale (likely to be the case for migration counts) to an additive one (as required in certain statistical analyses, see below).

The annual indices discussed so far may still contain extraneous variability related to year-to-year differences in the timing of missing data and in weather conditions that affect relationships among the *count population*, the *migrating population*, and the *monitored population* (Section 2). Indices for years in which coverage was missed in the tails and the peak of the season will tend to be overestimated and underestimated, respectively. A season with more days than usual of wind directions that concentrate birds at a given site will produce a larger annual index even when actual size of the monitored population does not change. Analysis of covariance (ANCOVA) has been used to help identify and remove some of the additional extraneous influences on migration counts (Hussell, 1981, 1985; Hussell *et al.*, 1992; Dunn et al., *in press* a, b; Pyle *et al.*, 1994). In the method described initially by Hussell (1981), annual abundance indices are calculated from regressions designed to partition variability in log-transformed spring or fall daily migration counts among variables for date in the season, weather, moon phase, year, and site (when migration counts are combined from different nearby areas that provide independent samples of the same monitored population). The procedure detects consistent influences of each variable on the transformed daily counts (e.g., fewer birds are normally recorded early and late in the season than in the middle; higher than average temperature on a given day in spring often brings more birds than otherwise expected; some species consistently occur more abundantly at one site than another). The coefficients of the dummy variables for the year that are estimated in the regression are used to calculate annual indices, which represent the daily count that would have occurred that year at a hypothetical average site on a standardized date under average weather and lunar conditions.

Pyle *et al.* (1994) used a two-step procedure to adjust annual count totals for the effects of date, weather, and lunar variables. They first used regression models to estimate average effects of each environmental variable over all species and years, then adjusted the daily migration counts of each species for the effects of those variables, and summed the adjusted counts within years.

Results show gross similarities in weather effects on daily numbers across all species (Pyle *et al.*, 1993). However, the specific combination of weather factors affecting migration counts differs among individual species (Hussell, 1981; D.J.T. Hussell, unpubl.). In addition, effects of weather on daily counts can differ among monitoring stations only a few kilometers apart, probably as a result of geographical configuration. Both these points are supported by a study of weather effects on bird numbers at British bird observatories (Darby, 1985). There is therefore little hope of developing universal weather corrections.

The ANCOVA analysis approach offers numerous advantages. The models allow curvilinear fits of data (if higher order variables are included), and the effects of many variables can be considered simultaneously. Because the models take account of date effects, missing days should not substantially affect the accuracy of annual indices, although their precision will be reduced because of smaller sample sizes. The procedure allows simultaneous consideration of data from different sites (each site represented by appropriate dummy and interaction variables) so that conditions specific to each site are taken into account. An added benefit is the detailed elucidation of weather conditions that cause large numbers of each species to appear at particular sites.

Even at sites that do not show dramatic day-to-day variations in numbers, it would probably be prudent to examine the data for weather effects before adopting a simpler method of calculating annual indices, such as the arithmetic and geometric mean methods described earlier. Analysis of daily fall banding captures from two inland sites at Kalamazoo, Michigan (E. H. Dunn, D.J.T. Hussell, and R. J. Adams unpubl.) indicated that at least one weather variable (out of 13) had significant effects on daily numbers of migrants in 16 of the 17 species examined.

Removal of extraneous, weather-related variability can result in dramatic changes in annual indices. In an analysis of hawk counts, the coefficient of variation (c.v.) among annual indices was reduced in Broad-winged Hawk (*Buteo platypterus*) from 59% (when indices were derived from summed daily counts) to 26% (indices derived from regression analysis). C.v. among annual indices for Sharp-shinned Hawk (*Accipiter striatus*) fell from 69% to 12% (Hussell, 1985). In both cases, most of the change resulted from downward correction of a single very high annual index, indicating that weather conditions causing big counts did not occur evenly among years (as was suggested might be the case by Bednarz and Kerlinger, 1989). The reduction in c.v. can have a large effect on the number of years of data required to detect a significant trend. For example, trends of weather-adjusted indices had higher significance levels than trends in indices based on raw counts, in 70% of 39 species tested (Pyle *et al.*, 1994).

Despite the promise shown by incorporation of weather into calculation of annual indices, there are drawbacks. The method is more time-consuming and expensive than calculation of simpler indices, and requires greater analytic skills. The data analyst must be knowledgable enough to understand and observe the assumptions of linear regression, transforming data appropriately, deleting inadequate data sets and choosing appropriate variables to include in the analysis (Richardson, 1974; Rao, 1981; Hussell *et al.*, 1992). These requirements generally preclude grass-roots analysis by the people who collect the data. Sample size (number of daily counts) is an important consideration in determining allowable number of independent variables, and several years of data are likely to be needed before the process of identifying consistent weather effects can begin.

Further research should be done on which weather variables to include in ANCOVA analyses. Weather can cause migrants to be more concentrated at a sampling site (e.g., Murray, 1964; Kerlinger and Gauthreaux, 1984), causing a greater proportion of the *migrating population* to enter the *count population* (Section 2). In this case it is appropriate to identify and remove the weather-related variability to the extent possible, as discussed above. However, certain weather conditions may simply stimulate migration, concentrating birds temporally in migratory pulses (Richardson, 1978) but leaving spatial distribution unchanged. A temporal pulse represents an increase in the daily *migrating population* over what would otherwise be expected at the monitoring site on a particular day, but does not necessarily change the proportion of that migrating population represented in the *count population*. All else being equal, if a weather factor stimulates only a temporal pulse, the total of all *count populations* over the season should be the same whether birds are concentrated into a few days or spread out over a longer period. If this is so, then indices calculated without consideration of these weather variables should perform better than those based on models that do include them, especially if the mean values of those variables differ from year to year. Many weather factors may concentrate migrants both temporally and spatially, however, complicating the picture further. Several avenues exist to exploring these questions further, and the subject should be followed up.

In summary, the use of ANCOVA models to correct migration counts for weather effects is relatively recent and needs further development. To date, however, it offers the best opportunity for separating changes in migrant numbers that actually reflect population change from those that result from confounding factors.

4.2. Trends

Once annual indices have been calculated, further analysis is required to determine whether they show trends over time. Visual examination may strongly suggest a population change, but if annual fluctuation is high and/or sample sizes are low, the trend may not be statistically significant. While statistical details of trend analysis are beyond the scope of this review, we can identify the important issues involved and briefly mention methods that have been applied to migration counts. The choice of method for determining trends is not trivial, as different methods can generate conflicting trends from the same data (e.g., Titus and Fuller, 1990).

A simple way to detect whether a significant population change has occurred is to compare the mean value of annual indices from an early set of years to the mean for a later group (e.g., Hackman and Henny, 1971, [although no statistical test was carried out]; Hussell, 1981; Bednarz and Kerlinger, 1989; Hagan *et al.*, 1992). Significance of difference between the annual abundance indices for any pair of years can be checked by applying t-tests or nonparametric means tests to the daily counts. The latter test assumes daily migration counts are independent samples of the *monitored population*, an assumption violated to the extent that lingering transients from a previous days' *migrating population* are included in subsequent migration counts. This may be a problem at some monitoring sites but not others (Section 3.1).

More commonly, linear regression is used to fit a straight line to annual indices (Berthold *et al.*, 1986b; Titus *et al.*, 1989; Hussell *et al.*, 1992; Dunn *et al.*, *in press* a, b; Pyle *et al.*, 1994). An advantage of this method is that it gives a single number (slope of the regression line) that expresses the average annual rate of change in bird numbers, or (if annual indices are log-transformed) the average annual proportionate rate of change. One can test whether the slope of the fitted line is significantly different from zero. Useful refinements are possible, such as weighting each annual index according to completeness of data (e.g., Hussell *et al.*, 1992). In an alternate approach, Titus *et al.* (1990) used nonparametric rank trend analysis to detect trends in raptor counts, but that produces no estimate of slope.

Fitting a straight line to nonlinear data is done all too often, although statistically inappropriate. Many authors have avoided this problem by not fitting trend lines at all, sometimes smoothing out some variability by calculating five-year moving averages (Spofford, 1969; Busse, 1973; Abraszewska-Kowalczyk, 1974; Nagy, 1977; Mueller *et al.*, 1988; Hill and Hagan, 1991; Baumanis, 1990), or using LOWESS, a LOcally WEighted Scatterplot Smoother (Titus *et al.*, 1989; see Taub, 1990; James *et al.*, 1992, for more information on the method). It is possible, however, to include higher-order year terms in a regression to fit a curved line to annual indices (e.g., Pyle *et al.*, 1994). The latter method offers the advantage that significance and magnitude of differences can be determined for the estimated values of any pair of years (Hussell and Brown, 1992). We think this method promises to be more generally useful than others tried to date and should be investigated further.

Regardless of the method used to determine trends, there is value in publishing annual indices as well as trend statistics. Annual indices from different sets of data can often be compared directly, but trends are not comparable if they cover different time periods (e.g., Marchant, 1992). Indices also allow one to detect annual variation that might suggest possible causes of population change (for example cold winters foretell lower subsequent breeding populations in certain species, Hjört and Lindholm, 1978; Hagan *et al.*, 1992).

There are difficult statistical and biological issues that need to be resolved if trends from separate migration counts are to be pooled into regional or national trends. The ANCOVA method can be used to calculate pooled annual indices for sites that are believed to provide independent samples of the same *monitored population,* such as those at Long Point, Ontario, which are no more than 28 km apart (Hussell, 1981; Hussell *et al.*, 1992). In principal, the same approach could be used to pool data from more distant sites that sample the same *monitored population* at different points along its migratory path. Possibly this approach can be extended to pool data from regional networks of counts, but it is not known how far apart the sites can be before the assumption that they all sample the same monitored population is seriously violated.

If migration counts at distant sites sample individuals from overlapping or separate monitored populations, then calculation of pooled indices or trends should weight counts from each site in proportion to the sizes of their monitored populations. Unfortunately, the relative sizes of the monitored populations are usually unknown. Titus and Fuller (1990) combined data from six hawk migration counts in eastern North America in a route regression analysis, using the relative abundance of birds at each station as a weighting factor. However, this assumes that a large *count population* must reflect a large *monitored population.* In fact, local topography can concentrate birds into large count populations even though the monitored population may be relatively small (see Fig. 1). Instead of pooling data from distant sites in a single regional analysis, it may be more appropriate to analyze results from each one separately and then look for regional patterns in results.

4.3. Summary of Analysis Section

Analysis methods play an important role in the use of *migration counts* for population monitoring. Standardization of data collection will make these counts more consistent estimators of birds present (the *count population*), while analytical procedure offers a route to dealing with identifiable variation in the proportions of the daily *migrating populations* that are present in the daily *count populations.* One of the most important contributions of ANCOVA to calculation of annual indices may be reduction in the number of years it takes to detect statistical significance of long-term trends. Trend analysis identifies significant population change, but annual indices may provide information useful for interpretation that trend statistics alone can hide. Statistical problems with pooling migration count data from distant sites may preclude calculation of national or regional trends.

5. EVALUATION OF MIGRATION COUNTS FOR POPULATION MONITORING

Previous sections have outlined methods for collecting and analyzing migration count data. But how do we know that migration counts tell us anything useful about population change? Many papers showing long-term trends in migration counts do not include any attempt to validate them by comparison with other data (e.g., Spofford, 1969; Busse, 1973; Abraszewska-Kowalczyk, 1974; Nagy, 1977; Riddiford, 1983; Hussell, 1985). There is some evidence, however, that migration counts do reflect real population size.

5.1. Incidental Evidence that Migration Counts Reflect Population Size

The relative abundances of 70 species observed at a migration station in the Swiss Alps matched their relative rankings in estimated population size (Jenni and Naef-Daenzer, 1986), and numbers of European migrants caught at oases in the Sinai were significantly correlated to estimated total population size (Safriel and Lavee, 1991). These results are perhaps unsurprising, however, as the population sizes involved vary by orders of magnitude.

Data from different migration stations frequently detect similar trends. One might argue that another migration count is not independent evidence of population change. Nonetheless, widespread agreement among trends from migration stations suggests at least that migration counts are not wholly unstable and peculiar to site. For example, in 14 species of hawks counted over a 16-year period at six widely-separated sites in eastern North America, there was consistency in about 60% of the comparisons of trends based on simple indices (Titus and Fuller, 1990). For 13 hawk species also counted over 16 years at two sites in Ontario and Minnesota, trends in indices derived from ANCOVA analyses were significantly related (Hussell and Brown, 1992). Trends from the Ontario site were also significantly correlated with the pooled trend for six stations calculated by Titus and Fuller (1990; including the Minnesota site), despite the fact that the time periods involved were not identical (13 years of overlap).

On the other hand, there can be important differences in trends among nearby migration stations. Trends in eight species from two banding sites in Michigan less than 1 km apart were not significantly correlated, although the lack of correspondence depended largely on striking differences between sites in only one species (E. H. Dunn, D.J.T. Hussell, and R. J. Adams, unpubl.).

Numerous papers contain nonstatistical comparisons of migration count trends to population trends from independent sources that are obvious enough to have been discernable by bird-watchers, and some authors demonstrate simple correlation between bird numbers and plausible causal events. Counts of Peregrine Falcons in Wisconsin, for example, declined and recovered in parallel with periods of use and banning of DDT (Mueller *et al.*, 1988). Bednarz *et al.* (1990) demonstrated similar correspondence in other raptors, although not all species acted as expected. White-throat (*Sylvia communis*) numbers at a Swedish bird observatory fluctuated in parallel with Sahel drought conditions (Hjört and Lindholm, 1978; Svensson, 1985), while Wren (*Troglodytes troglodytes*) numbers were negatively correlated to cold winters (Hjört and Lindholm, 1978). Similar negative correlation was found between migration counts of kinglets (*Regulus satrapa* and *R. calendula*) and cold winters in North America (Hagan *et al.*, 1992). Nuthatch (*Sitta europaea*) numbers observed at one migration concentration point fluctuated in parallel to the number of breeding pairs nearby (Gatter, 1974). A variety of warblers known to increase during spruce budworm outbreaks fluctuated simultaneously in migration counts (Jones, 1986; Stewart, 1987, [although these authors did not note the connection] Hill and Hagan, 1991; Hagan *et al.*, 1992; Hussell *et al.*, 1992). Other examples of general correspondence between migration counts and independent data on population trends include Sharrock (1969), Langslow (1978), Cowley (1979), and Berthold *et al.* (1986b).

5.2. Correspondence of Migration Count Trends to Trends in Breeding Populations

Stronger support for migration monitoring as a means of tracking population change comes from agreement in trends to those from independent monitoring programs, conducted at other times of year and with entirely different methods. Although no unimpeachable sources of population data exist, the more agreement found between independent monitoring programs, the stronger the evidence that *both* methods reflect population change. Comparisons of this nature may also help in deciding which methods of data collection and analysis are most effective (Berthold, 1976).

One of the earliest attempts at validation of migration counts by comparison to data from another monitoring program was by Svensson (1978), who compared sums of untransformed daily spring migration counts from Ottenby Bird Observatory (southern Sweden) to results from the following summer's plot-based breeding bird censuses in southern Sweden. He found less agreement between programs for trans-Saharan migrants than for shorter-distance migrants. We rearranged his results into two groups: species whose breeding range north of the bird observatory is more-or-less limited to the area covered by breeding season counts in Sweden, and those that breed in a much broader area. For 10 of the 11 species whose breeding range is limited to southern and central Sweden, there were positive correlations between annual migration count indices and breeding population indices (significant in three species; average correlation coefficient = 0.44). The average correlation for eight species that breed throughout Scandinavia was 0.01 (five of eight negative, one significantly so). Thus, a measure of correspondence between monitoring programs was evident when migration counts could be compared to breeding counts from the probable area of origin of a substantial portion of the migrant population. Correlations were weak, but migration data were untransformed and covered only six years. Comparable data extracted from a later paper with 13 years of data (Svensson *et al.*, 1986) showed significant positive correlation between migration and breeding season counts in 7 of 12 migrants breeding in southern Sweden, and 0 of 10 more widespread breeders.

In another European comparison, trends from the German migration monitoring program showed little correspondence to breeding population trends compiled from five nearby countries (Marchant, 1992). Data from the various programs were for different periods of years, however, and the breeding season trends from different countries were also poorly correlated to one another.

A preliminary analysis of visual counts from the Netherlands produced annual migration indices for four species, from 13 sites with 35+ daily counts each autumn (via regression analysis taking site, year and site-year interactions into account). Results from three of the species had annual indices significantly correlated to those from breeding season surveys in Finland (Ward Hagemeijer, personal communication).

Validation studies in North America have focused on comparisons of count results to the Breeding Bird Survey (BBS; Table I). All studies in this table refer to songbirds, except Hussell and Brown (1992), who dealt with hawks.

If one simply looks at the column in Table I that indicates degree of correspondence, the evidence for agreement appears positive. Notes in the table, however, draw attention to weaknesses in many of the studies. For example, some of the comparisons include species that are year-round residents (Note A, Table I). Counts in the migration season of residents probably include a high proportion of individuals on breeding territory, and such counts are probably fairly repeatable day-

TABLE I
Studies that Test for Correspondence of Trends in Migration Season Counts to Trends in BBS

Source	Count Type	Index calculation method[a]	Location	No. of species	No. of years	Count season	Degree of correspondence	Notes[b]
V. Kleen, C. Moore, and S. Droege, *pers. comm.*	One-day count	Log N/effort	Illinois	12	18	Spring	Trends correlate[c]	A, 1
V. Kleen, C. Moore, and S. Droege, *pers. comm.*	One-day count	Log N/effort	Illinois	41	18	Spring	No correlation of trends[c]	B, 1 & 3
Cyr and Larivée, 1993	Birders' checklists	Sighting frequency	Quebec	74	19	Spring and fall	4 (38)% opposite[d]	B, 1
G. Geupel and N. Nur, *pers. comm.*	Daily banding	Log N	California	35	13	Fall	Trends correlate[c] 3 (49%) opposite[d]	A, 1
Hagan *et al.*, 1992	Daily banding	Log N/effort	Massachusetts	38	19	Fall	Indices correlate in 63% of species[e]	A & B, 1 & 2
Hagan *et al.*, 1992	Daily banding	Log N/effort	Pennsylvania	39	19	Fall	Indices correlate in 26% of species[e]	A & B, 1 & 2
Data from D. Dawson, *pers. comm.*	Daily banding	N/effort	6 sites, eastern U.S.	104	19	Fall	Trends correlate[c]	B, 2
Data from West, 1992	One-day count	Log N/effort	Delaware	99	22	Spring	28% opposite[d]	B, 1 & 2
Data from West, 1992	One-day count	Log N/effort	Delaware	29	22	Spring	54% opposite[d]	D, 2
Pyle *et al.*, 1994	Daily banding and counts	ANCOVA	California	24	25	Spring and fall	0 (25)% opposite[d]	C, 2

(*continued*)

TABLE I (*Continued*)

Source	Count Type	Index calculation method[a]	Location	No. of species	No. of years	Count season	Degree of correspondence	Notes[b]
Hussell et al., 1992 and unpubl.	Daily banding and counts	ANCOVA	Ontario	45	21	Spring and fall	Trends correlate[c] *0 (27)% opposite*[d]	C, 3
Hussell and Brown, 1992 and unpubl.	Daily hawk count	ANCOVA	Ontario	8	15	Spring	Trends correlate[c] *0 (38)% opposite*[d]	C, 3
E. H. Dunn, D.J.T. Hussell, and R. J. Adams, unpubl.	Daily banding	ANCOVA	Michigan	11	13	Fall	Trends correlate[c] 0 (18)% opposite[d]	D, 3

[a]N is daily migration count, "/effort" indicates that daily counts were adjusted for effort. Sighting frequency indicates that the annual index was percent of checklists on which the species was recorded.

[b]Notes:

A. some (or many) of species in comparison are year-round residents at or close to sample site.
B. many (or all) of species are migrants that breed at or close to sample site as well as farther away.
C. all (or nearly all) species are migrants that only breed far from the sample site or, if sample site is within breeding range, species does not breed nearby.
D. comparison limited to species that only breed far from the study site.
1. BBS from same region as migration count.
2. BBS covers much broader region than probable portion of breeding range sampled by migration route.
3. BBS area limited to probable portion of breeding range sampled by migration count.

[c]Significant positive correlation ($P < 0.05$ for r or r_s, depending on study) between trends from the two programs (one test for all species combined).

[d]Percent of species with opposite signs of trends, when trends were significant in both programs ($P < 0.10$ except in Cyr and Larivée, 1993, in which $P < 0.05$). Figure in parentheses shows percent of all species compared in which trend signs were opposite (regardless of significance of trends).

[e]Significant correlation ($P < 0.05$) between annual indices in migration counts and BBS (one test per species). Lack of correlation may only reflect lack of long-term trend in the indices for a given species.

to-day regardless of weather or date. These counts might therefore have a high probability of agreement with BBS results. For species that are seen only on migration to a distant breeding ground, however, the probability of a stable count day-to-day is much lower (e.g., Fig. 2). This is probably the reason why trends in the one-day Illinois Spring Bird Count corresponded to Illinois BBS trends for resident species, while there was no correspondence between migration count trends for a selection of migrants and BBS in Illinois and northward (Table I).

Other validation studies include species that, although migratory, breed in the sample area as well as in more distant areas (Note B, Table I). It is possible that migration season counts at some sites could accurately reflect change in the locally-breeding segment of the population (as noted above), while inadequately sampling the portion of the population headed further north. If so, correspondence in trends between migration counts and distant BBS routes may come about not because counts of migrants track population change, but because there is range-wide population change that the monitoring station detects through its counts of local birds. It is not always easy to determine how significant a problem this might be, because some of the studies do not indicate which species were included in each comparison to BBS.

Ability to monitor locally breeding species is a valuable function and may be sufficient justification for migration season monitoring. Nonetheless the unique value of migration counts is their potential to track population change in species that do not breed locally; i.e., those that are seen only on migration. To determine whether this is possible, evaluation should focus on species that do not breed at or near the sample site(s) (See Note D, Table I). Moreover, BBS data used for comparison with migration counts should ideally come from a substantial portion of the probable breeding range, not merely from the immediate vicinity of the migration count. At the same time, however, BBS data should not cover a vast area far larger than the likely source of the *monitored population* (see notes 1 and 2, Table I). The most relevant validations, therefore, are those few in Table I with notations C or (preferably) D, and 3. These are discussed here in somewhat more detail.

West (1992) presented long-term trend data for effort-adjusted one-day counts in Delaware and from BBS for eastern North America (128 species), which we used to do a detailed comparison. Of 29 migratory species that breed only to the north of Delaware, 13 had strong ($P < 0.10$) trends in both the one-day counts and in BBS, but seven of these (54%) had trends that were opposite in sign. This suggests that species recorded only on migration were poorly tracked by one-day counts,

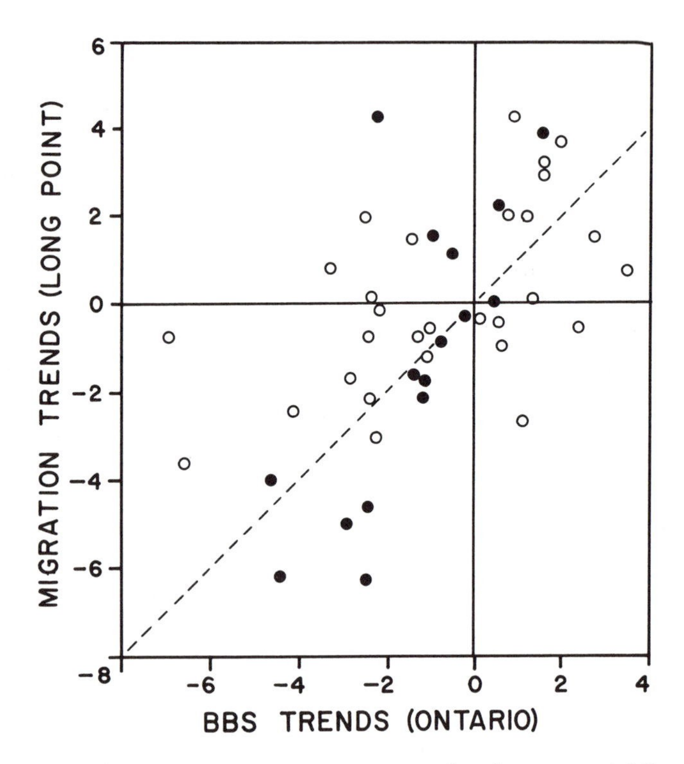

FIGURE 3. Rates of change (percent per year) in combined spring and fall migration indices from Long Point Bird Observatory DET's, versus median trends in the Breeding Bird Survey in Ontario, 1967–1987. Each point represents one species. Solid circles = species judged to be well-monitored by BBS (see text); open circles = all others. Spearman rank correlation coefficient = 0.60 for all species ($N = 45$, $P < 0.01$), and 0.75 for well-monitored species ($N = 16$, $P < 0.001$). Dashed line indicates one-to-one correspondence. From Hussell *et al.* (1992) and D.J.T. Hussell (unpublished).

although trends for residents were more in agreement with BBS (Table I). Kleen *et al.* (Table I) also found correspondence between one-day spring counts and BBS in resident species but not in migrants. Power analysis of migration counts suggests 5+ days of sampling per week over the entire migration season are required if significant population change is to be detected within a reasonable number of years (L. Thomas, *pers. comm.*). Checklist-type counts that pool unstandardized observations from many sites on many days during the season may perform better than one-day counts (e.g., Cyr and Larivée, 1993; Table I), but have not yet been rigorously evaluated for their ability to monitor populations of migrants that are not local breeders.

Trends from the Southeast Farallon Island (SEFI) station of Point Reyes Bird Observatory were compared to those from BBS for the entire western portion of North America by Pyle *et al.* (1994). Although some of the species analyzed breed on the nearby mainland, the exposed oceanic island itself is depauperate in breeding landbirds. Direction of trend (without consideration of trend significance) was the same in 78% of species in spring (N of species compared = 18) and 75% of species in fall (N = 24). There were only a few species that had significant trends in both SEFI and BBS (two in spring and six in fall), and none of these pairs of trends was opposite in sign.

Long Point Bird Observatory (LPBO) also has few breeding species as a result of its location (sandy peninsula in Lake Erie). LPBO trends for 45 species based on Daily Estimated Totals from three nearby sites correlated significantly to trends in Ontario BBS (Fig. 3). Signs of trends were opposite in 27% of the 45 species, but once again, there were no opposite signs among the 10 species in which both BBS and LPBO had significant trends (Table I). Many of the species examined have breeding ranges that extend well to the north of BBS coverage, such that the migration counts may have sampled substantially different portions of the breeding range than did BBS. When the analysis was restricted to the 16 species whose entire range north of LPBO is well-covered by BBS, correlation between the programs was strengthened and only 3 of 19 species (19%) had trends in opposite directions (none significant).

Hussell and Brown (1992) used regression analysis (including weather covariates) to produce annual indices for eight raptors migrating through southern Ontario in spring. Trends in these indices and in Ontario BBS were significantly correlated, despite the fact that the time periods covered did not correspond exactly. An updated version of their analysis using appropriate time periods also showed a significant positive correlation (Fig. 4).

Trends in log-transformed, effort-adjusted daily banding totals from a station in southern Michigan were calculated for 11 species that breed only to the north of the study site, using regression analysis to adjust for weather (E. H. Dunn, D.J.T. Hussell and R. J. Adams, unpubl.). This site does not concentrate migrants, unlike the sites just discussed (SEFI, LPBO and hawk counts). There was significant correlation between the trends in numbers banded and Michigan BBS trends (Fig. 5). This is one of the "cleanest" comparisons to date (see notes to Table I), and gives some of the strongest evidence that migration counts and BBS record the same phenomena.

Strong positive correlation between trends is only one criterion of agreement between independent data sets on population change. In

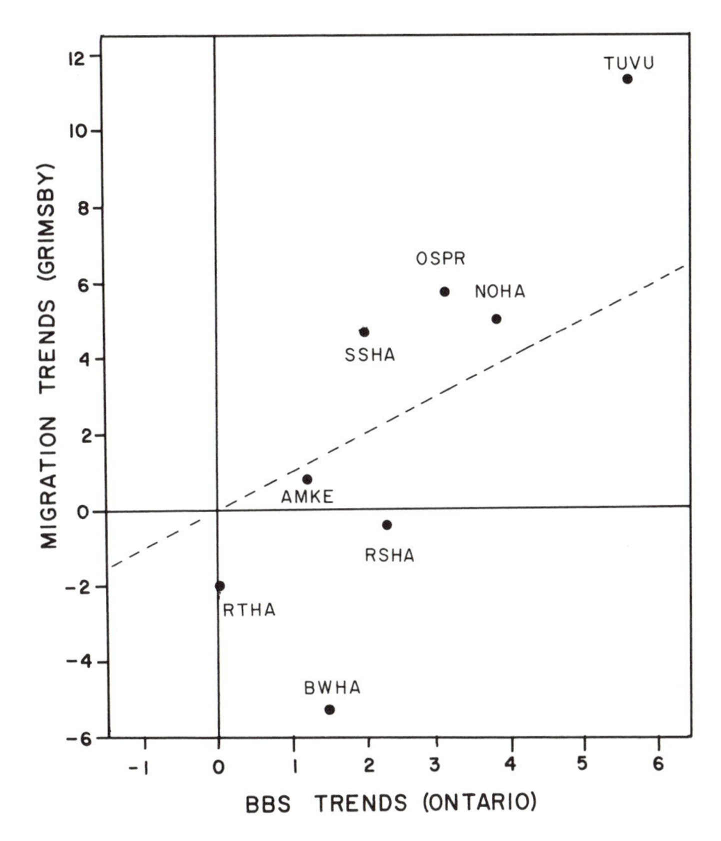

FIGURE 4. Rates of change (percent per year) in spring migration count indices for raptors at Grimsby, Ontario and median BBS trends for Ontario, 1975–1990. Spearman rank correlation $r_s = 0.810$ ($N = 8$, $P = 0.015$). Dashed line indicates one-to-one correspondence. TUVU = Turkey Vulture (*Cathartes aura*), OSPR = Osprey (*Pandion haliaetus*), NOHA = Northern Harrier (*Circus cyaneus*), SSHA = Sharp-shinned Hawk (*Accipiter striatus*), AMKE = American Kestrel (*Falco sparverius*), RSHA = Red-shouldered Hawk (*Buteo lineatus*), RTHA = Red-tailed Hawk (*B. jamaicensis*), BWHA = Broad-winged Hawk (*B. platypterus*).

addition there should be one-to-one correspondence in the magnitudes of the trends, as is suggested by Figs. 3–5. Where there is disagreement, the cause could lie with either or both programs, and neither should be taken as the standard for correctness.

Certain migration count sites sample unusually high proportions of young birds in autumn (Ralph, 1971; Dunn and Nol, 1980; Ehnbom et *al.*, 1993), but this may not affect ability of migration counts to monitor populations. Hussell *et al.* (1992) showed that although spring and fall

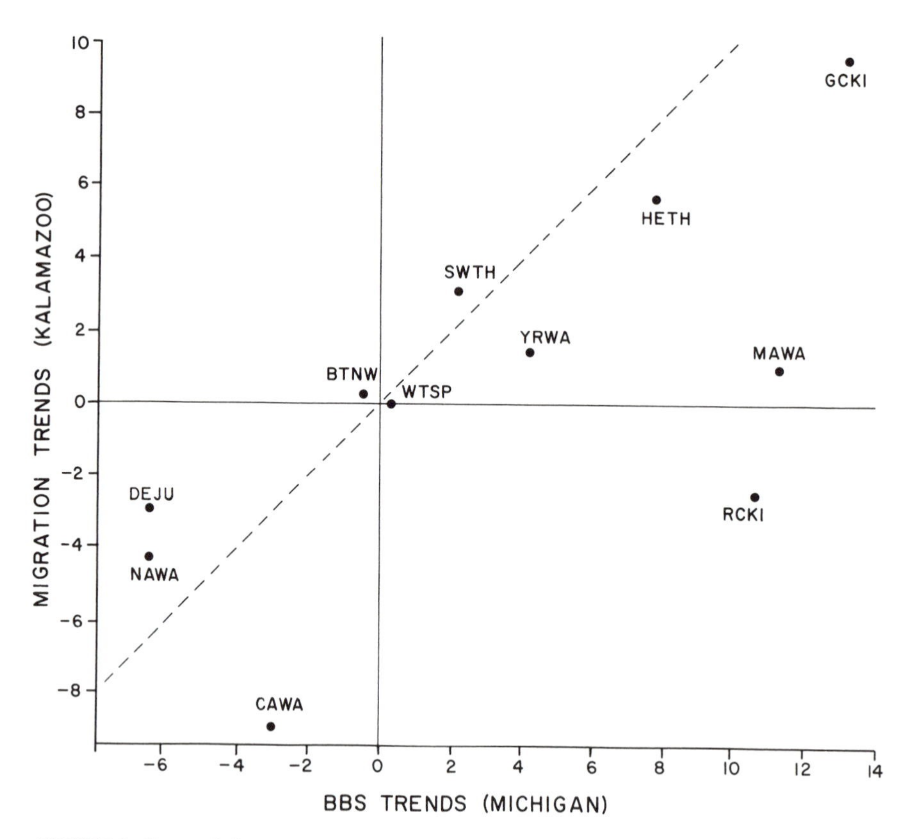

FIGURE 5. Rates of change (percent per year) in fall banding indices from two banding stations in Kalamazoo, Michigan 1979–90 (E. H. Dunn, D.J.T. Hussell, and R. J. Adams, unpublished), and mean BBS trends from that state, in migratory species that breed only north of the banding site. Spearman rank correlation = 0.75 (N = 11, $P < 0.01$). Dashed line indicates one-to-one correspondence. Codes for species: GCKI = Golden-crowned Kinglet (*Regulus satrapa*), RCKI = Ruby-crowned Kinglet (*R. calendula*), HETH = Hermit Thrush (*Catharus guttatus*), SWTH = Swainson's Thrush (*C. ustulatus*), NAWA = Nashville Warbler (*Vermivora ruficapilla*), MAWA = Magnolia Warbler (*Dendroica magnolia*), YRWA = Yellow-rumped Warbler (*D. coronata*), BTNW = Black-throated Green Warbler (*D. virens*), CAWA = Canada Warbler (*Wilsonia canadensis*), DEJU = Dark-eyed Junco (*Junco hyemalis*) and WTSP = White-throated Sparrow (*Zonotrichia albicollis*).

migration indices were rather weakly correlated ($r_s = 0.462$, $n = 42$, $P < 0.01$), trends from spring and fall counts combined correlated well with BBS (Fig. 3). Pyle et al. (1994) found BBS trends corresponded well both with fall and spring migration count trends, and other data in Table I also suggest counts from both seasons have monitoring potential.

5.3. Conclusions of Evaluation Section

The most appropriate comparisons of trends from migration counts and BBS (in terms of choice of species, analysis methods and area of BBS coverage) provide good evidence that migration counting can monitor species that are present at the count site only as migrants. Cases of positive correspondence with BBS include migration counts based on visual observation and banding, nocturnal and diurnal migrants, and sites that do and do not concentrate migrants.

Some of the weaker validation tests in Table I might be strengthened by more focused analyses (i.e., by selecting species in such a way as to upgrade codes in the final column of the table). More evaluation studies should be carried out, particularly for nearby locations to determine if all sites are equally suitable for sampling a given *monitored population*. Data from the southern U.S. should prove especially valuable, as there will be a larger set of species that are present at count sites only as migrants but whose breeding ranges are well-covered by BBS.

6. SYNTHESIS: THE PROS AND CONS OF USING MIGRATION COUNTS FOR POPULATION MONITORING

Assuming throughout this section that migration counts are, indeed, indicative of population change, what benefits might they offer? The chief one is that migration counts could track changes in species inaccessible to other population monitoring programs. There are 36 landbird species that are candidates for migration monitoring in which half or more of the entire North American breeding range is in remote areas north of road accessibility (Groups 1 and 2, Table II). Of these, 18 winter largely south of the U.S., beyond the reach of current winter monitoring programs such as Christmas Bird Counts (Group 1). Canadians may wish to use migration monitoring to track population change in 40 additional species, in which half or more of the breeding range within Canada also lies in inaccessible northern areas (Groups 3 and 4, Table II). Many of the birds listed in Table II deserve particular attention because of shrinking winter habitat (Diamond, 1991). Migration counts should be useful as well for several groups of birds that are not necessarily inaccessible in winter or summer, but which use habitats poorly covered by CBC or BBS (e.g., grassland and interior forest birds), or species that are rare, secretive or otherwise hard to detect on BBS routes (such as raptors, Bednarz and Kerlinger, 1989).

Migration counts could serve an important role in confirming

TABLE II
Landbird Species for which Migration Monitoring Could Be Particularly Helpful in Tracking Population Trends because of Inadequate Coverage by Other Monitoring Programs

Group 1. Species in which 50+% of North American breeding range is north of BBS coverage and 50+% of winter range is south of U.S.		Group 2. Species in which 50+% of North American breeding range is north of BBS coverage but substantial part of winter range is in U.S. and Canada.	
Peregrine Falcon	*Falco peregrinus*	Rough-legged Hawk	*Buteo lagopus*
Merlin	*Falco columbarius*	Winter Wren	*Troglodytes troglodytes*
Least Flycatcher	*Empidonax minimus*	Ruby-crowned Kinglet	*Regulus calendula*
Alder Flycatcher	*Empidonax alnorum*	Varied Thrush	*Ixoreus naevius*
Yellow-bellied Flycatcher	*Empidonax flaviventris*	Northern Shrike	*Lanius excubitor*
Swainson's Thrush	*Catharus ustulatus*	Water Pipit	*Anthus spinoletta*
Gray-cheeked Thrush	*Catharus minimus*	Yellow-rumped Warbler	*Dendroica coronata*
Philadelphia Vireo	*Vireo philadelphicus*	Le Conte's Sparrow	*Ammodramus leconteii*
Tennessee Warbler	*Vermivora peregrina*	American Tree Sparrow	*Spizella arborea*
Orange-crowned Warbler	*Vireo celata*	Harris' Sparrow	*Zonotrichia querula*
Cape May Warbler	*Dendroica tigrina*	White-throated Sparrow	*Zonotrichia albicollis*
Magnolia Warbler	*Dendroica magnolia*	White-crowned Sparrow	*Zonotrichia leucophrys*
Bay-breasted Warbler	*Dendroica castanea*	Golden-crowned Sparrow	*Zonotrichia atricapilla*
Blackpoll Warbler	*Dendroica striata*	Fox Sparrow	*Passerella iliaca*
Palm Warbler	*Dendroica palmarum*	Smith's Longspur	*Calcarius pictus*
Connecticut Warbler	*Oporornis agilis*	Lapland Longspur	*Calcarius lapponicus*
Wilson's Warbler	*Wilsonia pusilla*	Snow Bunting	*Plectrophenax nivalis*
Northern Waterthrush	*Seiurus noveboracensis*	Rusty Blackbird	*Euphagus carolinus*

(*continued*)

TABLE II (*Continued*)

Group 3. Species (additional to those in Group 1) in which 50+% of Canadian (but not North American) breeding range is north of BBS coverage and 50+% of winter range is south of U.S.		Group 4. Species (additional to those in Group 2) in which 50+% of Canadian (but not North American) breeding range is north of BBS coverage but substantial part of winter range is in U.S. and Canada.	
Osprey	*Pandion haleaetus*	Bald Eagle	*Haliaeetus leucocephalus*
Common Nighthawk	*Chordeiles minor*	Northern Harrier	*Circus cyaneus*
Eastern Kingbird	*Tyrannus tyrannus*	Sharp-shinned Hawk	*Accipiter striatus*
Olive-sided Flycatcher	*Contopus borealis*	Red-tailed Hawk	*Buteo jamaicensis*
Western Wood Pewee	*Contopus sordidulus*	Short-eared Owl	*Asio flammeus*
Say's Phoebe	*Sayornis saya*	Long-eared Owl	*Asio otus*
Tree Swallow	*Tachycineta bicolor*	Northern Flicker	*Colaptes auratus*
Violet-green Swallow	*Tachycineta thalassina*	Yellow-bellied Sapsucker	*Sphyrapicus varius*
Bank Swallow	*Riparia riparia*	Eastern Phoebe	*Sayornis phoebe*
Cliff Swallow	*Hirundo pyrrhonota*	Horned Lark	*Eremophila alpestris*
Barn Swallow	*Hirundo rustica*	American Crow	*Corvus brachyrhynchos*
Solitary Vireo	*Vireo solitarius*	Golden-crowned Kinglet	*Regulus satrapa*
Red-eyed Vireo	*Vireo olivaceus*	Townsend's Solitaire	*Myadestes townsendi*
Black-and-white Warbler	*Mniotilta varia*	Hermit Thrush	*Catharus guttatus*
Black-throated Green Warbler	*Dendroica virens*	American Robin	*Turdus migratorius*
Yellow Warbler	*Dendroica petechia*	Savannah Sparrow	*Passerculus sandwichensis*
Ovenbird	*Seirus aurocapillus*	Song Sparrow	*Melospiza melodia*
American Redstart	*Setophaga ruticilla*	Chipping Sparrow	*Spizella passerina*
Lincoln's Sparrow	*Melospiza lincolnii*	Dark-eyed Junco	*Junco hyemalis*
Western Tanager	*Piranga ludoviciana*	Swamp Sparrow	*Melospiza georgiana*

trends from other sources. Monitoring schemes do not always agree in the trends they document (e.g., Bednarz *et al.*, 1990; Dunn, 1991; Marchant, 1992), and no one type of monitoring survey can give a definitive answer for all species (Berthold, 1976). When a variety of independent data sources agree, therefore, the argument for validity of a widespread population change is greatly enhanced. The best evidence that BBS measures broad scale population change, for instance, is correspondence between BBS trends and those in migration counts and other independent data sets (Butcher, 1986; Holmes and Sherry, 1988; Butcher *et al.*, 1990; Temple and Carey, 1990b; references in Table I).

Compiling data from several monitoring programs provides more power for interpretation of trends. For example, Hagan (1993) used data from migration counts, BBS and Christmas Bird Counts (CBC) to document precipitous declines in northeastern U.S. breeding populations of Rufous-sided Towhee (*Pipilo erythrophthalmus*), and to suggest that wintering populations in the southeastern U.S. declined less strongly due to mixing of migrants with local residents whose populations were more stable. Migration counts can provide demographic data such as age proportions in fall which may shed light on annual reproductive success or on the season in which maximum mortality occurs. Mueller *et al.* (1977), for example, used age data from invasions of Northern Goshawk (*Accipiter gentilis*) to implicate reproductive failure as a factor in dispersal.

Migration counts may provide data that overcome some of the shortcomings specific to breeding season counts. Habitat-specific surveys could miss large-scale population changes resulting from shrinkage in the total amount of habitat available (Ginn, 1969), while migration counts would detect such changes. Migration counts should be less biased than are breeding season counts by birds' breeding status (and, therefore, singing behavior). Lastly, migration counts (at least where birds are concentrated by topographic features) should be less affected by habitat change at and around the count site than are site-specific breeding bird surveys.

Another positive aspect of migration counts is their potential for long-term continuity. Migration concentration points attract counters on an ongoing basis. Many bird observatories and hawk counts are financed and operated primarily by volunteers or nonprofit organizations, so they are not subject to the vagaries of government support. Volunteers with different skill levels and time commitment can often be accommodated.

Against these advantages of migration counts, we must weigh their shortcomings. Many of the potential drawbacks have been considered in this paper, and evidence has been presented that the one most frequently cited (i.e., variation in counts due to weather) is by no means fatal. Nonetheless, there are certain limitations to migration counts. It is probably not valid to generate continent-wide population trends by pooling migration data from different sites (see Section 4.2). All migration sampling sites may not be equally suitable, in which case there could be a shortage of sites in geographic areas where migration counts are needed for specific target species.

Intensive coverage of the migration season is required for the most effective detection of long-term population change, as is appropriate data analysis. Both may be beyond the capabilities of unaided volunteers, and finances must be found for personnel to fill in the gaps.

It may be difficult to outline the geographic locations where a specific monitored population breeds or winters. Band recoveries are often few and accumulate slowly, but quicker answers may be possible with special studies (e.g., of isotope ratios in feathers; Chamberlain *et al.*, 1994).

Lastly, migration counts are only suitable for tracking population trends in *monitored populations* that are wholly migratory. This excludes irruptive species, because the proportion of the population irrupting varies and destinations change from year to year. Migration counts of partially migratory *monitored populations* could only reflect population change if a constant proportion migrated each year. Even for species that are regular migrants, population monitoring may not be possible at sites located within the wintering range, because migration counts may be augmented or diminished by annual shifts in wintering distribution (e.g., Niles *et al.*, 1969; Terrill and Ohmart, 1984; Terrill and Crawford, 1988; Gwinner *et al.*, 1992). Nevertheless, migration counts in all these cases may provide valuable information on movements.

Overall, the potential benefits of migration counts for population monitoring appear substantial, while there are possible solutions for many of its drawbacks. As with any monitoring method, the shortcomings of migration counts should be kept in mind when interpreting results.

7. THE FUTURE OF MIGRATION MONITORING IN NORTH AMERICA

The Monitoring Working Group of Partners in Flight recognized that BBS was inadequate for monitoring inaccessible northern populations of landbirds (examples in Table II). It suggested that to fill this gap, a series of migration monitoring stations should be established across the northern edge of adequate BBS coverage, which corresponds to the northern edge of the more inhabited regions of Canada (Butcher *et al.*, 1993). As little evaluation of migration counts had been done at that time, however, the group recommended that before proceeding further, a workshop should be held to evaluate the potential of the proposed series of stations and to look at ways of validating the population indices derived from migration counts.

That workshop was held in September 1993, organized by the U.S. Fish and Wildlife Service (now National Biological Service) and the Canadian Wildlife Service. The final report in effect endorsed the Partners In Flight proposal for an east-west series of intensive migration monitoring stations, coupled with a more extensive program (perhaps harnessing the potential of general birders) to provide complementary data (Blancher *et al.*, 1993). The recommendations called for centralized establishment of standards and analysis of data, and close ties to the volunteer groups and individuals who collect the data. An administrative structure was proposed, and the first committees were appointed in early 1994.

An east-west chain of stations was recommended as a means of sampling all migration routes. The appropriate distance between sites is unknown, however, as we currently lack good information on the boundaries of the *monitored populations* of each species sampled at any given site. Stations near the northern limit of BBS coverage should be well positioned to monitor specific remote-breeding populations of species whose ranges are not covered, or are only partly covered by BBS. Southern stations (e.g., along the Gulf Coast) could also track far northern breeders, but the birds sampled may be drawn from much broader portions of the breeding range. Moreover, southern stations could not monitor northern breeders that winter at or north of the count sites (Section 6). They could, however, provide complementary data to BBS for many species that breed in the U.S.

Although migration monitoring sites anywhere in North America could contribute data, there are arguments to be made for defining a core group that is well-located to track particular target species. Efforts to establish permanent stations could be focused first on the core chain(s), rather than dispersing resources among many stations. Several existing stations along the Canada/U.S. border could potentially form part a northern chain specializing on northern-nesting species and populations (Fig. 6), although additional stations are needed to the east and west. There are several hawk-watching sites in northern areas that might also be included in a migration monitoring network (e.g., Cedar

FIGURE 6. Locations of existing migration monitoring stations in southern Canada and adjacent U.S. (solid circles). The dot-dash line shows the northern and western limits of contiguous Breeding Bird Survey routes in Canada (derived from Erskine, 1990). Dotted line shows approximate northern limit of trees. Locations are as follows: 1 = Beaverhill Lake, Alberta, 2 = Last Mountain Lake, Saskatchewan, 3 = Delta, Manitoba, 4 = Thunder Cape, Ontario, 5 = Whitefish Point, Michigan, 6 = Long Point, Ontario.

Grove, Wisconsin; Duluth, Minnesota; Grimsby, Ontario). Hawk counts were not considered in the Migration Monitoring Workshop (Blancher *et al.*, 1993), but most stations already collect standardized data according to rules of the Hawk Migration Association of North America, and their data can be analyzed with procedures that are similar to those used for counts of nocturnal migrants.

Centralized data analysis will be an important step in making migration counts more useful for population monitoring. Personnel at most stations do not have the time, expertise or funding to do appropriate analyses, and a uniform analytical procedure will ensure that results will be directly comparable. Analytical methods need further development, but this does not preclude centralized analysis in the meantime that uses the best methods currently available. Indeed, a centralized data analysis group will likely be in the forefront of further developments.

Much work remains to be done on evaluating the abilities of specific migration count programs to monitor population change. Many of the existing evaluations are difficult to interpret (Table I), and could profitably be redone with more restrictive criteria. A high priority is to determine whether regional checklist projects can detect population change in species that are recorded only during migration. If so, they have the potential to fill gaps in species or geographical coverage that cannot be provided by intensive, site-specific monitoring.

Lastly, cooperation and communication among the groups and individuals collecting migration data will be crucial in ensuring that migration counting becomes more generally useful as a population monitoring method. Assurance of stable funding for coordination of the program and analysis and reporting of results will also be important.

The process laid out for accomplishing all these steps (Blancher *et al.*, 1993) is currently being implemented. Activities of the next few years will be important in determining the degree to which migration counting becomes an effective continental program for monitoring bird populations.

ACKNOWLEDGEMENTS. This paper is based on a report written by the first author under a contract from the Canadian Wildlife Service to the Long Point Bird Observatory, which was improved by discussion ith Deanna Dawson, Sam Droege, George Fairfield, Linda Leddy, John Richardson, Chris Rimmer, and Jim Stasz. Helpful comments on early revisions were made by Peter Blancher, Michael Bradstreet, Greg Butcher, the late Bev Collier, Tony Gaston, Paul Kerlinger, and Jon McCracken, while thorough reviews of the penultimate draft were undertaken by Frances James and Nadav Nur. We thank the following for enabling us to use unpublished data or results: Mary Ellen Hebb (hawk counts from Grimsby, Ontario, compiled by volunteers for the Niagara Peninsula Hawk Watch), Raymond J. Adams, Kalamazoo Nature Center (mist-net capture data shown in Fig. 5) and Long Point Bird Observatory (data in Fig. 2). Lucille Brown, Ontario Ministry of Natural Resources, assisted with calculation of trends in Fig. 4. BBS trends were provided by Bruce Peterjohn, National Biological Service and Brian Collins, Canadian Wildlife Service. This paper is Ontario Ministry of Natural Resources, Southern Terrestrial Ecosystems Research contribution No. 94-04.

REFERENCES

Abraszewska-Kowalczyk, A., 1974, [Dynamics of number in some migrants observed at Polish Baltic coast in the years 1962–1970], (Polish, English summary), *Notatki Ornitologiczne* **15:**77–104.

Alerstam, T., 1978, Analysis and a theory of visible bird migration, *Oikos* **30:**273–349.

Askins, R. A., J. F. Lynch and R. Greenburg, 1990, Population declines in migratory birds in eastern North America, in: *Current Ornithology, Vol. 7,* (D. M. Power, ed.), *Plenum* Press, NY, pp. 1–57.

Askins, R. A., 1993, Population trends in grassland, shrubland, and forest birds in eastern North America, in: *Current Ornithology, Vol. 11,* (D. M. Power, ed.), Plenum Press, NY, pp. 1–34.

Bairlein, F., 1992, Recent prospects on trans-Saharan migration of songbirds, *Ibis* **134** Suppl. **1:**41–46.

Baker, R. R., 1993, The function of post-fledging exploration: a pilot study of three passerines ringed in Britain, *Ornis Scand.* **24:**71–79.

Bastian, A., and P. Berthold, 1991, [Are nocturnal migrants stationary or do they show directed migratory movements during daytime at their central European resting areas?], (German, English summary), *J. Orn.* **132:**325–327.

Baumanis, J., 1990, Long-term dynamics of some selected species of land birds during autumn migration in Pape, Latvia, *Baltic Birds* **5**(1):28–30.

Bednarz, J. C., and P. Kerlinger, 1989, Monitoring hawk populations by counting migrants, *Nat. Wild. Fed. Sci. & Tech. Series* **13:**328–342.

Bednarz, J. C., D. Klem, Jr., L. J. Goodrich, and S. E. Senner, 1990, Migration counts of raptors at Hawk Mountain, Pennsylvania as indicators of population trends, 1934–1986, *Auk* **107:**96–109.

Bennett, H. R., 1952, Fall migration of birds at Chicago, *Wilson Bull.* **64:**197–220.

Bergstrom, E. A., and W. H. Drury, Jr., 1956, Migration sampling by trapping: a brief review, *Bird-Banding* **27:**107–120.

Berthold, P., 1976, [Census methods in ornithology: overview and critical review.], (German, English Summary), *J. Orn.* **117:**1–69.

Berthold, P., and R. Schlenker, 1975, [The "Mettnau-Reit-Illnitz-Programm"—a long-term bird trapping program of the Vogelwarte Radolfzell with multiple goals], (German, English summary), *Vogelwarte* **28:**97–123.

Berthold, P., G. Fliege, U. Querner and R. Schlenker, 1986a, [Successful termination of the "Mettnau-Reit-Illnitz-Programm" of the Vogelwarte Radolfzell: a survey of the technical data and of following programs] (German, English summary), *Vogelwarte* **33:**208–219.

Berthold, P., G. Fliege, U. Querner and H. Winkler, 1986b, [Change in songbird populations in central Europe: analysis of trapping data], (German, English summary), *J. Orn.* **127:**397–437.

Berthold, P., G. Fliege, G. Heine, U. Querner and R. Schlenker, 1991, Autumn migration, resting behaviour, biometry and moult of small birds in central Europe, *Vogelwarte* **36:**1–224.

Berthold, P., A. Kaiser, U. Querner and R. Schlenker, 1993, [Analysis of trapping figures at Mettnau station, S. Germany, with respect to the population development in small birds: a 20 years summary], (German, English summary), *J. Orn.* **134:**283–299.

Bingman, V., 1980, Inland morning flight behavior of nocturnal passerine migrants in eastern New York, USA, *Auk* **97:**465–472.

Blancher, P., A. Cyr, S. Droege, D. Hussell and L. Thomas, 1993, Results of a U.S./Canada workshop on monitoring of landbirds during migration and recommendations towards a North American Migration Monitoring Program (MMP). (Unpubl. report available from P. Blancher at same address as first author, or from S. Droege, NBS, 12100 Beech Forest Dr., Laurel, MD 20708). 26 pp.

Blokpoel, H., and W. J. Richardson, 1978, Weather and spring migration of snow geese across southern Manitoba, *Oikos* **30:**350–363.

Busse, P., 1973, [Dynamics of numbers in some migrants caught at Polish Baltic coast 1961–1970], (Polish, English summary), *Notatki Ornitologiczne* **14:**1–38.

Busse, P., 1979, 20 years of Operation Baltic, *Ring* **101:**90–93.

Butcher, G. S., 1986, Populations of black ducks and mallards in winter; 1950–1983, Unpubl. report to U.S. Fish Wildl. Serv., Washington, D.C.

Butcher, G. S., and C. E. McCulloch, 1990, Influence of observer effort on the number of individual birds recorded on Christmas Bird Counts, in: *Survey Designs and Statistical Methods for the Estimation of Avian Population Trends*, (J. R. Sauer and S. Droege, eds.), *USF&WS Biol. Rept.* **90**(1), pp. 120–129.

Butcher, G. S., M. R. Fuller, L. S. McAllister, and P. H. Geissler, 1990, An evaluation of the Christmas Bird Count for monitoring population trends of selected species, *Wildl. Soc. Bull.* **18:**129–134.

Butcher, G. S., B. Peterjohn and C. J. Ralph, 1993, Overview of national bird population monitoring programs and databases, in: *Status and Management of Neotropical Migratory Birds* (D. M. Finch and P. W. Stangel, eds.), *U.S. Forest Serv. Gen. Tech. Rept.* **RM-229,** Fort Collins, CO, pp. 192–203.

Chamberlain, C. P., J. D. Blum, R. T. Holmes, S. Poulson, P. P. Mara and T. W. Sherry, 1994, Linking breeding and wintering grounds of long-distance migrants using stable and radiogenic isotope ratios, in: *Research Notes on Avian Biology 1994: Selected Contributions from the 21st International Ornithological Congress* (J. Dittami, W. Bock, M. Taborsky, R. van den Elzen and E. Vogel-Millesi, eds.), *J. Orn.* **135 (special issue):**23.

Cowley, E., 1979, Sand Martin population trends in Britain; 1965–78, *Bird Study* **26:** 113–117.

Cyr, A., and J. Larivée, 1993, A checklist approach for monitoring neotropical migrant birds: twenty-year trends in birds of Québec using ÉPOQ, in: *Status and Management of Neotropical Migratory Birds*, (D. M. Finch and P. W. Stangel, eds.), *U.S. Forest Serv. Gen. Tech. Rept.* **RM-229,** Fort Collins, CO, pp. 229–236.

Dalberg Peterson, F., 1976, Changes in numbers of migrants ringed at Danish bird observatories during the years 1966–75, *Dansk Orn. Foren. Tidsskr.* **70:**17–20.

Darby, K. V., 1985, Migration counts and local weather at British bird observatories: an examination by linear discriminant analysis, in: *Statistics in Ornithology*, (B.J.T. Morgan and P. M. North, eds.), *Lecture Notes in Statistics* **29.** Springer-Verlag, Berlin, pp. 37–64.

Dawson, D. K., 1990, Migration banding data: a source of information on bird population trends? in: *Survey Designs and Statistical Methods for the Estimation of Avian Population Trends*, (J. R. Sauer and S. Droege, eds.), *USF&WS Biol. Rept.* **90**(1), pp. 37–40.

DeSante, D. F., 1983, Annual variability in the abundance of migrant landbirds on southeast Farallon Island, California, *Auk* **100:**826–852.

Diamond, A. W., 1991, Assessment of the risks from tropical deforestation to Canadian songbirds, *Trans. 56th N. Amer. Wildl. & Nat. Res. Conf.*:177–194.

Duffy, K., and P. Kerlinger, 1992, Autumn owl migration at Cape May Point, New Jersey, *Wilson Bull.* **104:**312–320.

Dunn, E. H., 1991, Population trends in Canadian songbirds, *Bird Trends* **1**:2–11.

Dunn, E. H., and E. Nol, 1980, Age-related migratory behavior of warblers, *J. Field Ornithol.*, **51**:254–269.

Dunn, E. H., D.J.T. Hussell and R. J. Adams, *in press* a, Can annual productivity be monitored with mist-netting during migration? *U.S. Forest Service Gen. Tech. Rept.*

Dunn, E. H., D.J.T. Hussell and J. D. McCracken, *in press* b, A comparison of three count methods for monitoring songbird abundance during migration: banding, census and estimated totals, *U.S. Forest Service Gen. Tech. Rept.*

Eckert, K. R., 1990, Lakewood Pumping Station census of fall migration, *Loon* **62**:99–105.

Ehnbom, S., L. Karlsson, R. Ylvén, and Susanne Åkesson, 1993, A comparison of autumn migration strategies in Robins *Erithacus rubecula* at a coastal and an inland site in southern Sweden, *Ringing & Migr.* **14**:84–93.

Erskine, A. J., 1978, The first ten years of the co-operative Breeding Bird Survey in Canada, *Can Wildl. Serv. Rep. Series* **42**:1–59.

Erskine, A. J., 1990, Tables for a report on the co-operative Breeding Bird Survey in Canada through 1989. Unpubl. report to Canadian Wildlife Service, Sackville, NB.

Forsythe, B. J., and D. James, 1971, Springtime movements of transient nocturnally migrating landbirds in the Gulf Coastal Bend region of Texas, *Condor* **73**:193–207.

Fuller, M. R., and K. Titus, 1990, Sources of migrant hawk counts for monitoring raptor populations, in: *Survey Designs and Statistical Methods for the Estimation of Avian Population Trends*, (J. R. Sauer and S. Droege, eds.), *USF&WS Biol. Rept.* **90**(1), pp. 41–46.

Gatter, W., 1974, [Observations of an invasion of the Nuthatch (*Sitta europaea caesia*) at Randecker Maar, Schwabische Alp], (German, English summary), *Vogelwarte* **27**:203–209.

Gauthreaux, S. A., Jr., 1991, The flight behavior of migrating birds in changing wind fields: radar and visual analyses, *Amer. Zool.* **31**:187–204.

Ginn, H. B., 1969, The use of annual ringing and nest record card totals as indicators of bird population levels, *Bird Study* **16**:210–247.

Graber, R. R., and W. W. Cochran, 1960, Evaluation of an aural record of nocturnal migration, *Wilson Bull.* **72**:253–273.

Gwinner, E., H. Schwabl and F. Schwabl-Benzinger, 1992, The migratory time program of the Garden Warbler: is there compensation for interruptions?, *Ornis Scand.* **23**:264–270.

Hackman, C. D., and C. J. Henney, 1971, Hawk migration over White Marsh, Maryland, *Chesapeake Science* **12**:137–141.

Hagan, J. M. III, 1993, Decline of the Rufous-sided Towhee in the eastern United States, *Auk* **110**:863–874.

Hagan, J. M. III, T. L. Lloyd-Evans, J. L. Atwood and D. S. Wood, 1992, Long-term changes in migratory landbirds in the northeastern United States: evidence from migration capture data, in: *Ecology and Conservation of Neotropical Migrant Landbirds*, (J. M. Hagan and D. W. Johnston, eds.), Smithsonian Inst. Press, Washington, pp. 115–130.

Hall, G. A., and R. K. Bell, 1981, The diurnal migration of passerines along an Appalachian ridge, *Amer. Birds* **35**:135–138.

Hamilton, W. J., 1962, Evidence concerning the function of nocturnal call notes of migrating birds, *Condor* **64**:390–401.

Herremans, M., 1989, Habitat and sampling related bias in sex-ratio of trapped Blackcaps *Sylvia atricapilla*, *Ringing and Migr.* **10**:31–34.

Hill, N. P., and J. M. Hagan, III, 1991, Population trends of some northeastern North American landbirds: a half-century of data, *Wilson Bull.* **103**:165–182.

Hjört, C., and C.-G. Lindholm, 1978, Annual bird ringing totals and population fluctuations, *Oikos* **30**:387–392.

Holmes, R. T., and T. W. Sherry, 1988, Assessing population trends of New Hampshire forest birds: local vs. regional patterns, *Auk* **105**:756–768.

Holthuizjen, A. M. A. and L. Oosterhuis, 1985, Implications for migration counts from telemetry studies of Sharp-shinned Hawks (*Accipiter striatus*) at Cape May Point, New Jersey, Hawk Migration Assoc. N. Amer., *Proc. Hawk Migration Conf.* **4**:305–312.

Hussell, D. J. T., 1981, The use of migration counts for detecting population levels, *Stud. Avian Biol.* **6**:92–102.

Hussell, D. J. T., 1985, Analysis of hawk migration counts for monitoring population levels. Hawk Migration Assoc. N. Amer., *Proc. Hawk Migration Conf.* **4**:243–254.

Hussell, D. J. T., M. H. Mather, and P. H. Sinclair, 1992, Trends in numbers of tropical- and temperate-wintering migrant landbirds in migration at Long Point, Ontario, 1961–1988, in: *Ecology and Conservation of Neotropical Migrant Landbirds*, (J. M. Hagan and D. W. Johnston, eds.), Smithsonian Inst. Press, Washington, pp. 101–114.

Hussell, D. J. T., and L. Brown, 1992, Population changes in diurnally-migrating raptors at Duluth, Minnesota (1974–1989) and Grimsby, Ontario (1975–1990), Unpubl. report, Ministry of Natural Resources, Maple, Ontario, 67 pp.

Hutto, R. L., 1985, Seasonal changes in the habitat distribution of transient insectivorous birds in southeastern Arizona: competition mediated?, *Auk* **102**:120–132.

James, F. C., D. A. Wiedenfeld, and C. E. McCulloch, 1992, Trends in breeding populations of warblers: declines in the southern highlands and increases in the lowlands, in: *Ecology and Conservation of Neotropical Migrant Landbirds*, (J. M. Hagan and D. W. Johnston, eds.), Smithsonian Inst. Press, Washington, pp. 43–56.

Jean, A., and M. Razin, 1993, Monitoring migration in the Pyrenees: the case of the Wood Pigeon, *Columba palumbis*, *Bird Census News* **6**:83–89.

Jenni, L., and B. Naef-Daenzer, 1986, [Frequencies of migrating birds caught on an alpine pass in comparison with numbers of breeding birds in the area of origin], (German, English summary), *Ornithol. Beob.* **83**:95–110.

Jones, E. T., 1986, The passerine decline, *North Amer. Bird Bander* **11**:74–75.

Källender, H., and O. Rydén, 1974, Inter-observer differences in studies of visible migration at Falsterbo, *Ornis Scand.* **5**:53–62.

Kerlinger, P., and S. A. Gauthreaux, 1984, Flight behaviour of Sharp-shinned Hawks during migration. 1: over land, *Anim. Behav.* **32**:1021–1028.

Kuenzi, A. J., F. R. Moore, and T. R. Simons, 1991, Stopover of neotropical landbird migrants on East Ship Island following trans-Gulf migration, *Condor* **93**:869–883.

Lack, D., 1959, Migration across the sea, *Ibis* **101**:374–399.

Langslow, D. R., 1978, Recent increases of Blackcaps at bird observatories, *British Birds* **71**:345–354.

Laske, V. and A. Helbig, 1987, Influence of atmospheric turbidity on counts of diurnal migration and a method of correction, *Ardea* **75**:245–254.

Lowery, G. H., Jr., and R. J. Newman, 1955, Direct studies of nocturnal bird migration, in *Recent Studies in Avian Biology* (A. Wolfson, ed.), Univ. Illinois Press, Urbana, pp. 238–263.

Marchant, J. H., 1992, Recent trends in breeding populations of some common trans-Saharan migrant birds in northern Europe, *Ibis* **134** Suppl. **1**:113–119.

Martin, E. E., 1980, Diversity and abundance of spring migratory birds using habitat islands on the Great Plains, *Condor* **82**:430–439.

Martsching, P., 1986, Spring warbler migration at Brookside Park in Ames, *Iowa Bird Life* **56**:107–111.

Martsching, P., 1987, Fall warbler migration at Brookside Park in Ames, *Iowa Bird Life* **57**:112–117.

McCracken, J. D., D. J. T. Hussell and E. H. Dunn, 1993, A Manual for Monitoring Bird Migration, Long Point Bird Observatory, Port Rowan, Ontario, 65 pp.

Moore, F. R., P. Kerlinger, and T. R. Simons, 1990, Stopover on a Gulf Coast barrier island by spring trans-gulf migrants, *Wilson Bull.* **102**:487–500.

Mueller, H. C., D. D. Berger and G. Allez, 1977, The periodic invasions of Goshawks, *Auk* **94**:652–663.

Mueller, H. C., D. D. Berger and G. Allez, 1988, Population trends in migrating peregrines at Cedar Grove, Wisconsin, 1936–1985, in: *Peregrine Falcon Populations, their Management and Recovery*, (T. J. Cade, J. J. Enderson, C. G. Thelander and C. M. White, eds.), The Peregrine Fund, Inc., Boise, Idaho, pp. 497–506.

Murray, B. G., Jr., 1964, A review of Sharp-shinned Hawk migration along the northeastern coast of the United States, *Wilson Bull.* **76**:257–264.

Nagy, A. C., 1977, Population trend indices based on 40 years of autumn counts at Hawk Mountain Sanctuary in northeastern Pennsylvania, in: *Proceedings of World Conference on Birds of Prey*, 1975, ICBP, Vienna, pp. 243–253.

Niles, D. M., D. A. Rohwer, J. A. Jackson and J. D. Robins, 1969, An observation of midwinter nocturnal movement and tower mortality of Tree Sparrows, *Bird-Banding* **40**:322–323.

Nisbet, I. C. T., W. H. Drury, and J. Baird, 1963, Weight-loss during migration. Part I. Deposition and consumption of fat by the Blackpoll Warbler *Dendroica striata*, *Bird-Banding* **34**:107–138.

Nisbet, I. C. T., and W. H. Drury, Jr., 1969, A migration wave observed by moon-watching and at banding stations, *Bird-Banding* **40**:243–252.

Österlöf, S., and B.-O. Stolt, 1982, Population trends indicated by birds ringed in Sweden, *Ornis Scand.* **13**:135–140.

Pardieck, K., and R. W. Waide, 1992, Mesh size as a factor in avian community studies using mist nets, *J. Field Ornithol.* **63**:250–255.

Payevsky, V. A., 1990, Population dynamics of birds during 1960–1986 according to trapping data on the Courish Spit of the Baltic Sea. *Baltic Birds* **5**(2):69–73.

Pyle, P., N. Nur, R. P. Henderson, and D. F. DeSante, 1993, The effects of weather and lunar cycle on nocturnal migration of landbirds at Southeast Farallon Island, California, *Condor* **95**:343–361.

Pyle, P., N. Nur and D. F. DeSante, 1994, Trends in nocturnal migrant landbird populations at Southeast Farallon Island, California, 1968–1992, *Stud. Avian Biol.* **15**: 58–74.

Ralph, C. J., 1971, An age differential of migrants in coastal California, *Condor* **73**: 243–246.

Ralph, C. J., 1981, Age ratios and their possible use in determining autumn routes of passerine migrants, *Wilson Bull.* **93**:164–188.

Ralph, C. J., G. R. Geupel, P. Pyle, T. E. Martin, and D. F. DeSante, 1993, *Handbook of Field Methods for Monitoring Landbirds*, U.S. Forest Service Gen. Tech. Rep. PSW-GTR-144, 41 pp.

Rao, P. V., 1981, Summarizing remarks: estimating relative abundance (Part II), *Stud. Avian Biol.* **6**:110–111.

Rappole, J. H., and K. Ballard, 1987, Postbreeding movements of selected species of birds in Athens, Georgia, *Wilson Bull.* **99**:475–480.

Richardson, W. J., 1974, Multivariate approaches to forecasting day-to-day variations in the amount of bird migration, in: *Proceedings of Conference on Biological Aspects of* the *Bird/Aircraft Collision Problem*, (S. A. Gauthreaux, ed.), Clemson University, Clemson, South Carolina, pp. 309–329.

Richardson, W. J., 1978, Timing and amount of bird migration in relation to weather: a review, *Oikos* **30**:224–272.

Riddiford, N., 1983, Recent declines of Grasshopper Warblers *Locustella naevia* at British bird observatories, *Bird Study* **30**:143–148.

Riddiford, N., 1985, Grounded migrants versus radar: a case-study, *Bird Study* **32**: 116–121.

Robbins, C. S., D. Bystrak and P. H. Geissler, 1986, The Breeding Bird Survey: its first 15 years, 1965–1979, *USF&WS Resour. Publ.* **157**:1–196.

Robbins, C. S., J. R. Sauer, R. S. Greenberg and S. Droege, 1989, Population declines in North American birds that migrate to the neotropics, *Proc. Nat. Acad. Sci.* **86**: 7658–7662.

Russell, R. W., P. Dunne, C. Sutton and P. Kerlinger, 1991, A visual study of migrating owls at Cape May Point, New Jersey, *Condor* **93**:55–61.

Safriel, U. N., and D. Lavee, 1991, Relative abundance of migrants at a stopping–over site and the abundance in their breeding ranges, *Bird Study* **38**:71–72.

Sauer, J. R., and S. Droege (eds.), 1990, *Survey Designs and Statistical Methods for the Estimation of Avian Population Trends*, USF&WS Biol. Rept. **90**(1), 166 pp.

Scheider, F. G., and D. W. Crumb, 1985, Spectacular flight of Purple Finches at Derby Hill, *Kingbird* **35**:115.

Sharrock, J. T. R., 1969, Grey Wagtail passage and population fluctuations in 1956–67, *Bird Study* **16**:17–34.

Slack, R. S., C. B. Slack, R. N. Roberts, and D. E. Emord, 1987, Spring migration of Long-eared Owls and Northern Saw-whet Owls at Nine Mile Point, New York, *Wilson Bull.* **99**:480–485.

Spofford, W. R., 1969, Hawk Mountain counts as population indices in northeastern America, in: *Peregrine Falcon Populations, Their Biology and Decline*, (J. J. Hickey, ed.), University of Wisconsin, Madison, pp. 323–332.

Stedman, S. J., 1990, A synopsis of four decades of fall raptor migration study in Tennessee, *Newsletter H.M.A.N.A.* **15**(2):32–37.

Stewart, P. A., 1987, Decline in numbers of wood warblers in spring and autumn migration through Ohio, *North Amer. Bird Bander* **12**:58–60.

Svazas, S., 1991, Species composition and abundance of nocturnal autumnal bird migrants in the continental part of Lithuania, *Acta Ornithol. Lituanica* **4**:52–62.

Svensson, S. E., 1978, Efficiency of two methods for monitoring bird population levels: breeding bird censuses contra counts of migrating birds, *Oikos* **30**:373–386.

Svensson, S. E., 1985, Effects of changes in tropical environments on the North European avifauna, *Ornis Fennica* **62**:56–63.

Svensson, S., C. Hjört, J. Petterson and G. Roos, 1986, Bird population monitoring: a comparison between annual breeding and migration counts in Sweden, *Vår Fågelvarld* Suppl. **11**:215–224.

Taub, S. R., 1990, Smoothed scatterplot analysis of long-term breeding bird census data, in: *Survey Designs and Statistical Methods for the Estimation of Avian Population Trends*, (J. R. Sauer and S. Droege, eds.), *USF&WS Biol. Rept.* **90**(1), pp. 80–83.

Temple, S. A., and J. R. Carey, 1990a, Description of the Wisconsin Checklist Project, in: *Survey Designs and Statistical Methods for the Estimation of Avian Population Trends*, (J. R. Sauer and S. Droege, eds.), *USF&WS Biol. Rept.* **90**(1), pp. 14–17.

Temple, S. A., and J. R. Carey, 1990b, Using checklist records to reveal trends in bird populations, in: *Survey Designs and Statistical Methods for the Estimation of Avian*

Population Trends, (J. R. Sauer and S. Droege, eds.), *USF&WS Biol. Rept.* **90**(1), pp. 98–104.

Terrill, S. B., and R. D. Ohmart, 1984, Facultative extension of fall migration by Yellow-rumped Warblers (*Dendroica coronata*), *Auk* **101**:427–438.

Terrill, S. B., and R. L. Crawford, 1988, Additional evidence of nocturnal migration by Yellow-rumped Warblers in winter, *Condor* **90**:261–263.

Thompson, J. J., 1993, Modelling the local abundance of shorebirds staging on migration, *Theoret. Pop. Biol.* **44**:299–315.

Titus, K., and J. A. Mosher, 1982, The influence of seasonality and selected weather variables on autumn migration of three species of hawks through the central Appalachians, *Wilson Bull.* **94**:176–184.

Titus, K., M. R. Fuller and J. L. Ruos, 1989, Considerations for monitoring raptor population trends based on counts of migrants, in: *Raptors in the Modern World*, (B.-U. Meyburg and R. D. Chancellor, eds.), WWGBP, Berlin, London and Paris, pp. 19–32.

Titus, K., and M. R. Fuller, 1990, Recent trends in counts of migrant hawks from northeastern North America, *J. Wildl. Manage.* **54**:463–470.

Titus, K., M. R. Fuller, and D. Jacobs, 1990, Detecting trends in hawk migration count data, in: *Survey Designs and Statistical Methods for the Estimation of Avian Population Trends*, (J. R. Sauer and S. Droege, eds.), *USF&WS Biol. Rept.* **90**(1), pp. 105–113.

Van Tighem, K. J., and G. R. Burns, 1984, The 1984 spring bird migration: Elk Island National Park, *Alberta Naturalist* **14**:131–134.

Weisbrod, A. R., C. J. Burnett, J. G. Turner and D. W. Warner, 1993, Migrating birds at a stopover site in the Saint Croix River Valley, *Wilson Bull.* **105**:265–284.

West, R. L., 1992, Trend analysis of Delaware spring bird counts, *Delmarva Ornithol.* **24**:19–38.

Wiedner, D. S., P. Kerlinger, D. A. Sibley, P. Holt, H. Hough and R. Crossley, 1992, Visible morning flight of neotropical landbird migrants at Cape May, New Jersey, *Auk* **109**:500–510.

Winker, K., D. W. Warner and A. R. Weisbrod, 1992, Daily mass gains among woodland migrants at an inland stopover site, *Auk* **109**:853–862.

Young, B. E., 1991, Annual molts and interruption of the fall migration for molting in Lazuli Buntings, *Condor* **93**:236–250.

Zalakevicius, M., 1991, Visible bird migration as a part of the whole migratory passage, *Acta Ornithol. Lituanica* **4**:27–38.

CHAPTER 3

PTILOCHRONOLOGY
A Review and Prospectus

THOMAS C. GRUBB, JR.

1. INTRODUCTION

Several years ago, I introduced the width of daily growth bars on feathers as a novel index of nutritional condition (Grubb, 1989). Originally applied to birds, termed ptilochronology, the concept of monitoring nutritional status by measuring induced anabolism has since been extended to other taxonomic groups (Grubb, 1994). Since 1989, concerns about the validity of ptilochronology have been raised and investigated, and the method has been applied to a wide variety of conceptual issues concerning avian behavior and ecology. In this chapter, I review these developments, point out uncertainties about the method that need to be addressed, and suggest additional subjects in ornithology that could benefit from application of the method.

Nutrition may be defined as the rate of ingestion of assimilable energy and nutrients, and **nutritional condition** as the state of body components controlled by nutrition and which, in turn, influence an animal's fitness. I have reviewed extant methods for indexing avian

---

THOMAS C. GRUBB, JR. • Behavioral Ecology Group, Department of Zoology, The Ohio State University, Columbus, Ohio 43210.

Current Ornithology, Volume 12, edited by Dennis M. Power. Plenum Press, New York, 1995.

nutritional condition, noting difficulties inherent in the use of traditional indices such as fat supply or body mass (Grubb, 1994).

It is becoming apparent that factors other than risk of starvation influence fat supply, so the body mass of animals can be viewed as reflecting a trade-off among selective pressures. It is well known (e.g., Hogstad, 1989; Ekman and Lilliendahl, 1993) that within the dominance-structured flocks of wintering parids, dominant birds have priority of access to food resources and sometimes kleptoparasitize subordinates. Therefore, nutrition must be considered to be influenced directly by dominance position. If fat supply, as measured by body mass, were a valid index of nutritional condition, dominant birds should be heavier than subordinates. However, just the reverse was found within flocks of Willow Tits (*Parus montanus*, Ekman and Lilliendahl, 1993); subordinates carried more fat than dominants. Moreover, when the dominants were removed, the subordinates significantly reduced their body size as they became dominant, a situation interpreted as dominance-mediated access to food. Because of their reduced access, subordinates carried a larger "insurance policy" against starvation. The subordinates incurred some other cost, such as increased risk of predation, not borne by the lower body-mass dominants. However, if food were to become gradually scarcer, subordinates would resist starvation longer than dominants.

As an index of nutritional condition, induced anabolism appears to avoid such ambiguities. A feather is pulled from a bird and it is then determined how rapidly its replacement is grown (Grubb, 1989). As natural selection favors regeneration as rapidly as possible, reduction in growth rate reflects some nutritional constraint. Quantity of assimilable energy, rather than of any specific dietary nutrient, can limit induced feather growth (Grubb, 1989). Later, I will refer to Hogstad's (1992) study of induced feather growth. His study, also involving flocks of Scandinavian Willow Tits, showed a direct relationship between induced feather growth and dominance position within a flock. Although there is no direct comparison of body mass and feather growth rate in the same birds, it does appear that induced feather growth is the more valid indicator.

The technique of ptilochronology is detailed in Grubb (1989). A feather is characterized by a series of parallel dark and light bands oriented obliquely to the shaft, or rachis. From evidence to be detailed later, we are quite confident that a combination of one dark band and an adjacent light band (together termed a growth bar) constitutes 24-hours' growth of the feather. The method assumes a direct relationship between growth bar width and nutritional condition, and can be applied to growth bars on either an original feather grown during a molt, or on a feather induced at a known time by pulling the original. As shown later, most studies to date have focused on induced feathers because the environmental conditions under which they are grown can be closely monitored. In practice, a bird is caught and a single rectrix, usually the outermost right, is pulled and stored. The bird is then released. Several weeks later, after the induced feather has been fully regenerated, the bird is recaptured, and the induced feather pulled. The average width of 10 growth bars is determined for both original and induced feathers, with the former value used to standardize for absolute bird size prior to subsequent use of induced growth as an index of nutritional condition.

In 1989, I raised questions on the limitations of ptilochronology. Additional concerns about the method and its interpretation have been expressed by others (Murphy and King, 1991; White *et al.*, 1991). I will first summarize the results of these inquiries before dealing with the method applied to issues in avian biology.

2. CAUSES OF GROWTH BARS

A growth bar consists of a swath of keratin across the vane of a feather. Such swaths are circular in cross section in the developing feather; they become "bars" once that portion of the feather clears the sheath and unfurls. The darker band within each growth bar is laid down during the day and the lighter band at night (Wood, 1950). Wood suggested that the dark and light bands comprising a growth bar are due to differences in optical properties brought about by differential pigment incorporation into the keratin matrix of the feather during the day and night. Pigment differences between dark and light bands are quite noticeable when the rectrices of some species (e.g., Northern Cardinal, *Cardinalis cardinalis*) are inspected under transmitted light only. However, one can also see growth bars in pure white feathers (e.g., rectrices of Herring Gulls, *Larus argentatus*), suggesting that pigment is not necessary.

Preliminary evidence suggests that whatever their cause, growth bars are not due to differences in the external anatomy of feather barbs or barbules. On a scanning electron micrograph of a Tufted Titmouse (*Parus bicolor*) rectrix, there was no evident discontinuity in structure across a known boundary between a dark and a light band (T. C. Grubb, Jr., unpublished).

It is important to distinguish between growth bars and fault bars. Fault bars appear as thin places on a feather due to incomplete growth of barbules connecting adjacent barbs. While fault bars are oriented in the same manner as growth bars with reference to the rachis, they do not appear to have the same cause. Fault bars appear sporadically, or not at all, indicating some short-term, acutely stressful period in a bird's life. King and Murphy (1984) found that White-crowned Sparrows (*Zonotrichia leucophrys*) produce a fault bar in response to the stress of being handled rather than dietary deficiency.

3. UBIQUITY OF GROWTH BARS

Inspection of live specimens and study skins of dozens of species suggests that production of growth bars is inherent in the normal growth of a feather. Growth bars are most readily detectable on freshly grown, monochromatic (i.e., nonpied) flight feathers, but their appearance can be masked by several factors. Growth bars are difficult or impossible to discern on massive feathers such as the remiges and rectrices of large birds (e.g., cranes, eagles); it is often possible to detect growth bars in smaller feathers. For example, we are currently investigating age-specific parental effort in Canada Geese (*Branta canadensis*) by examining growth bar width on greater secondary coverts (S. L. Earnst and T. C. Grubb, Jr., unpublished).

Growth bars seem to be obscured by nonmelanin pigments (e.g., rectrices of Red-tailed Hawk, *Buteo jamaicensis*) and by heavily barred pigment patterns (e.g., rectrices of House Wren, *Troglodytes aedon*).

Growth bars are most discernable on newly grown feathers and become obscured with age due to fading and/or abrasion. Because of these effects of age, to imply that growth bars on freshly grown induced feathers are more discernable than on freshly molted original feathers (e.g., Brodin, 1993) may not be correct.

Our studies of Leach's Storm-petrel (*Oceanodroma leucorhoa*) illustrate another potential difficulty in interpreting growth bars (R. A. Mauck and T. C. Grubb, Jr., unpublished). Among the species we have examined, only this petrel has not grown an induced feather marked by a regular alternation of dark and light bands. Instead, on the rectrices we have induced during the breeding season, we see a complicated and irregular progression of darker and lighter bands. By contrast, the original feather, grown during the previous postbreeding molting period, is marked by a much more regular and typical alternation of dark and light bands. Irregular progression of activity and rest periods during the breeding season may cause the unusual banding on induced feathers. This petrel commutes long distances between breeding colony and oceanic feeding grounds and is active day and night (Grubb, 1972). The absence of a regular diurnal rhythm of activity and rest during the breeding season may be reflected in the jumbled appearance of growth bars. Recent studies using on-board recording devices have shown that an irregular progression of rest and activity is the norm during the breeding season in another procellariiform, the Wandering Albatross (*Diomedea exulans;* Afanasyev and Prince, 1993). Even though the five albatrosses monitored were more active during the day, they also spent 59% of the dark hours in flight.

4. ONE GROWTH BAR DENOTES 24 HOURS OF FEATHER GROWTH

Recent evidence reinforces the statement by Michener and Michener (1938) that one growth bar denotes 24 hours worth of feather growth. We have taken daily measurements of feathers induced in captive birds. For Carolina Chickadees (*Parus carolinensis*), Tufted Titmice, and White-breasted Nuthatches (*Sitta carolinensis*), growth increments calculated from daily measurements of the outermost right rectrix agree with the measured width of growth bars on the same feather after it had been pulled (T. C. Grubb, Jr. and A. P. Marshall, unpublished).

Brodin (1993) has provided powerful confirmation to the 24-hour feather growth. He fed sunflower seeds dosed with radioactive sulfur to captive Willow Tits while the birds were growing an induced feather. The radioactivity was incorporated into the keratin laid down in the growing feather and could be detected by exposing photographic film to the radioactivity. By dosing the same bird on two separate days, measuring on the developed film the distance between radioactive segments, and dividing that distance by the time between doses, Brodin determined precisely the average daily growth of the feather. When that measurement was compared with the average width of growth bars, the match was exact. There seems little question that, except in a few unusual cases such as the petrel mentioned earlier, one growth bar does equal 24-hour growth of a feather.

5. GROWTH BAR WIDTH IS SENSITIVE TO NUTRITIONAL CONDITION

Two studies support the fundamental assumption of ptilochronology that rate of feather growth is a function of a bird's nutritional condition. Grubb and Cimprich (1990) tested the prediction that birds with access to supplementary food would grow an induced feather with wider growth bars than birds without additional food. Examining a total of 10 age/sex categories across four species (Downy Woodpecker, *Picoides pubescens;* Carolina Chickadee, Tufted Titmouse, White-breasted Nuthatch) wintering in Ohio deciduous woodlands, they found that in every case, the rate of induced feather growth adjusted for variation in bird size was greater in the supplemented group. In five comparisons, the difference was significant, and in a sixth comparison, involving male Carolina Chickadees, it just missed significance ($P = 0.06$).

A second experiment (Grubb, 1991) examined the reciprocal prediction from the nutritional-index assumption, that depriving a bird of food would cause a reduction in the width of growth bars on an induced feather. Captive Carolina Chickadees grew an induced rectrix while on a diet that was at or below the ad libitum rate of food consumption. Corrected for bird size, growth bar width of birds on 80% of the ad libitum diet was significantly narrower than that of birds on either 90% or 100% of ad libitum (Fig. 1). The latter two groups did not differ statistically. Thus, for a limited number of species evidence supports ptilochronology.

6. OTHER POSSIBLE FACTORS CONTROLLING GROWTH BAR WIDTH

If variation in growth bar width is to be interpreted correctly as indicating variation in nutritional condition, other potential causal factors must be known and controlled. Several projects have addressed whether

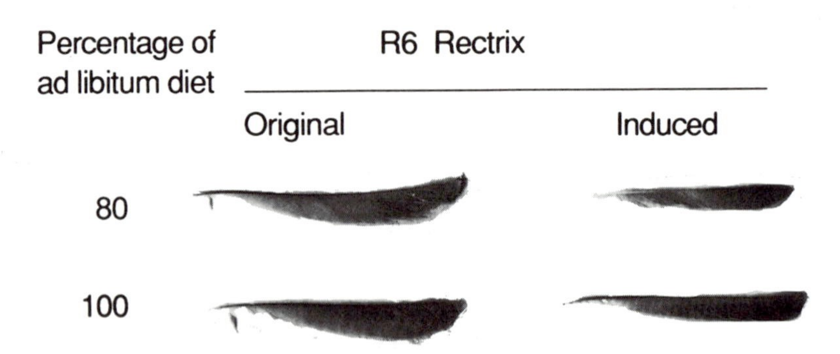

FIGURE 1. Original and induced outermost right rectrix grown by two Carolina Chickadees, one maintained on 80% and the other on 100% of the ad libitum diet determined on a per-bird basis.

any environmental factor other than net energy intake might cause variation in daily feather growth outside of the normal molting period.

6.1. Ambient Temperature and/or Wind Chill

Feathers grow from the papilla of the feather follicle (Watson, 1963). It is possible that exposure of the pygostyle (where rectrices originate) to cold temperatures and prevailing winds could cool this area and reduce feather growth rate. By affecting growth rate of an induced feather, such a relationship could be a major confounding factor in studies of temperate-zone birds in winter. Therefore, we undertook a controlled laboratory investigation of the matter.

Male White-breasted Nuthatches were captured in their territories, had their outer right rectrix pulled, and were then housed individually in wind-tunnel cages within chest-type freezers (Zuberbier and Grubb, 1992). Three birds were randomly assigned to each of nine treatment groups comprising all possible combinations of −15, −5, and +5 °C temperature, and 0.0, 0.5 and 1.0 m sec^{-1} wind speed. Under conditions of ad libitum food and water (or crushed ice), the birds lived in the wind tunnels for the several weeks required to grow out one rectrix.

There was no statistically significant effect of wind, temperature or the wind—temperature interaction on average growth bar width. Results in one nuthatch species indicate that any reduction in induced feather growth under conditions below the thermoneutral zone is unlikely to be caused by peripheral physiological effects operating at the level of the follicle. We ascribe to other causes the variation in feather growth during our winter studies of avian communities (see below).

6.2. Follicle History

Ptilochronology has considerable potential for long-term studies such as those monitoring habitat quality. In any protocol for long-term monitoring, an appealing design feature would employ daily growth bar widths on successive feathers pulled from the same follicle, thus controlling for any inter-follicular variation in feather growth. We performed a laboratory experiment to ascertain whether follicle history, independent of other factors, could influence the growth rate of a series of feathers grown within one molt cycle from the same follicle (Grubb and Pravosudov, 1994).

We arranged for various House Sparrows (*Passer domesticus*) to grow a first, second, or third induced rectrix from the outermost right follicle of the tail at the same time that each of the birds grew a first induced rectrix from the outermost left follicle. Results supported the hypothesis that follicle history is not an important factor controlling daily feather growth. Growth bar widths on the first, second, and third induced rectrices from one follicle did not differ significantly either from each other or from the growth bar width of the first induced rectrix being grown at the same time from another follicle. The interval between sequential feather inductions was seven weeks, so it would be worth knowing whether shorter time intervals between successive inductions would begin to affect the rate of induced growth.

6.3. Endogenous Cycles

These three studies suggest that induced feather growth is affected by factors that cycle annually. Growth bar width on induced rectrices varied significantly with season in a population of Northern Cardinals that had year-round access to supplementary food (Grubb *et al.*, 1991). During the winter, daily growth was slowest, and the finished feathers were shortest and lightest in mass.

Laboratory studies of birds on ad libitum food indicate that total feather length, and not daily growth, is controlled by cyclical endogenous factors. Female American Tree Sparrows (*Spizella arborea*) exposed to artificially long daylength (LD 20:4) grew out an induced rectrix that was significantly shorter in total length than that of short-day (LD 8:16) controls. However, the two treatment groups did not differ in growth bar width. These results suggested to White and Kennedy (1992) that the number of days taken by a follicle to grow out a feather could be controlled hormonally, while the amount of growth per day might be controlled by the bird's nutritional condition. Additional support for this interpretation comes from the follicle history study of House Sparrows on ad libitum food (Grubb and Pravosudov, 1994). We found that growth bar width of first-induced feathers did not vary statistically across induction dates. However, between the November and February induction dates there was a steady and significant increase in the total length of the induced feather. It is not known how hormones or other factors might determine the number of days that a follicle remains active.

7. APPLICATIONS

Induced anabolism, as an index of nutritional condition, represents a new tool for testing hypotheses of ultimate causation (Table 1).

TABLE I
Examples of Conceptual Areas where Predictions Centered on Differences in Nutritional Condition Could Be Tested by Ptilochronology

Age	Kinship
Sex	Deme
Dominance	Species
Dispersal Strategy	Niche
Foraging strategy	Habitat
Caching strategy	Migratory pathway
Dialect group	Guild
Information center	Community
Breeding experience	Ecotype
Parental effort	Parasite load
Support from helpers	Exposure to toxicants

Here, I review research employing the technique in a variety of conceptual areas.

7.1. Territory Size

For two species, we have indications that territory size is related to nutritional condition. One hypothesis explaining the reduction in numbers of Loggerhead Shrikes (*Lanius ludovicianus*) across North America has centered on changes to the landscape caused by shifting agricultural practices. Clearing of fence rows and field corners has been thought to reduce the density of hunting perches below the level required for shrike maintenance metabolism and reproduction (Yosef, 1992). Below some minimum density of perches, a shrike would be required to fly such long distances between perches across unhuntable "dead space" within its territory that defense would no longer be economically feasible in terms of net energy intake. Yosef and Grubb (1992) supported this hypothesis by finding evidence for two predictions. A rather crude measure of net energy intake, prey captures per hour divided by percent time flying, and daily feather growth of an induced rectrix were both significantly negatively related to territory size. Apparently, birds on larger territories were forced to invest more energy in commuting and defense for each unit of energy found, and the growth bars indicated that they suffered a real cost in reduced nutritional condition. Similarly, during the 1990 breeding season, R. Dettmers (unpublished) found a significant inverse correlation (r^2 =

0.539) between territory size and growth bar width in male Hooded Warblers (*Wilsonia citrina*) breeding in southern Ohio (Fig. 2). As territory size increased from 0.2 to 2.0 ha, the rate of daily induced feather growth decreased by 50%. It is interesting that in birds holding the smaller territories, daily feather growth on the induced feather grown during the breeding season actually exceeded daily growth during the previous prebasic molt. Was molting more nutritionally stressful than breeding in such birds? Dettmers' attempt to replicate these results during 1991 failed completely because all warblers failed to grow an induced rectrix (R. Dettmers, unpublished). The summer of 1991 was marked by the worst drought in recorded Ohio history, so a lack of moisture that year may have reduced the warblers' insect prey base below the threshold required to permit any feather regeneration.

Both the shrike and warbler studies were only correlational, so there is a real possibility that some birds of poor quality would have grown out their induced rectrix slowly even had they been on small territories. Required is a controlled manipulative test of the prediction that supplementary food would cause smaller territories and wider growth bars. Such an experiment has not yet been attempted, but Yosef

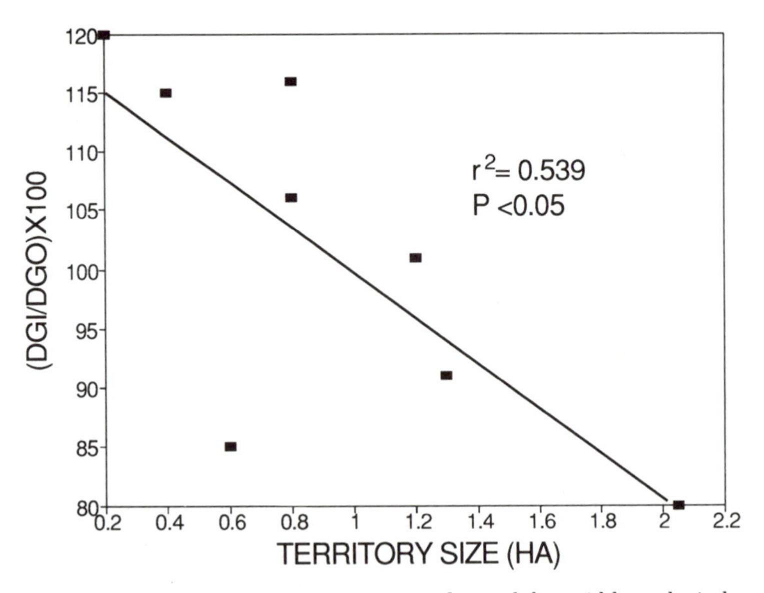

FIGURE 2. Relationship between territory size and growth bar width on the induced R6 rectrix of breeding male Hooded Warblers. DGI/DGO, the growth bar width of the induced feather divided by the growth bar width of the original feather, is a ratio employed to control for variation in body size.

(1992) showed that augmenting the number of hunting perches caused breeding shrikes to reduce territory size and to increase both number of fledglings and fledging success (fledglings per egg). Not incidentally for this endangered species, the local density of breeding shrikes was also increased by the manipulation. Additional pairs colonized areas that became available as the territory size of the resident pairs shrank.

7.2. Reproductive Effort

Breitwisch (1989) concluded that the concept of reproductive effort is essentially the same as Trivers' (1972) earlier concept of parental investment, namely the commitment of resources to the raising of an individual offspring at the cost of committing resources to other offspring, either present or future. The intensity of present reproductive behavior, or parental input (*sensu* Evans (1990), could thus reflect a trade-off between present offspring production and the ability to produce additional offspring in the future. White *et al.* (1991) used ptilochronology to show that raising additional offspring adversely affected the nutritional condition of female Starlings (*Sturnus vulgaris*). The researchers moved young nestlings among nest boxes to create broods of two, four, or six, and induced females to grow out one rectrix while raising the experimental young. Birds with the largest broods worked the hardest; feeding visits per hour varied positively and significantly with brood size. Feather growth indicated that the added work had a nutritional cost; females raising the largest broods had significantly narrower growth bars and the overall mass of the completed rectrix was significantly lightest in that group. Interestingly, body mass completely failed as an indicator of nutritional stress. At the point when the birds were recaptured to pull the induced rectrix, the body mass of the females that had raised six nestlings was actually slightly greater than that of the other two treatment groups, but the difference was not significant.

R. A. Mauck and I (Mauck and Grubb, in press) are using ptilochronology as part of our investigation of age- and sex-specific reproductive effort in Leach's Storm-petrel. The results we now have suggest that parental investment in this species is sex-specific, but not age-specific. During the 1991 and 1992 seasons, we induced feathers on petrels ranging from 5 to 21 years of age at the start of the chick-rearing period. Because of the difficulty in reading growth bars in this species that I described above, we used total length of the growing feather as an index of nutrition. Since we pulled the induced feather before it was fully grown, and since we knew the number of days it had been growing, we used the number of days of growth as a covariate in the analysis of total length. Other covariates in our general linear model were total length of the original feather (to standardize for body size) and age of the adult. Sex was treated as a categorical variable. When the effects of length of the original feather (partial $F = 7.4$; $P < 0.0001$) and number of days since feather induction (partial $F = 78.5$; $P < 0.0001$) were removed, there was no statistical effect of age on growth of the original rectrix (partial $F = 0.50$; $P = 0.48$). The effect of sex was significant (partial $F = 8.12$; $P = 0.005$), with males growing their induced rectrix more slowly (1.74 ± 0.08 SD mm day^{-1}, n = 62) than females (1.79 ± 0.09 mm day^{-1}, n = 56) over an average growth period of 49.9 ± 3.0 SD days. While these data suggest that the nutritional condition of the males was less than that of the females, it is possible that males initiated feather regeneration later than did females. Such delayed initiation of growth, rather than any difference in rate of regeneration, could have accounted for the shorter total length. However, while possible, and also indicative of sex-specific nutritional condition, the mechanism for such a sex-specific delay in the start of feather regeneration is difficult to imagine.

Results from both the starling and petrel studies indicate the need to compare the subsequent fate of experimental and control birds. We need to relate apparent reductions in nutritional condition during one breeding season to reductions in survivorship and/or reproductive output in succeeding breeding seasons.

7.3. Nutritional Condition of Fledglings

At least one study is probing the relationship between the survivorship of fledglings and their nutritional condition during the weeks immediately preceding independence. Preliminary results (R. L. Mumme, G. E. Woolfenden, and T. C. Grubb, Jr., in prep.) indicate that nutritional condition in recently-fledged Florida Scrub Jays (*Aphelocoma c. coerulescens*) is improved by the experimental addition of food to their homeranges. In this species, most rectrix growth occurs after fledging, so we predicted that if added food improves nutritional status of postfledging jays, supplemented fledglings should grow rectrices at a faster daily rate (wider growth bars) than unsupplemented controls.

Four of 10 territories were food-supplemented daily from 0–30 days post-fledging; the other six territories were retained as unmanipulated controls. The prediction that growth bar width would be greater in supplemented territories was tested with a general linear model in which food supplementation was treated as a categorical variable and Julian hatch day and number of fledglings were treated as covariates. To ensure statistical independence of replicates, the territory, not the individual bird, was used as the primary sampling unit. With the small sample presently in hand, the analysis has not quite shown a significant difference between treatment and control territories ($P = 0.10$; Fig. 3), even though a biological effect is strongly suggested. Power analysis ($\beta = 0.05$; proportionate difference = 0.34; one-tailed $\alpha = 0.05$) indicates that experimental and control sample sizes of seven would be required to show the difference statistically. We are now accumulating records from additional territories.

7.4. Habitat Selection

I turn now to several studies that have applied ptilochronology to several aspects of habitat quality, social dominance in winter flocks, and caching. Because it appears to be sensitive to subtle variation in nutritional condition, the rate of feather growth has considerable potential as an indicator of habitat quality for birds. One group of possible

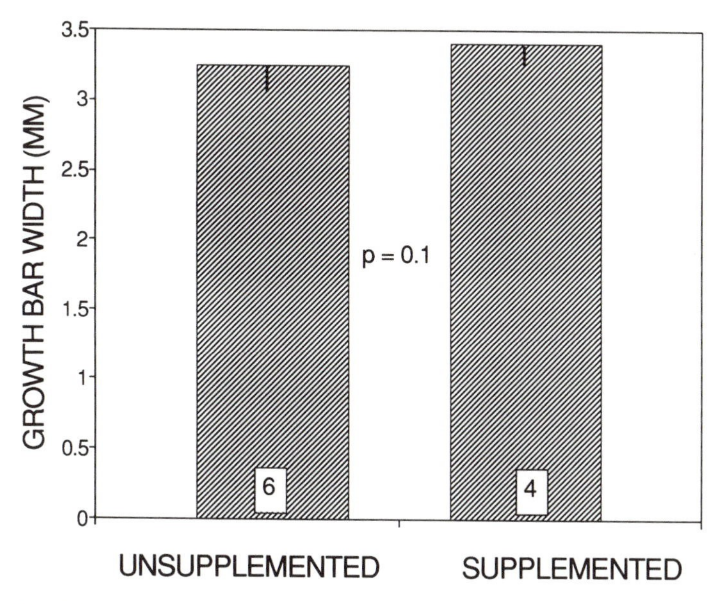

FIGURE 3. Adjusted mean growth bar width on the original R6 rectrix of unsupplemented and food-supplemented fledgling Florida Scrub Jays. Standard deviations and sample sizes (number of territories) are shown in the bars.

applications might involve testing hypotheses about the ultimate causation of habitat selection. Another might relate to quantity and quality of habitat in relation to the conservation of avian species. As an example of the latter application, Grubb and Yosef (1994) have recently completed an analysis of habitat quality as it affects nutritional condition of Loggerhead Shrikes.

We identified four habitat types in which this species resides permanently in southcentral Florida, built-up urban, palmetto scrub, citrus, and fenced pasture. We analyzed characteristics of an original rectrix (grown during the autumnal prebasic molt), and confined our analysis to adult birds under the assumption that, unlike recently dispersed juveniles, adults live-trapped in winter were likely to have previously bred and molted in the same location (Yosef, 1992). As we collected feathers from adult birds of both sexes during each of two winters, our general linear statistical model employed sex and year as categorical variables while treating tarsus length as a covariate to control for bird size.

Both growth bar width (Fig. 4) and total feather mass were signifi-

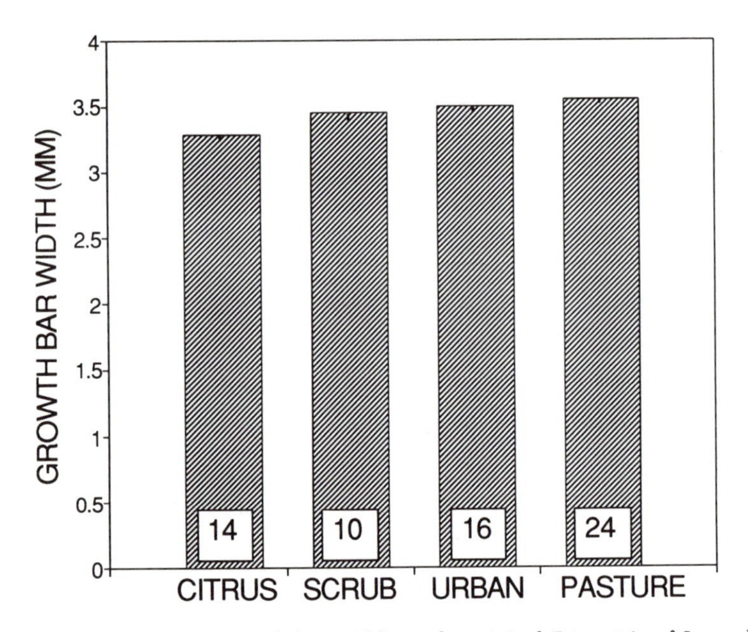

FIGURE 4. Adjusted mean growth bar width on the original R4 rectrix of Loggerhead Shrikes wintering in four habitat types in southcentral Florida. Standard deviations and sample sizes are shown in the bars.

cantly related to habitat type. Shrikes in CITRUS grew their rectrix less each day (partial $F = 5.90$; $P = 0.001$, $n = 64$) and less massively *in toto* (partial $F = 5.19$; $P = 0.003$) than did the average shrike across the four habitats. By contrast, shrikes in PASTURE grew a rectrix that was longer, heavier, and had wider growth bars than the average for the four habitats.

Taken together, these results suggest that in each of two years something about pastures and citrus groves, respectively, caused increased and reduced nutritional condition in molting resident shrikes. Pasture habitats contain the highest densities of shrikes in central Florida, suggesting that a combination of short grass and abundant fence-post hunting perches provides optimal foraging habitat and consequent excellent nutrition for the species.

In Florida, citrus is routinely sprayed during the autumn with mitacide/insecticide compounds known to be toxic to nontarget organisms. Such compounds could have had a chronic, sublethal effect on shrikes sufficiently intense to retard feather growth. Alternatively, application of pesticides in the groves could have depressed the shrikes' food supply, reducing the birds' nutritional condition sufficiently to be detectable by ptilochronology. It seems possible, also, that reduced nutritional condition could have been caused by both direct and indirect effects of citrus grove management practices. In cases such as this where feather growth indicates substandard habitat quality, it will be important to collect corroborating evidence such as feeding rates, survivorship, and fecundity.

In another application of ptilochronology to habitat quality in a species at risk, D. C. Rudolph and R. N. Conner (unpubl.) obtained evidence suggesting that nutritional condition in a Texas population of Red-cockaded Woodpeckers (*Picoides borealis*) was habitat-specific. These workers pulled an original rectrix from adults living either in mixed stands of loblolly (*Pinus taeda*) and shortleaf (*Pinus echinata*) pine or in pure stands of longleaf pine (*Pinus australis*). In a three-way ANOVA, the difference between the adjusted average width of six growth bars from birds in the two habitat types approached significance ($P = 0.12$). *T*-test comparisons revealed that average growth bar width of males did not differ between habitats ($P = 0.72$), but the same comparison in females approached significance ($P = 0.12$). Males may have been buffered against the full impact of the poorer habitat because their social dominance over females allowed them to monopolize higher quality feeding areas. Such an interpretation is supported by experimental results obtained for the congeneric Downy Woodpecker, also a species in which males dominate females socially. Wintering female downies grew an induced feather faster when given supplemental food, but there was no analogous increase for males of that species (Grubb and Cimprich, 1990). Interestingly, in the case of the Red-cockaded Woodpeckers, the apparent habitat-specific variation in nutritional condition of adult females indexed by feather growth was paralleled by a difference in provisioning rate of nestlings; those in longleaf pine were fed at a much higher rate.

7.5. Social Dominance

Two studies have used ptilochronology to examine possible consequences of social dominance for a bird's nutritional condition, one within species and one across species. Norwegian Willow Tits start the winter in monospecific flocks consisting of a dominant male and female that have bred in the area the previous summer, and two male–female pairs of juveniles that have dispersed from their natal areas and immigrated into the dominant pair's territory. Within a flock of six, all males are socially dominant to all females at feeders and everywhere else in the flock home range. There is also a rigid dominance hierarchy within each sex, with the adult most dominant. Thus, the dominance hierarchy within a flock is adult male > one juvenile male > the other juvenile male > adult female > one juvenile female > the other juvenile. In a nonmanipulative study, Hogstad (1992) used ptilochronology to suggest that the behavior of the adult male toward his mate increased her nutritional condition. Hogstad noted that when the alpha male was close (<5 m) to the alpha female, the proportion of total time she spent foraging (0.70) and scanning (0.30) were greater and less, respectively, than when the alpha male was not close to her (>5 m). Hogstad interpreted these results to indicate that when the alpha male was close by, the female no longer needed to be quite so vigilant against supplanting attacks for food by the juvenile males, both of whom were dominant to her in one-on-one interactions, so she could devote more of her time to foraging for food. Comparison of growth bar widths tended to reinforce this conclusion. The adult female ranked lower in social dominance than the juvenile males, but she grew an induced rectrix just as rapidly. As expected, growth bar width in the alpha male and juvenile females was significantly wider and narrower, respectively, than that in the adult female and juvenile males. These results were interpreted as indicating that the increased foraging time provided to the alpha female by her mate's guarding behavior allowed her to remain in better nutritional condition. Time and energy spent by the alpha male in mate defense would be selected for if the alpha female's enhanced condition resulted in an earlier start to egg-laying the following spring. There is now good evidence that early breeding is adaptive because it produces early dispersers that have a better-than-average chance of finding a position within a flock the following winter (Nilsson, 1987).

Hogstad's interpretation would be buttressed by a demonstration that an alpha female deprived of the protection of an alpha male would suffer a reduction in nutritional condition, as indicated by narrower growth bars. Unfortunately, such a manipulation may be difficult to perform because it is known that when an alpha member of a flock of Willow Tits disappears, the beta member of the same sex almost immediately divorces its present mate to pair with the alpha bird of the opposite sex (Hogstad, 1989). Thus, if the alpha male were to be removed, the female might well continue to receive mate protection, but now from the same male that had been defended against while it was in the beta position.

Ptilochronology has also been employed to search for nutritional consequences of social dominance between species. Among the well-studied heterospecific winter foraging flocks of bark-foraging birds, there is usually a clear-cut social dominance hierarchy among species. In Ohio woodlots, for example, the dominance hierarchy consists of Downy Woodpeckers > White-breasted Nuthatches > Tufted Titmice > Carolina Chickadees. Cimprich and I (Cimprich, 1992; Cimprich and Grubb, 1994) have simultaneously tested two contrasting predictions about nutritional condition within such flocks. One prediction was deduced from competition theory and the other from mixed-species flocking theory.

Competition theory applied to mixed-species flocks holds that, when in the presence of socially dominant heterospecifics, socially subordinate birds should be displaced to parts of their niche not occupied by the dominant (Morse, 1974). Under such circumstances, one would predict that the nutritional condition of the subordinates would suffer. By contrast, mixed-species flocking theory holds that in the flocks of larger size made possible by foraging with heterospecifics in addition to conspecifics, a bird should be able to devote less time to being vigilant for predators and more time to looking for food (Barnard and Thompson, 1985). Thus, birds of socially subordinate species should realize better nutritional condition in the company of heterospecifics even if those heterospecifics are socially dominant.

Cimprich and I tested these predictions by manipulating the species composition of bark-foraging birds within forest fragments. Here in the Midwest, woodland remnants persist as habitat islands within an ocean of soybeans and corn. Between the fall dispersal of juveniles, ending in October, and the spring dispersal of young birds that have failed to secure a breeding territory where they have overwintered, there is a period of several months during which birds of several species remain completely sedentary. In particular, during this period, birds that disappear or are removed by a researcher from one of these island woodlands are not replaced. This fact allowed us to remove one species from experimental woodlots and then to compare with controls the behavioral, nutritional, and survivorship responses of the remaining species. We concentrated on one congeneric pair of species, the socially subordinate Carolina Chickadee and the socially dominant Tufted Titmouse. At the beginning of each of several winters, we transported all titmice from island woodlands to distant release sites, and pulled a rectrix from the chickadees that we then released back into the woodland fragment. By recording behavior and retrapping birds at the end of the winter, we were able to amass the responses of the chickadee to loss of the socially dominant flockmate.

For both female ($P = 0.09$) and male ($P = 0.07$) chickadees, there was a nearly-significant *increase* in daily growth of an induced rectrix after removal of the titmice. For females, the increase was six percent; for males it was ten percent.

A similar increase in the absence of titmice also occurred in mass of the induced feather; this increase reached statistical significance in females ($P = 0.01$). There was a statistically significant interaction between titmouse removal and availability of American beech (*Fagus americana*) mast, an important, yet sporadic, winter food source. In mast years in our study area, chickadees having found a "nut" often are unable to open it before being supplanted by a titmouse. During 1990–91, a year in which beech mast was to be easily found by birds probing the leaf litter, female chickadees in the presence and absence of titmice, respectively, produced the lowest (3.71 ± 0.05 mm) and highest (4.00 ± 0.06 mm) recorded daily growth of an induced rectrix. During two other winters when beech mast was scarce, daily induced feather growth in female chickadees was uniformly low in both treatment and control woodlots.

Our general conclusion from analysing the responses of chickadees to the removal of titmice has been that under normal circumstances, chickadees pay a nutritional price for foraging with the dominant congener. While it is possible that the loss in nutritional condition incurred in the presence of titmice is more than compensated for by a reduction in risk of hawk predation provided by the extra vigilance capability of the titmice, we have not been able to demonstrate any change in within-winter mortality of chickadees after titmouse removal. We are now midway through the reciprocal experiment in which we are assessing the responses of the socially dominant titmouse to the removal of chickadees (Pravosudov and Grubb, in prep.). It will be instructive to see whether the asymmetry in social dominance produces an asymmetry in relative daily feather growth in the absence of the congener.

7.6. Caching

Research on caching behavior has so far employed ptilochronology in two ways, to confirm the adaptive significance of caching and to investigate the distribution of cached items among potential consumers. Although it would be quite surprising if birds that expend large amounts of energy in caching food realize no return for their effort, it has been worthwhile to confirm the nutritional benefits of the behavior. During the summer and autumn months, Gray Jays (*Perisoreus canadensis*) in northern Alaska cache prodigious amounts of food, everything from blueberries to bits of dead moose. During October, Waite (1990) provided Alaskan jays in six flock territories with pellets of dried dogfood, which the birds promptly cached in surrounding trees. Four territories were retained as controls. About a month later, he caught all the birds and removed one rectrix from each before releasing them back into their territories. Thus, induced feather growth commenced several weeks after the supplementing period, by which time the only extra food present was that which the birds had cached. While regenerating the rectrix, the birds existed under conditions of such short daylength that they were active for as little as 4 hours during each 24-hour interval. In February, the jays were recaptured and their induced rectrix pulled.

Feather growth indicated that the cached supplemental food improved the jays' nutritional condition. Using the territory as the primary sampling unit, Waite calculated that growth bar width, rectrix mass and rectrix length were 9.2%, 10.6%, and 7.1% greater, respectively, in the supplemented birds than in the controls. All these differences were statistically significant. Furthermore, in all the supplemented territories every member of the flock of jays grew an induced rectrix, while in three of the four control territories, at least one jay failed completely to grow a replacement. Waite's results strongly indicate that over the several weeks following a period of food storage, birds can realize a nutritional advantage from the sequestered food.

The time window over which stored food is known to augment nutritional condition has recently been extended. Nilsson *et al.* (1994)

also selected territories to be supplemented or left as controls, but their system was the mosaic of territorial pairs of Eurasian Nuthatches (*Sitta europaea*) wintering within island woodlots in the farm country of southern Sweden. Having allowed pairs to cache unlimited numbers of sunflower seeds over a period of 30 days during autumn, they refrained from pulling the original feather of controls and experimentals until fully 40–56 days after the remaining supplementary food had been removed. Even with such a delay, the induced retrices of supplemented birds were longer, heavier and had wider growth bars than those of controls. Though the strength of the inferences to be drawn from these results is limited because the bird rather than the woodlot was treated as the primary sampling unit, they do suggest that the nuthatches were prudent hoarders, using their caches only when required to during energetically stressful periods of the winter. Behavioral records of cache retrieval supported this idea. In three females whose foraging behavior was monitored closely, there was a statistically significant negative relationship between ambient temperature and rate of cache retrieval and consumption. For instance, at +8 °C, each bird retrieved about two seeds per hour, while at −8 °C, the consumption rate had risen to eight seeds per hour, a four-fold increase. This relationship appeared to be independent of the number of cached seeds remaining because hourly consumption rate over the course of the winter was independent of the number of days since feeders had been removed. The ability to adjust the use of a sequestered food supply in response to environmental conditions extending over a period of weeks or months would seem to be a strong selective force favoring winter territoriality.

Over the past decade or so some workers have portrayed the caching strategy of scatter hoarders as an optimal trade-off between benefits and costs (e.g., Hurley and Robertson, 1987; Waite, 1991). A scatter-hoarder would use its time and energy available for caching most efficiently if it concentrated its caches in the immediate vicinity of the food items' source. However, it has been argued that the chance of a naive competitor finding such food items by area-restricted search is an inverse function of how densely they have been cached. This line of thinking has lead to the proposition that in each circumstance there should be some optimal cache density that produces the lowest ratio of cache loss to caching effort. Until quite recently, this body of theory has not been fully accessible to testing because of the problem of determining the consumption rates of caches by any given cacher and its competitors for the caches. Brodin's (1993) ingenious use of "radio-ptilochronology" appears to provide a method for testing the theory empirically.

Brodin showed that if a bird is permitted to consume a food item containing radioactive sulfur (S^{35}) while growing an induced feather, the radioactivity is quickly incorporated into that day's growth bar. In preliminary results now being augmented, Brodin showed that if he allowed only one member of a pair of Willow Tits to cache radioactive sunflower seeds, radioactivity showed up not only on the induced rectrix of the cacher, but also on the induced feather being grown at the same time by its mate. Furthermore, the result was symmetrical; both the male and female of a pair consumed seeds cached by the mate. It appears that all necessary information can be obtained to calculate from the intensity of the image on an autoradiograph the number of radioactive seeds consumed per day by any given bird. Thus, Brodin seems to have provided us with a powerful tool for assessing the validity of optimal caching theory. The radio-labeling technique could also tell us the extent to which long-term cachers like Waite's jays and Nilsson's nuthatches feed their nestlings with caches made the previous autumn.

8. PROSPECTUS

We now have considerable support for the validity and sensitivity of ptilochronology as an index of nutritional condition, and we are seeing the method employed to address a variety of conceptual issues. Here, I consider lingering uncertainties about the method that require additional attention, before ending this review with some suggestions for how ptilochronology might be employed in the future.

Additional information on the basic mechanics and physiology of feather growth would be welcome. What, exactly, causes dark bands and light bands? Is the causal mechanism for feather length really independent from the causal mechanism for daily growth, as suggested by the results of White and Kennedy (1992)? If so, how does a follicle "know" when to stop producing a feather? How are growth bars related to pigment bars? For example, is one of the dark pigment bars on a House Wren (*Troglodytes aedon*) rectrix laid down every 24 hours? On a rectrix of a Ring-necked Pheasant (*Phasianus colchicus*) or female American Kestral (*Falco sparverius*) is some constant multiple of the 24-hour period represented by feather growth between successive dark pigment bars? If such a relationship exists, then rate of feather growth could be employed as a nutritional index even in species with feathers too heavily pigmented for growth bars to be discerned.

Reduced feather growth appears to be a reliable indicator of nutritional condition, but how precise is the relation between the two? Does reduced feather growth of a rectrix reflect the cost of regenerating that one feather, or does it indicate a general energy-conservation syndrome where many physiological systems are slowed simultaneously? For example, when Waite's unsupplemented Gray Jays were regenerating a rectrix at a slower rate than cache-supplemented birds, were they also producing red blood cells at a slower rate and were their claws growing more slowly? In short, is induced feather growth just one of a number of potential indices of induced anabolism that one could use to calibrate nutritional condition (Grubb, 1994)?

We need more controlled experiments detailing the relationship between extent of dietary deficiency and reduction in feather growth rate. First, we need more assurance that the relationship between the two variables found in Carolina Chickadees (Grubb, 1991) is universal. Second, the degree of the relationship needs to be quantified more precisely. Ideally, we would like to be able to say that a certain reduction in growth bar width, say of Loggerhead Shrikes in citrus groves, corresponds to a certain percentage shortfall in the net intake of assimilable energy. Comparison of daily induced feather growth suggests that supplemented Eurasian Nuthatches wintering in Sweden were in better nutritional condition than supplemented White-breasted Nuthatches wintering in Ohio (Fig. 5). Were they? If so, why? Waite's finding that some cache-unsupplemented jays completely failed to regenerate a rectrix suggests the presence of a threshold value of nutritional condition necessary for feather regeneration. What is this value? How and why does it vary across species and/or environments?

How important are potentially confounding causes of variation in feather growth that have not been examined? Most avian species can only be aged as juvenile or adult. Does senility affect feather regeneration in very old birds? Despite our present thinking to the contrary, are there any populations of wild birds in which feather growth is limited by some specific nutrient other than assimilable energy?

As a practical matter, how close in time can one pull sequential feathers from the same follicle without affecting feather growth? I have noticed in some cases that a rectrix induced shortly before the normal molting period will not be molted and replaced in the normal fashion. Instead, it will be retained and carried for months, and quite possibly to the next molt, in company with an otherwise full complement of new rectrices. While only anecdotal, this observation indicates that some influence of follicle history on feather growth does occur.

How valid is feather growth rate as an indicator of relative fitness? We are in need of direct evidence that birds with a lower rate of feather

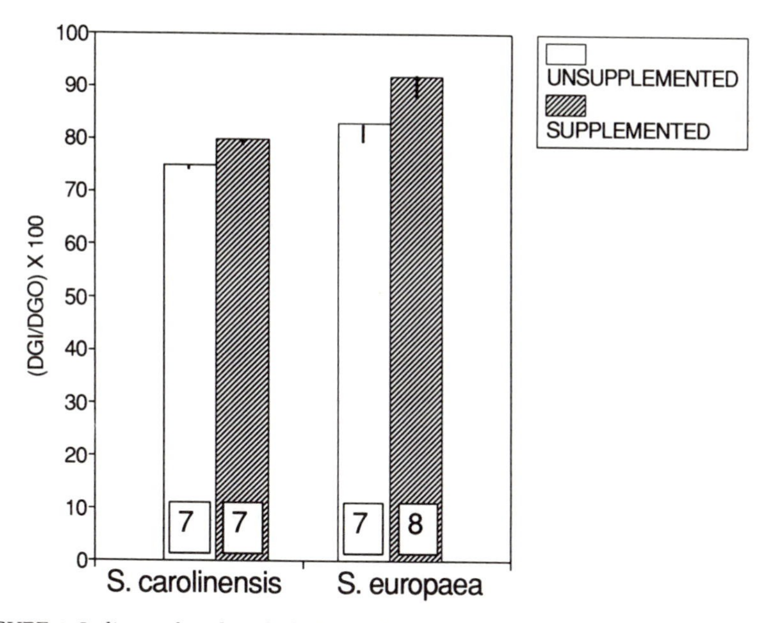

FIGURE 5. Indices of induced daily feather growth of unsupplemented and food-supplemented White-breasted and Eurasian Nuthatches. Standard deviations and sample sizes are shown in the bars.

regeneration also suffer a reduction in survivorship, fecundity, or both. At this point, our only evidence for such relationships is indirect. Both the width of growth bars on an induced rectrix (Hogstad, 1992) and survivorship (Koivula and Orell, 1988) increased with increasing social status within flocks of wintering Willow Tits, but the two studies were performed in different years and in different countries. As detailed earlier, White *et al.* (1991) found an inverse relationship between manipulated brood size and the width of growth bars on an induced feather grown by female Starlings feeding nestlings. In a different study of the same species, survivorship was also inversely related to brood size (Clobert *et al.*, 1987).

Notwithstanding the caveats raised in the paragraphs above, and undoubtedly others as well, it does appear that ptilochronology will be a useful tool in many projects where nutritional condition and its assumed relationship with fitness are of interest. The regeneration of a feather may be a type of induced anabolism uniquely suitable for manipulative study. First, a fully-grown feather is held within the follicle solely by dead connective tissue, so inducing regeneration apparently results in little or no discomfort to the bird. Second, the loss of one tail feather probably has little of no effect on a bird's ability to fly, or on the insulation or water-repellency of the feather coat. Thus, the removal of a feather has little impact on fitness independent of anabolic demand. Third, even after the induced feather has been fully grown, it can be divided into daily growth increments, so multiple captures are not required to determine rate of anabolism. Fourth, in the majority of bird species, an induced feather is retained until the next molt, so a record of nutritional condition can be obtained weeks or months after the original feather has been pulled.

Finally, ptilochronology seems particularly suitable as a conservation tool for detecting small changes in the quality of avian habitats. Reduced rates of feather growth could provide an early warning of habitat degradation and could alert us to which habitats would be most likely to become population sinks.

ACKNOWLEDGMENTS. I thank my students and colleagues for advice and discussion, for their company during many happy days of field work, and for permission to use unpublished material. My research has been supported by The Ohio State University, The Frank M. Chapman Fund of The American Museum of Natural History, and National Science Foundation.

REFERENCES

Afanasyev, V., and Prince, P. A., 1993, A miniature storing activity recorder for seabird species, *Ornis Scand.* **24**:243–246.

Barnard, C. J., and Thompson, D. B. A., 1985, *Gulls and Plovers: The Ecology and Behaviour of Mixed-Species Foraging Groups*, Croom Helm, London, England.

Breitwisch, R., 1989, Mortality patterns, sex ratios, and parental investment in monogamous birds, in: *Current Ornithology*, vol. 6 (D. M. Power, ed.), Plenum Press, New York, pp. 1–50.

Brodin, A., 1993, Radio-ptilochronology-tracing radioactively labelled food in feathers, *Ornis Scand.* **24**:167–173.

Cimprich, D. A., 1992, Costs and benefits for the Carolina Chickadee (*Parus carolinensis*) of flocking in winter with a socially dominant congener, the Tufted Titmouse (*Parus bicolor*), Unpubl. M.S. thesis, The Ohio State University, Columbus, Ohio.

Cimprich, D. A., and T. C. Grubb, Jr., 1994, Costs and benefits for the Carolina Chickadee (*Parus carolinensis*) of flocking in winter with a socially dominant congener, the Tufted Titmouse (*Parus bicolor*), *Ecology* (in press).

Clobert, J., Bauchau, V., Dhondt, A. A., and Vansteen Wegen, C., 1987, Survival of breeding female Starlings in relation to brood size, *Acta Oecologica Oecol. Gener.* **8**: 427–433.

Ekman, J. B., and Lilliendahl, K., 1993, Using priority to food access: Fattening strategies in dominance-structured Willow Tit (*Parus montanus*) flocks, *Behav. Ecol.* **4:** 232–238.

Evans, R. M., 1990, The relationship between parental input and investment, *Anim. Behav.* **39:**797–798.

Grubb, T. C., Jr., 1972, Smell and foraging in shearwaters and petrels, *Nature* **237:** 404–405.

Grubb, T. C., Jr., 1989, Ptilochronology: Feather growth bars as indicators of nutritional status, *Auk* **106:**314–320.

Grubb, T. C., Jr., 1991, A deficient diet narrows growth bars on induced feathers, *Auk* **108:**725–727.

Grubb, T. C., Jr., 1994, On induced anabolism, induced caching, and induced construction as unambiguous indices of nutritional condition. *Western Found. Vert. Zool.* (in press).

Grubb, T. C., Jr., and Cimprich, D. A., 1990, Supplementary food improves the nutritional condition of wintering woodland birds: evidence from ptilochronology. *Ornis Scand.* **21:**277–281.

Grubb, T. C., Jr., and Pravosudov, V. V., 1994, Ptilochronology: Follicle history fails to influence growth of an induced feather. *Condor* (in press).

Grubb, T. C., Jr., Waite, T. A., and Wiseman, A. J., 1991, Ptilochronology: induced feather growth in Northern Cardinals varies with age, sex, ambient temperature, and day length, *Wilson Bull.* **103:**435–445.

Grubb, T. C., Jr., and Yosef, R., 1994, Habitat-specific nutritional condition in Loggerhead Shrikes (*Lanius ludovicianus*): Evidence from ptilochronology, *Auk* (in press).

Hogstad, O., 1989, Social organization and dominance behavior in some *Parus* species. *Wilson Bull.* **101:**254–262.

Hogstad, O., 1992, Mate protection in alpha pairs of wintering Willow Tits, *Parus montanus, Anim. Behav.* **43:**323–328.

Hurley, T. A., and Robertson, R. J., 1987, Scatterhoarding by territorial red squirrels: a test of the optimal density model, *Can. J. Zool.* **65:**1247–1252.

King, J. R., and Murphy, M. E., 1984, Fault bars in the feathers of White-crowned Sparrows *Zonotrichia leucophrys gambeli:* Dietary deficiency or stress of captivity and handling, *Auk* **101:**168–169.

Koivula, K., and Orell, M., 1988, Social rank and winter survival in the Willow Tit, *Parus montanus, Ornis Fenn.* **65:**114–120.

Mauck, R. A., and Grubb, T. C., Jr. In press. Parent petrels shunt added reproductive costs to offspring. *Anim. Behav.*

Michener, H., and Michener, J. R., 1938, Bars in flight feathers, *Condor* **40:**149–150.

Morse, D. H., 1974, Niche breadth as a function of social dominance, *Amer. Natur.* **108:**818–830.

Murphy, M. E., and King, J. R., 1991, Ptilochronology: A critical evaluation of assumptions and utility, *Auk* **109:**673–676.

Nilsson, J-.A., 1987, Establishment of juvenile Marsh Tits in winter flocks: An experimental study. *Anim. Behav.* **38:**586–595.

Nilsson, J-.A., Kallander, H., and Persson, O., 1994, A prudent hoarder: Effects of long-term hoarding in the European Nuthatch, *Ornis Scand.* (in press).

Trivers, R., 1972, Parental investment and sexual selection, in: *Sexual Selection and the Descent of Man,* (B. Campbell, ed.), Aldine, Chicago.

Waite, T. A., 1990, Effects of caching supplemental food on induced feather regeneration in wintering Gray Jays *Perisoreus canadensis:* A ptilochronology study, *Ornis Scand.* **21:**122–128.

Waite, T. A., 1991, Economics and consequences of scatterhoarding in gray jays (*Perisoreus canadensis*), Unpubl. Ph.D. dissertation, The Ohio State University, Columbus, Ohio.

Watson, G. E., 1963, Feather replacement in birds, *Science* **139**:50–51.

White, D. W., and Kennedy, E. D., 1992, Growth of induced feathers in photostimulated American Tree Sparrows, *Condor* **94**:543–545.

White, D. W., Kennedy, E. D., and Stouffer, P. C., 1991, Feather regrowth in female European Starlings rearing broods of different sizes, *Auk* **108**:889–895.

Wood, H. B., 1950, Growth bars in feathers, *Auk* **67**:486–491.

Yosef, R., 1992, Territoriality, nutritional condition, and conservation in Loggerhead Shrikes (*Lanius ludovicianus*), Unpubl. Ph.D. Dissertation, The Ohio State University, Columbus, Ohio.

Yosef, R., and Grubb, T. C., Jr., 1992, Territory size influences nutritional condition in nonbreeding Loggerhead Shrikes (*Lanius ludovicianus*): A ptilochronology approach, *Conserv. Biol.* **3**:447–449.

Zuberbier, G. M., and Grubb, T. C., Jr., 1992, Ptilochronology: Wind and cold temperatures fail to slow induced feather growth in captive White-breasted *Nuthatches Sitta carolinensis* maintained on ad libitum food, *Ornis Scand.* **23**:139–142.

CHAPTER 4

INDIVIDUAL VOICE DISCRIMINATION IN BIRDS

MARCEL M. LAMBRECHTS
and ANDRE A. DHONDT

1. INTRODUCTION

Animals discriminate among familiar and unfamiliar conspecifics probably to avoid wasting time and energy during social interactions. Individual recognition may reduce the chance of expensive fights among territorial neighbors during resource defense, or allow mates, kin, or parents and offspring to find each other where familiar individuals intermingle with unfamiliar ones. Because individual recognition may involve a time-consuming learning process, or because it may be costly in some other way, individual recognition will probably only evolve when benefits outweigh these costs (e.g., Beer, 1970; Wilson, 1975; Falls, 1982, Beecher and Stoddard, 1990).

Vocalizations possess both species-specific characteristics and individually distinctive features (e.g., Borror, 1959; Marler, 1960; Schleidt, 1976; Becker, 1982; Falls, 1982; Nelson, 1989; Elfström, 1990). Marler (1960) predicted that vocal features used to recognize species

MARCEL M. LAMBRECHTS • CNRS-CEFE, B. P. 5051, 34000 Montpellier Cedex 1, France. ANDRE A. DHONDT • Cornell Laboratory of Ornithology, 159 Sapsucker Woods Road, Ithaca, New York, 14850.

Current Ornithology, Volume 12, edited by Dennis M. Power. Plenum Press, New York, 1995.

should differ from those used to recognize individuals within species. Species recognition by vocalizations may require that vocal characteristics show little or no variation among conspecific individuals. By contrast, individual recognition would be most efficient when vocal characteristics differ a lot among individuals with little variation within individuals. While species recognition is not necessarily the result of a learning process, it is generally assumed that individual recognition is completely learned.

It is well established that birds recognize the calls of mates, and use vocalizations in parent–offspring recognition, kin recognition, or neighbor–stranger discrimination (Beer, 1970; Falls, 1982, for reviews of studies before 1982) (e.g., Beecher, 1982; Jouventin, 1982; McArthur, 1982; Stoddard and Beecher, 1983; Beecher et al., 1985; Sieber, 1985; Beecher *et al.*, 1986; Medvin and Beecher, 1986; Payne *et al.*, 1988; Saunders and Wooller, 1988; Van Elsacker *et al.*, 1988; Ydenberg *et al.*, 1988; Beecher, 1990; Beecher and Stoddard, 1990; Elfström, 1990; Galeotti and Pavan, 1991; Payne *et al.*, 1991; Speirs and Davis, 1991; Weary *et al.*, 1992; Chaiken, 1992; Kent, 1992; Galeotti and Pavan, 1993; Falls, 1992; McGregor, 1993). Studies of mate recognition and neighbor–stranger recognition do provide direct evidence for individual vocal recognition. Studies of parent-offspring discrimination did not test whether parents discriminate among their offspring or whether offspring recognize each parents individually (see Falls, 1982).

Knowing exactly which vocal cues are important for individual discrimination can give valuable information about the speed, the timing, and the potential significance of individual recognition. Different vocal features can potentially be used to discriminate among individuals (e.g., Schleidt, 1976; Falls, 1982; Beecher, 1989; Brémond *et al.*, 1990; Elfström, 1990; Robisson, 1990; Horning *et al.*, 1993; Weary *et al.*, 1990; Chaiken, 1992; Robisson, 1992; Bauer and Nagl, 1992). It does not mean, however, that all the individually distinctive traits ("signature" traits) are used. In comparison to calls, bird song often shows a higher level of structure complexity. Therefore, relatively more signature traits can potentially be used in individual recognition by song. In many songbirds, individuals sing a number of distinct themes (called song, phrase or syllable types) of the conspecific song to form a song repertoire. The songs constituting the repertoire, the rate of song switching, and the rendition of songs all may differ considerably among individual birds (e.g., Kroodsma and Miller, 1982; Weary *et al.*, 1990). A few experimental studies were able to determine vocal features that are important for individual recognition. White-throated Sparrows (*Zonotrichia albicollis*) recognize neighbors by the frequency (pitch) of the first three notes of a song (Brooks and Falls, 1975). Field Sparrows (*Spizella pusilla*) increase response strength when the neighbor's songs are lowered by about 400 Hz (Nelson, 1989), suggesting that song frequency characteristics are used in neighbor recognition. Meadow Pipits (*Anthus pratensis*) discriminate among individual neighbors when presented with playback of the introductory phrase of a song (Motif I), but not when presented with the terminating part of a song (Motif II) (Elfström, 1990). Other than these studies, little is known about processes of individual discrimination by song alone.

The aim of this paper is not to give a complete review of individual recognition of vocalizations by birds. Here, we will review and discuss hypotheses about processes of individual recognition recently set forth in the bird song literature, with special emphasis on work related to individual "voice" recognition.

2. NEIGHBOR–STRANGER DISCRIMINATION AND EVIDENCE FOR INDIVIDUAL RECOGNITION BY SONG

Neighbor–stranger discrimination is almost exclusively reported in noncolonial birds (Table I; but see Speirs and Davis, 1991, for a colonial bird) (see Falls, 1982, Ydenberg *et al.*, 1988). In most experiments testing neighbor–stranger discrimination, territorial birds were presented with a single playback stimulus (one rendition of a song) at the center or the edge of the territory, and the response quantified before, during, and/or after the playback period (e.g., interval between the start of playback and the first response, approach, minimal distance from the stimulus, call rate, song rate, number of flights). Differences in the extent of neighbor–stranger discrimination between the different studies could to some extend be due to the number of speakers used during the experiment (one or two playback stimuli in the territory; e.g., Searcy *et al.*, 1981; Stoddard *et al.*, 1990), the origin of the nonneighbor song (e.g., from a distant nonneighbor, or not [Kroodsma, 1976]), the order of delivery of songs of neighbors and strangers (e.g., Falls and d'Agincourt, 1981; Hansen, 1984), the stimuli used during the experiment (e.g., playback with or without a stuffed bird; Belcher and Thompson, 1969, *v.* Emlen, 1971), events to which subjects were exposed before the start of the experiment (e.g., a neighbor entered the territory of the subject before the experiment; Godard, 1993), the time of the year when the experiments were carried out (e.g., subjects were not habituated to neighbor's song at the beginning of the territorial season; Harris and Lemon, 1976), the density of the population in which the

TABLE I
Playback Studies of Neighbor–Stranger Discrimination (NSD) and [a]Individual Neighbor Recognition (INR) in Territorial Non-cooperative Songbirds in Relation to the Song Repertoire Size and the Playback Stimulus Used.

	Song repertoire size	Playback stimulus	Evidence for NSD	Evidence for INR	Reference
Seiurus aurocapillus	1 song	1 song	yes	—	Weeden and Falls, 1959
Zonotrichia albicollis	1 song	1 song	yes	—	Falls, 1969
Z. albicollis		1 song	yes	—	Lemon and Harris, 1974
Z. albicolis		1 song	yes	yes	Brooks and Falls, 1975; Falls and Brooks, 1975
Passerina cyanea	1 song	1 song	yes	—	Emlen, 1971
		1 song	no	—	Belcher and Thompson, 1969
Spizella pusilla	1 song	1 song	yes	—	Goldman, 1973
Geothlypis trichas	1 song	1 song	yes	yes	Wunderle, 1978
Zonotrichia leucophrys	1 song	1 song	yes	—	Baker *et al.*, 1981
Anthus pratensis	1 song 3–6 syllables	1 song	yes	yes	Elfström, 1990
Parus atricapillus	1 song(?)	?	yes	?	Ratcliffe, 1990
Hylocichla fuscescens	1–3 songs	1–3 songs	yes	—	Weary *et al.*, 1987
Emberiza citrinella	1–4 songs	3 songs	yes	—	Hansen, 1984
Fringilla coelebs	1–6 songs	1 song	weak	—	Pickstock and Krebs, 1980
Parus major	1–8 songs	1 song	yes	—	Krebs, 1971
P. major		1 song	weak	—	Järvi *et al.*, 1977
P. major		1 song	yes	—	Falls *et al.*, 1982
P. major		1 song	yes	yes	McGregor and Avery, 1986; McGregor, 1993

Agelaius phoenicus	1–8 songs	1 song	yes	—	Yasukawa *et al.*, 1982
A. phoenicus	2 songs	1 song[a]	no	—	Beletsky, 1983
Setophaga ruticilla	2–8 songs	2–5 songs	yes	—	Weary *et al.*, 1992
Wilsonia citrina	2–9 songs	1 song	—	yes	Godard, 1991
W. citrina		1 song	yes	yes	Godard, 1993
Pipilo erythrophthalmus	3–8 songs	1 song[b]	yes	—	Richards, 1979
Melospiza georgiana	4–5 songs	1 song	yes	—	Searcy *et al.*, 1981
Sturnella neglecta	5–10 songs	1 song[c]	yes	—	Falls, 1982
S. neglecta		2 songs	yes	—	Falls and d'Agincourt, 1981
Melospiza melodia	6–12 songs	1 song	weak	—	Harris and Lemon, 1976
M. melodia		1 song	weak	—	Kroodsma, 1976
M. melodia		1 song	"very weak"	—	Searcy *et al.*, 1981
M. melodia		1 song[c]	yes	—	Beecher *et al.*, 1990
M. melodia		1 song	yes	yes	Stoddard *et al.*, 1991
Parus bicolor	8–12 songs	1 song	yes	no	Schroeder and Wiley, 1983
Dendroica petechia	20 songs	4–6 songs	yes	—	Weary *et al.*, 1992
Zosterops lateralis chlorocephala	49–60 syllables	?	no	—	Slater, 1991
Sturnella magna	50–100 songs	2 songs	weak	—	Falls and d'Agincourt, 1981
Erithacus rubecula	100–250 phrases	36–48 phrases	yes	yes	Brindley, 1991
E. rubecula		?	yes	—	Brémond, 1968
Ficedula hypoleuca	"many" songs	?*a*	yes	?	Lampe and Slagsvold, 1994
Icteria virens	"very large"	?	yes	yes	Ritchison, 1988

[a]: Test with females.
[b]: Two songs including one control song.
[c]Two-speaker design.
? Unknown.
— Not tested.
Weak: Significant response in only a few response measures, but a weak tendency for a stronger to strangers.

experiments were carried out (e.g., Harris and Lemon, 1976), loudspeaker placement (e.g., center versus edge effect; Beletsky, 1983; Giraldeau and Ydenberg, 1987; Stoddard *et al.*, 1990), the history of the bird (e.g., whether the subject learned much songs of neighbors before, or not; McGregor and Avery, 1986) and/or the response measure considered (e.g., matched counter-singing versus other response measures; Falls *et al.*, 1982; Stoddard *et al.*, 1992). In the majority of the studies birds responded more strongly to songs of strangers than to songs of neighbors, even if neighbor–stranger discrimination was weak (e.g., Harris and Lemon, 1976; Pickstock and Krebs, 1980; Järvi *et al.*, 1977) (see Falls, 1982, for the distinction between neighbor–stranger discrimination and kin or dialect recognition).

Neighbor–stranger discrimination was often interpreted as individual recognition, although relatively few studies demonstrated individual neighbor recognition (Table I). Falls and Brooks (1975), and others (Table I), showed that territorial birds respond strongly to the song of a stranger, and weakly to the song of a neighbor when the song comes from the neighbor's normal territory boundary. However, when a neighbor's song is played from the "wrong" territorial boundary (i.e., a different boundary from where the bird normally sings), the subject responds to the song as if it was sung by a stranger. Therefore, birds make a clear association between a particular territory and a particular song, which suggests that they use song to identify individuals rather than neighbors as a class.

Neighbor–stranger recognition has often been attributed to the phenomenon of habituation (e.g., Verner and Milligan, 1971; Petrinovich and Peeke, 1973; Falls, 1982) to neighbors that sing from their normal territory boundary (e.g., Kroodsma, 1976; Harris and Lemon, 1976; Falls, 1982). However, some studies indicate that individual neighbor recognition involves associative learning (e.g., "operant conditioning"), rather than habituation (Wiley and Wiley, 1977; Richards, 1979; Yasukawa *et al.*, 1982; Godard, 1991).

3. PROCESSES OF INDIVIDUAL RECOGNITION

3.1. Composition of Song Repertoires

A number of studies have focused on the consequences of learning on intraspecific variation in individual song repertoires and the distribution of distinct songs within and among populations (see McGregor and Krebs, 1982; Kroodsma and Miller, 1982; Lemon *et al.*, 1985; Slater,

1989). These studies showed that each individual song repertoire may consist of a unique combination of similar and/or dissimilar songs (so-called song types) (e.g., Harris and Lemon, 1972; Hansen, 1981; McGregor and Krebs, 1982; Lemon *et al.*, 1985), or that the order of songs delivered in a bout may be individually distinctive (e.g., Martin, 1990; Weary *et al.*, 1990). Because songs are learned (copied) from territorial neighbors, and the individual repertoire composition may differ among habitats or geographic regions, the song repertoire composition provides information about the origin of individuals, and therefore allows neighbor–stranger discrimination and recognition of individuals (e.g., immigrants, Hansen, 1984). Also, some species include in their songs individually distinctive imitations of vocalizations or songs of other species (e.g., Richards, 1979; Helb *et al.*, 1985; Eens *et al.*, 1992). Rufous-sided Towhees (*Pipilo erythrophthalmus*), for instance, seem to use individually distinctive heterospecific imitations to recognize neighbors (Richards, 1979).

The first field studies reported neighbor-stranger discrimination especially in species with a single song or a small song repertoire (less than 10 song types per individual) (see Falls, 1982; Kroodsma, 1982; Table 1). Neighbor-stranger discrimination was weak or not existing in some other species (e.g., Harris and Lemon, 1976; Kroodsma, 1976; Falls and d'Agincourt, 1981; Searcy *et al.*, 1981; Slater, 1991). In experimental studies comparing pairs of related species exposed to similar tests, neighbor-stranger discrimination was weaker in species with larger song repertoires than in species with smaller ones (e.g., Falls and d'Agincourt, 1981; Searcy *et al.*, 1981; Falls, 1982). This is an unexpected finding because individually distinctive traits are probably more likely to be found in larger song repertoires (Kroodsma, 1982). Some authors argued that, in larger individual song repertoires, different themes of the species-specific song are likely to be more similar in structure, and the performance time of each theme is shorter than in small-song repertoires. Therefore, individual recognition may be more difficult in species with larger repertoires if birds would be limited in their capabilities to perceive, learn, or memorize songs (e.g., Kroodsma, 1976; Wiley and Wiley, 1977; Krebs and Kroodsma, 1980; Nottebohm *et al.*, 1981; Falls, 1982; McGregor and Avery, 1986; McGregor, 1989; Stoddard *et al.*, 1991; McGregor, 1993).

Recently, Weary *et al.* (1992) did not find a clear negative relationship between neighbor–stranger discrimination and song repertoire size among 20 species for which data were available (see also Table I). For instance, European Robins (*Erithacus rubecula*) show neighbor–stranger discrimination despite having large and complex song repertoires (up to 250 different phrase types per individual) (Brémond, 1968; Brindley, 1991).

Godard (1991) showed recently that migratory Hooded Warblers (*Wilsonia citrina*) have a long-term memory of individual neighbors. Furthermore, birds can recognize many more song types than they actually sing (Shy *et al.*, 1986; McGregor and Avery, 1986; Stoddard *et al.*, 1992). For instance, in Song Sparrows (*Melospiza melodia*) having a relatively small song repertoire, captive individuals can learn many more song types than the total number of song types produced by all territorial neighbors in the field (Stoddard *et al.*, 1992). This means that Song Sparrows are potentially able to recognize the complete repertoire composition of all neighbors. It does not reject the hypothesis, however, that individual recognition takes more time in species with larger and more complex song repertoires.

In most playback experiments testing neighbor-stranger recognition, birds were presented with a single rendition of a song during a trial, not a complete song repertoire (Table I). This means that birds can use one distinct song to identify neighbors (e.g., Stoddard *et al.*, 1990), and suggests that the performance of the complete song repertoire is not always needed for individual recognition. Some studies showed neighbor-stranger discrimination when birds were presented with a small repertoire (Hansen, 1984; Weary *et al.*, 1987) or a series of distinct songs during a trial (Weary *et al.*, 1992). These experiments did not examine, however, whether the test birds used particular songs in the repertoire (e.g., familiar songs) or whether they used the complete song repertoire composition to identify neighbors.

3.2. Rendition of Songs

Although birds copy (learn) songs of conspecifics, these copies are not exact (e.g., Lemon, 1975; Kroodsma and Miller, 1982; Slater, 1989), perhaps because of auditory problems during the song learning process or improvization. Thus, the rendition of a song varies among individuals, even if songs of different individuals are classified as same types (e.g., Stoddard *et al.*, 1991; Podos *et al.*, 1992; Beecher *et al.*, 1994).

Neighbor–stranger discrimination and individual recognition could be related to the level of song sharing among neighbors and strangers. Neighbor recognition may be more difficult when neighbors and strangers share similar songs. The "deceptive mimicry" hypothesis assumes that, when a newcomer in an area would imitate the songs of a settled neighbor (intraspecific mimicry), the newcomer may have a higher chance to acquire a territory if other birds would confuse the identity of the settled bird and the newcomer during matched countersinging (e.g., see Payne, 1981, 1982; Craig and Jenkins, 1982; Schroeder and Wiley, 1983; Beecher *et al.*, 1994b). However, this hypothesis is not supported by field observations in different species (Payne, 1983; McGregor and Krebs, 1984a; Baptista, 1985).

Neighbor-stranger discrimination does exist in species having a single song (Table I) and where neighbors share similar songs (Falls, personal communication). This would mean that species having single-song repertoires are sensitive to relatively subtle differences in songs.

In Tufted Titmice (*Parus bicolor*), in which neighbors share similar songs and matched counter-singing happens frequently, no evidence was found for neighbor recognition (Schroeder and Wiley, 1983). On the other hand, song sharing did not seem to influence neighbor–stranger discrimination in Redstarts (*Setophaga ruticilla*) (Weary *et al.*, 1992). In Song Sparrows, response strength diminished sooner when territorial males were presented with playback of one rendition of a song than when they were presented with a series of clearly dissimilar songs (different song types), or a series of similar songs (different renditions of a single song type) (Stoddard *et al.*, 1988). This suggests that Song Sparrows perceive small variations in the micro–structure of song which could be used in individual recognition. However, neither study could rule out the possibility that shared songs delays the learning of shared neighbor songs.

3.3. Individual "Voices"

It is reasonable to assume that differences in the structure of the vocal system among individuals (e.g., vocal folds, size of the vocal apparatus, lung structure, vocal tract) are responsible for individual differences in "voice" characteristics, as in human voice (Kersta, 1962; Tosi, 1979; Zemlin, 1988). Thus, all the songs in an individual song repertoire may share certain vocal features that are unique for each individual bird (e.g., Beecher, 1989; Weary *et al.*, 1990).

Individual "voice" recognition is well known in our own species. Humans can recognize the voice of a familiar individual even if this individual produces a sentence, or a song for the first time. Thus, if birds would use "voice" characteristics in individual recognition, they should be able to recognize familiar individuals when hearing a previously unused song. This is in contrast to the widely accepted belief that vocal features used for individual recognition should have been learned (e.g., Falls, 1982).

The first laboratory results consistent with individual "voice" recognition were recently reported by Weary and Krebs (1992). They recorded song repertoires (four song types each) of two Great Tits (*Parus major*). Then they trained five Great Tits, not familiar with those two repertoires, in an operant conditioning experiment. Three individuals were trained to respond positively (approaching a feeder) to two song types of one unfamiliar Great Tit and respond negatively (not approach the feeder) when two song types from the other Great Tit were played. For two other subjects this training was reversed. Feeder visits following a positive training song were rewarded by allowing the bird access to the food in a feeder. Feeder visits following the negative training songs were not rewarded. Once birds had learned to discriminate between the training songs, they were played the other, novel songs from the two training repertoires. Subjects visited the feeder more often after hearing songs from the 'positive' repertoire, than they did after hearing songs from the 'negative' one. These findings suggest that the test birds were able to discriminate among conspecifics using songs they had never heard before, thus supporting the possibility of individual "voice" recognition in standardized laboratory condition.

More recently, Beecher *et al.* (1994a) used operant conditioning techniques to examine if Song Sparrows perceive variation in "voice" characteristics. To avoid potential problems with "pseudoreplication" (Kroodsma, 1989; McGregor *et al.*, 1992), they used playback songs of a larger number of song types and of a larger number of different individuals than Weary and Krebs (1992; see above). In contrast to the findings for the Great Tit (see above), the captive Song Sparrows did not seem to recognize the "voices" of different individuals in laboratory conditions, and confused the "voices" of different individuals when they produced similar songs.

Although the existence of individual "voice" recognition cannot be rejected, the findings of Beecher *et al.* (1994a) indicate that birds recognize individuals more easily by vocal features other than "voice" characteristics. This is supported by results of field experiments. In Great Tits, the ability to recognize the songs of new neighbors is negatively related to the similarity with songs of old neighbors to which the birds were exposed before (McGregor and Avery, 1986; McGregor, 1989). These proactive memory constraints are not predicted if individual "voice recognition would be important in the wild. Furthermore, we may expect that individual "voices" are expressed in all notes constituting a song. Thus birds should be capable of using all notes constituting a song to identify individuals. By contrast, some studies suggest that some parts in a song are more important in individual recognition than other parts of the same song (see Brooks and Falls, 1975; Elfström, 1990; Horning *et al.*, 1993).

Most, if not all, studies of individual recognition examined vocal cues that can be quantified relatively easily with sound spectrograms. Weary and Krebs (1992) suggested that maximum frequency, song duration (number of phrases per song), or "drift" (changes in note rate within songs) may be involved in individual "voice" recognition because these three song characteristics differed statistically significantly among individual Great Tit song repertoires (e.g., Lambrechts and Dhondt, 1987; Weary *et al.*, 1990; Lambrechts, 1992). However, the two song repertoires used in the operant conditioning experiment of Weary and Krebs (1992) do not appear to differ in maximum frequency, all the songs that were used consisted of four phrases per song, and since drift is more pronounced at the end of a song (e.g., Lambrechts and Dhondt, 1987), different songs most probably did not differ in the amount of drift. Weary and Krebs (1992) found that the test birds responded somewhat differently to three different tape recorders used, although the tape recorders were of the same brand and type. If individual "voice" recognition existed in Great Tits, this result would indicate that birds may recognize individual "voices" with song features that are subtle and difficult to describe with simple sonagrams (e.g., timbre, Cynx *et al.*, 1990; Dhondt and Lambrechts, 1992).

4. HOW TO DISTINGUISH AMONG THE PROCESSES OF INDIVIDUAL RECOGNITION

Although three processes of individual recognition are proposed in literature on song (see above), the distinctions among these processes are not always clear. For instance, the decision whether two individuals with one song type have a different repertoire composition or whether they sing different versions of a single song type, may depend on how song types are classified. If we decide that the two birds sing different song types, then we would describe the difference between the two individuals as one of repertoire composition. If not, we would describe the difference as one of *rendition* within song type. Furthermore, individual differences in the rendition of a song type may be caused by differences in "voice" characteristics among individuals.

To avoid defining the process of individual recognition on the basis of arbitrary classifications of vocalizations (song types), we may define four processes. (1) Birds recognize unique features in each vocalization that are not shared between different vocalizations of an individual. (2) Birds recognize conspecifics using vocal features that are unique for the individual and that are found in a limited number of vocalizations within the individual repertoire. (3) Birds identify conspecifics with characteristics that are shared by all the vocalizations of an individual. (4) Different vocal features that are individually distinctive (e.g., "voice" characteristics, repertoire composition, song structure, sequence of song delivery) are used simultaneously to identify individuals.

Besides the general processes of individual recognition described above, each song may consist of a number of individually distinctive parameters, such as the number of notes per song, the note rate, and the duration and frequencies of notes constituting a song. Birds seem to be especially sensitive to small variations in frequency characteristics (e.g., pitch) of song (Dooling, 1980; Weary, 1990; Ratcliffe and Weisman, 1992). Individuals can be recognized using the frequency features of a single song (e.g., Brooks and Falls, 1975; Nelson, 1989; Loesche *et al.*, 1992). However, frequency (pitch) may also be individually distinctive at other levels, such as the song repertoire composition, the order of song delivery, and the individual "voice."

Different species may not use the same vocal characteristics for individual recognition. For instance, species with small song repertoires and short songs may learn the characteristics of each song in the repertoire, while species with very large song repertoires and very elaborate songs may only learn a part of the songs or repertoire (e.g., McGregor, 1991).

5. DESIGN AND EXECUTION OF EXPERIMENTS TESTING INDIVIDUAL "VOICE" RECOGNITION IN THE FIELD

To minimize errors in playback experiments, McGregor *et al.* (1992) mentioned a number of features about the design and execution of playback experiments, which should be taken into account before an experiment is carried out. Some of these features are especially important in experiments investigating individual "voice" recognition.

If individual "voice" characteristics can be identified, these could be altered and the altered calls played to subjects as tests. However, Great Tits and other birds may be capable of detecting minor variations in acoustical features, such as those caused by the tape recorder used (Weary and Krebs, 1992), so researchers should pay much attention to the material used to record and broadcast vocalizations.

Acoustic features that are shared by different vocalizations of the same bird do not necessarily reflect "voice" characteristics, if the structure of vocalizations are influenced by environmental factors and if the environment in which a bird vocalizes is not the same for different individuals (e.g., height of song post, vegetation in a territory, size of nest cavity, distance between sender and receiver). For instance, the vegetation structure may influence the microstructure of acoustical signals during propagation resulting in sound degradation (e.g., Richards, 1981; Wiley and Richards, 1982; Morton, 1982; McGregor and Krebs, 1984b). Because the different songs of a distant bird propagate through the same environment, the taperecorder will record different degraded songs of the same individual that share acoustic characteristics. Therefore, investigators should control for the origin of recordings used in playback experiments, for instance by using vocalizations of birds that were recorded in more standardized laboratory conditions.

Examining "voice" recognition in laboratory conditions may give a number of problems. First, birds may be presented with only a small sample of songs per individual song repertoire. For example, in one experiment Beecher *et al.* (1994a) only used one song per individual repertoire as the training stimulus, and Weary and Krebs (1992) used two songs per individual repertoire. This gives only a very small sample with which the listener can generalize and recognize the common features of the individual "voice." Secondly, subjects are usually presented with a single song of a few seconds during each trial. In natural populations, however, (1) listeners can make an association between particular territories and particular songs, (2) neighbors may repeat single songs in bouts of several minutes (redundancy), and (3) listeners can compare the "voices" of neighbors during matched countersinging. Therefore, individual "voice" recognition may be a less difficult task in the field than in the operant conditioning experiments mentioned above, assuming that environmental conditions do not alter the subtle "voice" characteristics during sound propagation in the field.

The only direct evidence of individual "voice" recognition is that vocalizations of familiar individuals can always be recognized, even if these vocalizations are produced for the first time. Individual "voice" recognition is difficult to test in field conditions because the vocalizations used in playback may have been learned by the test birds before the start of the experiments. For instance, playback experiments testing neighbor recognition can only be carried out after song territories have been determined and after the test songs of the neighbors were recorded. Therefore, tests of neighbor recognition in the field cannot provide direct evidence for individual "voice" recognition.

One way to test individual "voice" recognition in natural populations is to work with songs of strangers. "Habituation" is a well-known concept in studies of bird song and consists of a stimulus-specific decrease in response strength to repeated exposure to a particular rendition of a song (e.g., Verner and Milligan, 1971; Petrinovich and Peeke,

1973; Krebs, 1976; Falls, 1982). When birds are habituated to one song in a playback experiment they will normally increase response strength again when they are played a new dissimilar song. In playback experiments testing individual "voice" recognition, a test bird could be habituated to a song A1 of a stranger in one trial, and to song A1 of the same bird or to an artificial song A2 with altered "voice" characteristics of song A1 in a following trial. If the test birds would be able to distinguish between the different "voices" of song A1 and song A2, the test birds should habituate more quickly to the "voice" of song A1 than that of song A2 (Fig. 1). This experiment may test if birds are able to recognize a feature that was identified as a "voice" characteristic.

In another experiment, a subject could be presented with many song types from the same individual song repertoire to provide the subject with enough information to work out the common "voice" features. For instance, the subject could be habituated to songs A, B, . . . , Y, of the same song repertoire in a series of trials, and then presented with a novel song Z from the same bird and/or a song Z from a different bird.

Searcy *et al.* (1994) carried out a habituation experiment similar as those described above. Territorial Red-winged Blackbirds (*Agelaius*

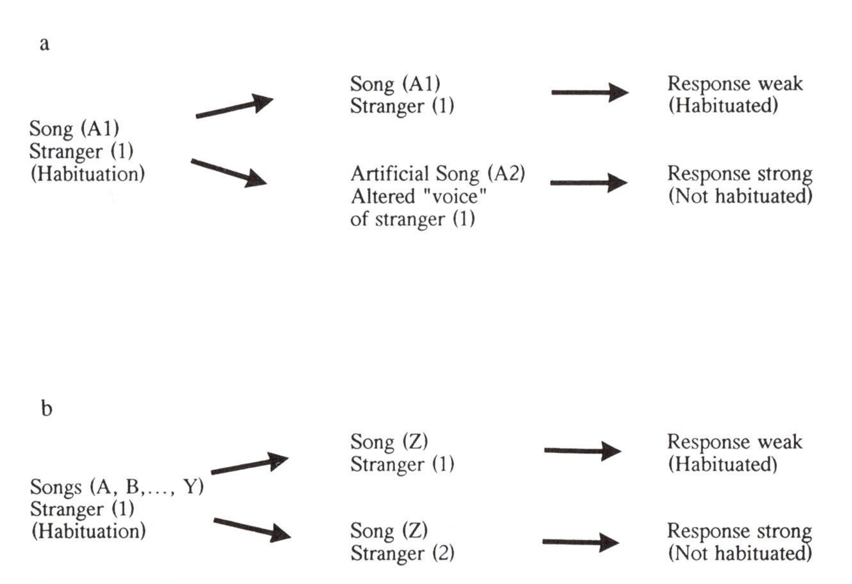

FIGURE 1. A playback experiment to test individual "voice" recognition in the field, and the predicted response if "voice" recognition exists.

phoenicus) decreased response strength with repeated presentation of a single song (habituation), and recovered response strength again after switching to a new song. Response strength recovered much stronger when the new song was dissimilar than when it resembled the song to which the birds were habituated. Individual "voice" characteristics did not seem to influence recovery of response strength, because responses did not recover more strongly when the second song was from a different male than when it was produced by the same male. However, Searcy *et al.* could not rule out that the subjects may have been habituated to vocal features other than those used in individual "voice" recognition. Furthermore, because the number of songs per individual song repertoire during the habituation phase were small, the subjects probably did not have an opportunity to learn the common "voice" features in this experiment.

6. THE POTENTIAL SIGNIFICANCE OF INDIVIDUAL "VOICE" RECOGNITION

European Starlings (*Sturnus vulgaris*) use contact calls of nestlings to discriminate between their own and strange chicks during the last week before fledging (Van Elsacker *et al.*, 1988). At the end of the nestling stage, young European Starlings also develop "distress" screams that are normally produced in response to a predator or an intruder entering the nest (Chaiken, 1992). Although they have no or only a few opportunities to hear the distress screams of their young in the nest, Starlings seem to be able to discriminate between the screams of their own nestlings and those of neighboring nestlings (Chaiken, 1992). One could, therefore, hypothesize that the parents recognized the novel screams if the parents learned to recognize their own young via the frequently produced contact calls, and if the different call types shared individually distinctive vocal characteristics.

Young birds are particularly vulnerable after fledging, and parents sometimes have to defend their offspring against predators. Because antipredator behavior is risky, we would expect that defense of fledglings is limited to a bird's young. After fledging, young starlings of different nests intermingle (Van Elsacker *et al.*, 1988). Thus, in these circumstances, individual "voice" recognition may help parents to recognize and defend their own screaming young against predators.

In territorial defense, neighbor–stranger discrimination and individual recognition by song may reduce the chance of time-consuming interactions among territorial neighbors, it may help to decide quickly whether a singing opponent poses a threat, and it may prevent that information about the outcome of former contests between neighbors is lost (e.g., Weeden and Falls, 1959; Falls, 1982; Ydenberg *et al.*, 1988; Getty, 1989). Neighbor recognition should therefore develop at the beginning of the territorial season when males establish their territories.

Some processes of neighbor discrimination demand that all song types of all neighbors should have been learned. This may take a lot of time, especially in populations with high densities or in species with large and complex song repertoires. Individual "voice" recognition does not require that the complete song repertoire of all neighbors is known, and therefore may increase the speed of neighbor recognition.

Although it is generally assumed that songbirds possess a song repertoire to increase their success in male–male competition or in mate attraction, the exact role of song repertoires is not yet understood (reviews in Craig and Jenkins, 1982; Kroodsma and Miller, 1982; Kroodsma and Byers, 1991). The "honest advertisement" hypothesis, for instance, assumes that song repertoire size gives reliable information of "male quality" (Baker *et al.*, 1986). Exchange of information about male quality may give advantages in territorial defense or female attraction (e.g., Baker *et al.*, 1986; Lambrechts, 1992). This hypothesis predicts that singers perform the complete song repertoire in a short period of time allowing listeners to learn the complete song repertoire of opponents as quickly as possible. If this hypothesis were true, individual "voice" recognition would not give additional advantages in comparison to other processes of individual recognition by song. Other functional hypotheses of song repertoires assume that individual "voice" recognition is not or weakly developed in birds (Falls, 1982). According to the "antihabituation" hypothesis, singers should switch song types frequently to counteract habituation in listeners (e.g., Krebs, 1976). If listeners would habituate to "voice" characteristics, switching song types would not be an effective way to prevent habituation. The "Beau Geste" hypothesis predicts that territorial birds switch song types and perches at the same time to mislead potential newcomers into believing that the habitat is more densely occupied than it really is (e.g., Krebs, 1977). This hypothesis can only work if newcomers do not recognize residents by "voice." Thus if the existence of individual "voice" recognition would be confirmed with well-designed experiments, it could have important consequences for the interpretation of the significance of song and song repertoires in birds.

7. SIGNATURE ADAPTATIONS AND PERCEPTUAL ADAPTATIONS FOR INDIVIDUAL RECOGNITION

Individually distinctive vocal (signature) traits may have been evolved via natural selection to facilitate individual recognition when needed (e.g., Beecher, 1982; Falls, 1982; Beecher *et al.*, 1985). This may explain why, in interspecific comparisons of parent–offspring recognition, the appearance of both individual recognition and signature traits are related to the level of coloniality of a species (e.g., Beecher, 1982; Jouventin, 1982; Stoddard and Beecher, 1983; Beecher *et al.*, 1985; Medvin and Beecher, 1986; Beecher *et al.*, 1986; Beecher and Stoddard, 1990). Loesche *et al.* (1991) showed that colonial Cliff Swallows (*Hirundo pyrrhonota*), non-colonial Barn Swallows (*H. rustica*), and Starlings discriminate more easily among chick begging calls of different Cliff Swallows than among chick begging calls of different Barn Swallows. Furthermore, the colonial Swallow species discriminated faster than the non-colonial Swallow species. The findings of Loesche *et al.* indicate that in some species, special signature and perceptual adaptations may have evolved to facilitate individual vocal recognition.

In bird song, the level of individual neighbor recognition clearly differs among species, which could be partly due to interspecific differences in song repertoire size (see above). Therefore, it was proposed that, independent of a lack of an ability to recognize individuals, interspecific differences in the level of neighbor recognition could be caused by interspecific differences in learning opportunities of songs (see above) or by a stronger proactive memory interference in species with larger song repertoires (McGregor and Avery, 1986; McGregor, 1991). Learning constraints alone are probably not responsible for these interspecific differences because individual recognition by song was recently found in species with very complex song repertoires. Different species may therefore have evolved adaptive differences in learning abilities (e.g., adaptive differences in the rate of habituation; Kroodsma, 1976; see also Kroodsma and Canady, 1985), assuming that individual "voice" recognition by song is weakly developed. Today, it is unclear whether neighbor recognition by song could be considered as just being the consequence of the inevitable existence of individual vocal features, or whether songbirds evolved special signature adaptations or perceptual adaptations for individual neighbor recognition by song. Some studies showed that features used in individual recognition are especially found at the beginning of a song, while features used in species recognition are found at the end of a song (e.g., Elfström, 1990).

This organization of the song could be considered as a signature adaptation, if territorial defense requires fast individual recognition of conspecific neighbors rather than fast species recognition (e.g., the chance of escalated conflicts is probably higher during conspecific interactions than during heterospecific ones).

In contrast to studies of bird calls, studies of bird song do not show relationships between the ability to recognize neighbor, and ecological or social characteristics of the species.

8. CONCLUDING REMARKS

Although in this paper we discussed the different potential processes of individual vocal recognition in bird song, similar kinds of processes of individual recognition probably exist in other bird vocalizations or other animal groups as well. For instance, all vocalizations in the individual vocal repertoire may be exposed to the same constraints of the vocal apparatus (e.g., body size, Ryan and Brenowitz, 1985; Tosi, 1979). Therefore, both calls and songs may share common "voice" features, and individual "voice" recognition may be possible using vocalizations other than song (see Chaiken, 1992).

The existing experimental evidence suggest that "voice" recognition is not or only weakly developed in birds, but this may have been partly due to the way playback experiments were carried out. Whether "voice" recognition gives important advantages during conspecific interactions, in comparison to other processes of individual recognition, may depend on (1) the vocal features characterizing the "voice" (which should be recognizable in field conditions), (2) the time span in which novel vocalizations are introduced (which should be long), (3) the minimal sample of vocalizations per individual repertoire that is needed to work out the common "voice" features of an individual (this sample should be relatively small in comparison to the total individual repertoire), and (4) the memory capacity of birds. Therefore, we suggest that field studies should examine the timing of the introduction of novel vocalizations. Psychophysical experiments can determine how many times a bird must hear novel vocalizations before these are memorized, and if the number of vocalizations that a bird has learned influences the capacity to learn novel vocalizations (i.e., proactive memory interference; McGregor and Avery, 1986; McGregor, 1989). Additional studies are necessary to identify "voice" characteristics, whether vocal features are individually distinctive and whether vocal features are shared by all the vocalizations in the individual vocal repertoire. Well-designed operant conditioning experiments or field experiments could examine if birds perceive signature traits and use common "voice" features and/or other vocal characteristics in individual recognition. All these studies may finally help to determine the relative importance of "voice" recognition and other processes of individual vocal recognition during intraspecific communication.

9. SUMMARY

Although individual recognition by vocalizations was observed in many bird studies, it is not clearly understood which vocal features are used for individual recognition. This paper reviews and discusses hypotheses on vocal cues used to recognize individuals by song, with emphasis on the new hypothesis of individual recognition that all vocalizations of an individual may share distinctive features, as in human voice. The potential significance of individual "voice" recognition in birds is discussed. Some suggestions for the design and execution of experiments testing "voice" recognition are made.

ACKNOWLEDGMENTS. We are grateful to J. Banbura, J. B. Falls, W. A. Searcy, P. K. Stoddard, and D. M. Weary, for helpful comments on the manuscript. M. D. Beecher and W. A. Searcy kindly provided unpublished data.

REFERENCES

Baker, M. C., Thompson, D. B., and Sherman, G. L., 1981, Neighbor/stranger song discrimination in the White-crowned Sparrow, *Condor* **83**:265–267.

Baker, M. C., Bjerke, T. K., Lampe, H., and Espmark, Y., 1986, Sexual response of female Great Tits to variation in sizes of males' song repertoires, *Am. Nat.* **128**:491–498.

Baptista, L. F., 1985, The functional significance of song sharing in the White-crowned Sparrow, *Can. J. Zool.* **63**:1741–1752.

Bauer, H.-G., and Nagl, W., 1992, Individual distinctiveness and possible function of song parts of Short-toed Treeceepers (*Certhia brachydactyla*). Evidence from multivariate song analysis, *Ethology* **91**:108–121.

Becker, P. H., 1982, The coding of species-specific characteristics in bird sounds, in: *Acoustic Communication in Birds*, Volume 2 (D. E. Kroodsma, and E. H. Miller, eds.), Academic Press, New York, pp. 213–252.

Beecher, M. D., 1982, Signature systems and kin recognition, *Amer. Zool.* **22**:477–490.

Beecher, M. D., 1989, Signalling systems for individual recognition: an information theory approach, *Anim. Behav.* **38**:248–261.

Beecher, M. D., 1990, The evolution of parent-offspring recognition in swallows, in:

Contemporary Issues in Comparative Psychology (D. A. Dewsbury, ed.), Sinauer, Mass., pp. 360–380.

Beecher, M. D., and Stoddard, P. K., 1990, The role of bird song and calls in individual recognition: contrasting field and laboratory perspectives, in: *Comparative Perception*, Vol. 2 (W. C. Stebbins, and M. A. Berkeley, eds.), New York, John Wiley, pp. 375–408.

Beecher, M. D., Stoddard, P. K., and Loesche, P., 1985, Recognition of parents' voices by young Cliff Swallows, *Auk* **102:**600–605.

Beecher, M. D., Medvin, M. B., Stoddard, P. K., and Loesche, P., 1986, Acoustic adaptations for parent-offspring recognition in Swallows, *Exp. Biol.* **45:**179–193.

Beecher, M. D., Campbell, S. E., and Burt, J. M., 1994a, Song perception in the Song Sparrow: birds classify by song type but not by singer, *Anim. Behav.* **47:**1343–1351.

Beecher, M. D., Campbell, S. E., and Stoddard, P. K., 1994b, Correlation of song learning and territory establishment strategies in the Song Sparrow, *Proc. Natl. Acad. Sci., U.S.A.*, **91:**1450–1454.

Beer, C. G., 1970, Individual recognition of voice in the social behavior of birds, *Adv. Study Behav.* **3:**27–74.

Belcher, J. W., and Thompson, W. L., 1969, Territorial defense and individual song recognition in the Indigo Bunting, *Jack-Pine Warbler* **47:**76–83.

Beletsky, L. D., 1983, An investigation of individual recognition by voice in female Redwinged Blackbirds, *Anim. Behav.* **31:**355–362.

Borror, D. J., 1959, Variation in the songs of the Rufous-sided Towhee, *Wilson Bull.* **71:**54–72.

Brémond, J. C., 1968, Recherches sur la sémantique et les éléments vecteurs d'information dans les signaux acoustiques du Rouge-Gorge (*Erithacus rubecula* L.), *Terre Vie* **2:**109–220.

Brémond, J. C., Aubin, T., Nyamsi, R. M., and Robisson, P., 1990, Le chant du Manchot Empereur (*Aptenodytes forsteri*): Recherche des paramètres utilisables pour la reconnaissance individuelle. *C.R. Acad. Sci. Ser. III Sci. Vie* **311:**31–35.

Brindley, E. L., 1991, Response of European Robins to playback of song: neighbour recognition and overlapping, *Anim. Behav.* **41:**503–512.

Brooks, R. J., and Falls, J. B., 1975, Individual recognition by song in White-throated Sparrows: III. Song features used in individual recognition, *Can. J. Zool.* **53:**1749–1761.

Chaiken, M., 1992, Individual recognition of nestling distress screams by European Starlings (*Sturnus vulgaris*), *Behaviour* **120:**138–150.

Craig, J. L., and Jenkins, P. F., 1982, The evolution of complexity in broadcast song in passerines, *J. Theor. Biol.* **95:**415–422.

Cynx, J., Williams, H., and Nottebohm, F., 1990, Timbre discrimination in Zebra Finch (*Taeniopygia guttata*) song syllables, *J. Comp. Psychol.* **104:**303–308.

Dhondt, A. A., and Lambrechts, M. M., 1992, Individual voice recognition in birds, *TREE* **7:**178–179.

Dooling, R. J., 1980, Behavior and psychophysics of hearing in birds, in: *Comparative Studies of Hearing in Vertebrates*, (A. N. Popper, and R. R. Fay, eds.), Springer-Verlag, Berlin, pp. 261–288.

Eens, M., Pinxten, R., Verheyen, R. F., 1992, Song learning in captive European Starlings, *Sturnus vulgaris, Anim. Behav.* **44:**1131–1143.

Elfström, S. T., 1990, Individual and species-specific song patterns of Rock and Meadow Pipits: physical characteristics and experiments, *Bioacoustics* **2:**277–301.

Emlen, S. T., 1971, The role of song in individual recognition in the Indigo Bunting, *Z. Tierpsychol.* **28**:241–246.

Falls, J. B., 1982, Individual recognition by sounds in birds, in: *Acoustic Communication in Birds*, Volume 2 (D. E. Kroodsma, and E. H. Miller, eds.), Academic Press, New York, pp. 237–278.

Falls, J. B., 1992, Playback: a historical perspective, in: *Playback and Studies of Animal Communication* (P. K. McGregor, ed.), Plenum Press, New York, pp. 11–33.

Falls, J. B., and Brooks, R. J., 1975, Individual recognition by song in White-throated Sparrows: II. Effects of location, *Can. J. Zool.* **53**:1412–1420.

Falls, J. B., and d'Agincourt, L. G., 1981, A comparison of neighbor-stranger discrimination in Eastern and Western Meadowlarks, *Can. J. Zool.* **59**:2380–2385.

Falls, J. B., Krebs, J. R., and McGregor, P. K., 1982, Song matching in the great tits (*Parus major*): the effects of similarity and familiarity, *Anim. Behav.* **30**:997–1009.

Galeotti, P., and Pavan, G., 1991, Individual recognition of male Towny Owls (*Strix aluco*) using spectrograms of their territorial calls, *Ethol. Ecol. Evol.* **3**:113–126.

Galeotti, P., and Pavan, G., 1993, Differential responses of territorial Tawny Owls *Strix aluco* to the hooting of neighbours and strangers, *Ibis* **135**:300–304.

Getty, T., 1989, Are dear enemies in a war of attrition?, *Anim. Behav.* **37**:337–339.

Giraldeau, L.-A., and Ydenberg, R., 1987, The center-edge effect: the result of a war of attrition between territorial contestants, *Auk* **104**:535–538.

Godard, R., 1991, Long-term memory of individual neighbours in a migratory songbird, *Nature* **350**:228–229.

Godard, R., 1993, Tit for tat among neighboring Hooded Warblers, *Behav. Ecol. Sociobiol.* **33**:45–50.

Goldman, P., 1973, Song recognition by Field Sparrows, *Auk* **90**:106–113.

Hansen, P., 1981, Coordinated singing in neighbouring Yellowhammers (*Emberiza citrinella*), *Natura Jutlandica* **19**:121–138.

Hansen, P., 1984, Neighbour-stranger song discrimination in territorial Yellowhammer *Emberiza citrinella* males, and a comparison with responses to own and alien song dialects, *Ornis scandinavia* **15**:240–247.

Harris, M. A., and Lemon, R. E., 1972, Songs of Song Sparrows (*Melospiza melodia*): individual variation and dialects, *Can. J. Zool.* **50**:301–309.

Harris, M. A., and Lemon, R. E., 1976, Response of male Song Sparrows *Melospiza melodia* to neighbouring and non-neighbouring individuals, *Ibis* **118**:421–424.

Helb, H.-W., Dowsett-Lemaire, F., Bergmann, H.-H., and Conrads, K., 1985, Mixed singing in European songbirds—a review, *Z. Tierpsychol.* **69**:27–41.

Horning, C. L., Beecher, M. D., Stoddard, P. K., and Campbell, S. E., 1993, Song perception in the Song Sparrow: Importance of different parts of the song in song type classification, *Ethology* **94**:46–58.

Järvi, T., Radesäter, T., and Jakobsson, S., 1977, Individual recognition and variation in the Great Tit (*Parus major*) two-syllable song. *Biophon* **5**:4–9.

Jouventin, P., 1982, Visual and vocal signals in Penguins, their evolution and adaptive characters, Berlin, Verlag Paul Parey.

Kent, J. P., 1992, The relationship between the hen and chick, *Gallus gallus domesticus*: the hen's recognition of the chick, *Anim. Behav.* **44**:996–998.

Kersta, L. G., 1962, Voiceprint identification, *Nature* **186**:1253–1257.

Krebs, J. R., 1971, Territory and breeding density in the Great Tit, *Parus major* L., *Ecology* **52**:2–22.

Krebs, J. R., 1976, Habituation and song repertoires in the Great Tit, *Behav. Ecol. Sociobiol.* **1**:215–227.

Krebs, J. R., 1977, The significance of song repertoires, the Beau Geste hypothesis, *Anim. Behav.* **25**:475–478.

Krebs, J. R., and Kroodsma, D. E., 1980, Repertoires and geographical variation in bird song, *Adv. in Study Behav.* **11**:143–177.

Kroodsma, D. E., 1976, The effect of large song repertoires on neighbor 'recognition' in male song sparrows, *Condor* **78**:97–99.

Kroodsma, D. E., 1982, Song repertoires: Problems in their definition and use, in: *Evolution and Ecology of Acoustic Communication in Birds*, Vol. 2, (D. E. Kroodsma, and E. H. Miller, eds.), Academic Press, London, pp. 125–146.

Kroodsma, D. E., 1989, Suggested experimental designs for song playbacks, *Anim. Behav.* **37**:600–609.

Kroodsma, D. E., and Byers, B. E., 1991, The function(s) of bird song, *Amer. Zool.* **31**:318–328.

Kroodsma, D. E., and Canady, R. A., 1985, Differences in repertoire size, singing behavior, and associated neuroatonomy among Marsh Wren populations have a genetic basis, *Auk* **102**:439–446.

Kroodsma, D. E., and Miller, E. H., 1982, *Acoustic Communication in Birds*, Volume 1 and 2, Academic Press, New York.

Lambrechts, M. M., 1992, Male quality and playback in the Great Tit, in: *Playback and Studies of Animal Communication* (P. K. McGregor, ed.), Plenum Press, New York, pp. 135–152.

Lambrechts, M., and Dhondt, A. A., 1987, Differences in singing performance between male great tits, *Ardea* **75**:43–52.

Lampe, H. M., and Slagsvold, T., 1994, Individual recognition based on male song in a female songbird, *J. Ornithol.* **135**:163.

Lemon, R. E., 1975, How birds develop song dialects, *Condor* **77**:385–406.

Lemon, R. E., Cotter, R., MacNally, R. C., Monette, S., 1985, Song repertoires and song sharing by American Redstarts, *Condor* **87**:457–470.

Loesche, P., Beecher, M. D., and Stoddard, P. K., 1992, Perception of Cliff Swallow calls by birds and humans, *J. Comp. Psychol.* **106**:239–247.

Loesche, P., Stoddard, P. K., Higgins, B. J., and Beecher, M. D., 1991, Signature versus perceptual adaptations for individual vocal recognition in Swallows, *Behaviour* **118**:15–25.

McArthur, P. D., 1982, Mechanisms and development of parent-young recognition in the Pinon Jay (*Bymnorhinus cyanocephalus*), *Anim. Behav.* **30**:62–74.

McGregor, P. K., 1989, Pro-active memory interference in neighbour recognition by a song bird. *Proc. Int. Orn. Cong.* **XIX**:1391–1397.

McGregor, P. K., 1991, The singer and the song: on the receiving end of bird song, *Biol. Rev.* **66**:57–81.

McGregor, P. K., 1993, Signalling in territorial systems: a context for individual identification, ranging and eavesdropping, *Phil. Trans. R. Soc. Lond.* B **340**:237–244.

McGregor, P. K., and Avery, M. I., 1986, The unsung songs of Great Tits (*Parus major*): learning neighbours' songs for discrimination, *Behav. Ecol. Sociobiol.* **18**:311–316.

McGregor, P. K., and Krebs, J. R., 1982, Song types in a population of Great Tits (*Parus major*): their distribution, abundance, and acquisition by individuals, *Behaviour* **79**:126–152.

McGregor, P. K., and Krebs, J. R., 1984a, Song learning and deceptive mimicry, *Anim. Behav.* **32**:280–287.

McGregor, P. K., and Krebs, J. R., 1984b, Sound degradation as a distance cue in great tit (*Parus major*) song, *Behav. Ecol. Sociobiol.* **16**:49–56.

McGregor, P. K., Catchpole, C. K., Dabelsteen, T., Falls, J. B., Fusani, L., Gerhardt, H. C., Gilbert, F., Horn, A. G., Klump, G. M., Kroodsma, D. E., Lambrechts, M. M., McComb, K. E., Nelson, D. A., Pepperberg, I. M., Ratcliffe, L., Searcy, W. A., and Weary, D. M., 1992, Design of playback experiments: the thornbridge hall NATO ARW consensus, in: *Playback and Studies of Animal Communication*, Plenum Press, New York, pp. 1–9.

Marler, P., 1960, Bird song and mate selection, in: *Animal Sounds and Communication*, (W. F. Lanyon, and W. Tavolga, eds.), A.I.B.S., Washington, D.C., pp. 348–367.

Martin, D. J., 1990, Songs of the Fox Sparrow. III. Ordering of songs, *Wilson Bull.* **102:**655–671.

Medvin, M. B., and Beecher, M. D., 1986, Parent-offspring recognition in the Barn Swallow, *Anim. Behav.* **34:**1627–1639.

Morton, E. S., 1982, Grading, discreteness, redundancy and motivation-structure rules, in: *Acoustic Communication in Birds*, Volume 1 (D. E. Kroodsma, and E. H. Miller, eds.), Academic Press, New York, pp. 183–212.

Nelson, D. A., 1989, Song frequency as a cue for recognition of species and individuals in the Field Sparrow (*Spizella pusilla*), *J. Comp. Psychol.* **103:**171–176.

Nottebohm, F., Kasparian, S., and Pandazis, C., 1981, Brain space for a learned task, *Brain Research* **213:**99–109.

Payne, R. B., 1981, Song learning and social interaction in Indigo Buntings. *Anim. Behav.* **29:**688–697.

Payne, R. B., 1982, Ecological consequences of song matching: breeding success and intraspecific song mimicry in Indigo Buntings, *Ecology* **63:**401–411.

Payne, R. B., 1983, The social context of song mimicry: song-matching dialects in Indigo Buntings (*Passerina cyanea*), *Anim. Behav.* **31:**788–805.

Payne, R. B., Payne, L. L., and Rowley, I., 1988, Kin and social relationships in Splendid Fairy-wrens: recognition by song in a cooperative bird, *Anim. Behav.* **36:**1341–1351.

Payne, R. B., Payne, L. L., Rowley, I., and Russell, E. M., 1991, Social recognition and response to song in cooperative Red-winged Fairy-wrens, *Auk* **108:**811–819.

Petrinovich, L., and Peeke, H. V. S., 1973, Habituation to territorial song in the White-crowned Sparrow (*Zonotrichia leucophrys*), *Behav. Biol.* **8:**743–748.

Pickstock, J. C., and Krebs, J. R., 1980, Neighbour-stranger discrimination in the Chaffinch (*Fringilla coelebs*), *J. Ornithol.* **121:**105–108.

Podos, J., Peters, S., Rudnicky, T., Marler, P., and Nowicki, S., 1992, The organization of song repertoires in song sparrows: themes and variations, *Ethology* **90:**89–106.

Ratcliffe, L. M., 1990, Neighbour/stranger discrimination of whistled songs in Black-capped Chickadees, *Proc. Int. Orn. Congr.* **XX** (Abstract 852).

Ratcliffe, L., and Weisman, R., 1992, Pitch processing strategies in birds: a comparison of laboratory and field studies, in: *Playback and Studies of Animal Communication*, Plenum Press, New York, pp. 211–223.

Richards, D. G., 1979, Recognition of neighbors by associative learning in Rufous-sided Towhees, *Auk* **96:**688–693.

Richards, D. G., 1981, Estimation of distance of singing conspecifics by the Carolina Wren, *Auk* **98:**127–133.

Ritchison, G., 1988, Responses of Yellow-breasted Chats to the song of neighboring and non-neighboring conspecifics, *J. Field Ornithol.* **59:**37–42.

Robisson, P., 1990, The importance of the temporal pattern of syllables and the syllable structure of display calls for individual recognition in the genus Aptenodytes, *Behav. Processes* **22:**157–163.

Robisson, P., 1992, Vocalizations in Aptenodytes Penguins: Application of the two-voice theory, *Auk* **109:**654–658.

Ryan, M. J., and Brenowitz, E. A., 1985, The role of body size, phylogeny, and ambient noise in the evolution of bird song, *Am. Nat.* **126:**87–100.

Saunders, D. A., and Wooller, R. D., 1988, Consistent individuality of voice in birds as a management tool, *Emu* **88:**25–32.

Schleidt, W. M., 1976, On individuality: the constituents of distinctiveness, in: *Perspectives in Ethology*, Vol. 2 (P. P. G. Bateson, and P. H. Klopfer, eds.), Plenum Press, New York, pp. 299–310.

Schroeder, D. J., and Wiley, R. H., 1983, Communication with shared song themes in Tufted Titmice, *Auk* **100:**414–424.

Searcy, W. A., McArthur, P. D., Peters, S. S., and Marler, P., 1981, Response of male song and Swamp Sparrows to neighbour, stranger and self songs, *Behaviour* **77:**152–166.

Searcy, W. A., Coffman, S., and Raikow, D. F., 1994, Habituation, recovery, and the similarity of song types within song repertoires in Red-winged Blackbirds (*Agelaius phoeniceus*) (Aves, Emberizidae), *Ethology* **98:**38–49.

Shy, E., McGregor, P. K., and Krebs, J. R., 1986, Discrimination of song types by male Great Tits, *Behav. Proc.* **13:**1–12.

Sieber, O. J., 1985, Individual recognition of parental calls by Bank Swallow chicks (*Riparia riparia*), *Anim. Behav.* **33:**107–116.

Slater, P. J. B., 1989, Bird song learning: causes and consequences, *Ethology, Ecology and Evolution* **1:**19–46.

Slater, P. J., 1991, The response of the Silvereyes to individual variation in their territorial song, *Emu* **91:**36–40.

Speirs, E. A. H., and Davis, L. S., 1991, Discrimination by Adelie Penguins, *Pygoscelis adeliae*, between the loud mutual calls of mates, neighbours and strangers, *Anim. Behav.* **41:**937–944.

Stoddard, P. K., and Beecher, M. D., 1983, Parental recognition of offspring in the Cliff Swallow, *Auk* **100:**795–799.

Stoddard, P. K., Beecher, M. D., and Willis, M. S., 1988, Response of territorial male Song Sparrows to song types and variations, *Behav. Ecol. Sociobiol.* **22:**125–130.

Stoddard, P. K., Beecher, M. D., Horning, C. L., and Willis, M. S., 1990, Strong neighbor-stranger discrimination in Song Sparrows, *Condor* **92:**1051–1056.

Stoddard, P. K., Beecher, M. D., Horning, C. L., and Campbell, S. E., 1991, Recognition of individual neighbors by song in the Song Sparrow, a species with song repertoires, *Behav. Ecol. Sociobiol.* **29:**211–215.

Stoddard, P. K., Beecher, M. D., Campbell, S. E., and Horning, C. L., 1992, Song-type matching in the Song Sparrow, *Can. J. Zool.* **70:**1440–1444.

Stoddard, P. K., Beecher, M. D., Loeshe, P., and Campbell, S. E., 1992, Does memory constrain individual recognition in a bird with song repertoires? *Behaviour* **122:**274–287.

Tosi, O., 1979, *Voice identification. Theory and legal applications*, University Park Press, Baltimore.

Van Elsacker, L., Pinxten, R., and Verheyen, R., 1988, Timing of offspring recognition in adult Starlings, *Behaviour* **107:**122–130.

Verner, J., and Milligan, M. M., 1971, Responses of male White-crowned Sparrows to playback of recorded songs, *Condor* **73:**56–64.

Weary, D. M., 1990, Categorization of song notes in Great Tits: which acoustic features are used and why?, *Anim. Behav.* **39:**450–457.

Weary, D. M., and Krebs, J. R., 1992, Great Tits classify songs by individual voice characteristics, *Anim. Behav.* **43**:283–287.
Weary, D. M., Lemon, R. E., and Date, E. M., 1987, Neighbour-stranger discrimination by song in the Veery, a species with song repertoires, *Can. J. Zool.* **65**:1206–1209.
Weary, D. M., Norris, K. J., and Falls, J. B., 1990, Song features birds use to identify individuals, *Auk* **107**:623–625.
Weary, D. M., Lemon, R. E., and Perreault, S., 1992, Song repertoires do not hinder neighbor-stranger recognition, *Behav. Ecol. Sociobiol.* **31**:441–447.
Weeden, J. S., and Falls, J. B., 1959, Differential response of male Ovenbirds to recorded songs of neighboring and more distant individuals, *Auk* **76**:343–351.
Wiley, R. H., and Richards, D. G., 1982, Adaptations for acoustic communication in birds: sound transmission and signal detection, in: *Evolution and Ecology of Acoustic Communication in Birds*, Vol. 1, (D. E. Kroodsma, and E. H. Miller, eds.), Academic Press, London, pp. 131–181.
Wiley, R. H., and Wiley, M. S., 1977, Recognition of neighbors' duets by Stripe-backed Wrens, *Campylorhynchus nuchalis, Behaviour* **62**:10–34.
Wilson, E. O., 1975, *Sociobiology: The New Synthesis*, Harvard Univ. Press, Cambridge, Massachusetts.
Wunderly, J. M., Jr., 1978, Differential response of territorial Yellowthroats to the songs of neighbors and non-neighbors, *Auk* **95**:389–395.
Yasukawa, K., Bick, E., Wagman, D. W., and Marler, P., 1982, Playback and speaker replacements on song-based neighbor, stranger and self-discrimination in male Red-winged Blackbirds, *Behav. Ecol. Sociobiol.* **10**:211–215.
Ydenberg, R. D., Giraldeau, L. A., and Falls, J. B., 1988, Neighbours, strangers and the asymmetric war of attrition. *Anim. Behav.* **36**:343–347.
Zemlin, W. R., 1988, *Speech and Hearing Science*, Prentice-Hall, Inc., New Jersey.

CHAPTER 5

THE EVOLUTION OF BIRD COLORATION AND PLUMAGE ELABORATION

A Review of Hypotheses

UDO M. SAVALLI

1. INTRODUCTION

Although physically homogeneous, birds come in a wide variety of colors, rivaling butterflies and tropical reef fishes. Such variation invites attention and many hypotheses have been developed to explain the evolution of avian color and pattern. In addition, plumage of birds have structural elaborations that go beyond its insulative and aerodynamic properties including various kinds of ruffs, elongated tails (retrices or coverts), crests, and plumes which I refer to collectively as "ornaments." Since many hypotheses for the evolution of color and pattern (especially conspicuous colors and patterns) involve communicative functions that apply also to the evolution of ornaments, I treat both color and ornaments in this review.

Many explanations for bird coloration and ornamentation have been

UDO M. SAVALLI • Department of Integrative Biology, University of California, Berkeley, California 94720. *Present affiliation:* Department of Biology, Allegheny College, Meadville, Pennsylvania 16335.

Current Ornithology, Volume 12, edited by Dennis M. Power. Plenum Press, New York, 1995.

proposed, but testing these hypotheses has not kept pace. Evidence for some of the hypotheses is lacking altogether; for the remainder, the supporting evidence varies from anecdotal observations to intraspecific variation, using natural variation, experimental manipulations of the color or ornament, and interspecific comparisons. No study to date had addressed all hypotheses relevant to the particular color pattern or ornament being investigated, and consequently, it is never made clear if the data support or refute the alternative hypotheses. It is my hope that this review will stimulate researchers of animal coloration to address all of the alternatives and not just the favored hypothesis.

The hypotheses differ in the range of color patterns and ornaments they address. Some make very specific predictions about color patterns and consequently can account for only a small amount of the variation in bird coloration. For example, several hypotheses apply only to eye-stripes, another only to eye color. Other hypotheses, such as sexual selection, make very general predictions, such as the pattern or ornament should be conspicuous or costly. Such hypotheses explain a large variety of traits as divergent as a widowbird's tail, a rooster's comb, a peacock's train, the red head of a woodpecker, or the black bib of a sparrow. While these hypotheses can account for some features of the trait, they cannot account for the enormous diversity in conspicuous traits—why some birds have red heads and others long tails even though the same basic process (such as sexual selection or aposomatism) may be at work. Recently, approaches are being developed that could predict specific patterns, although in different directions. One of these approaches suggests that the signal (the color pattern or ornament) is closely related to its message; by using information theory (how the signal and the message relate to each other) and an understanding of the messages relevant to a particular species, the optimal color pattern or ornament could be predicted (see section 3.3). The other approaches consider coloration designed to optimize perception by the receiver of the signal; predicting the optimal color or ornament would require knowledge of the perceptual systems of the receivers. Approaches to coloration are still in their infancy and should provide exciting new insights.

The quantification of color and pattern has received little attention, especially by ornithologists, but is critical to the aforementioned approaches. In particular, such vague terms as "conspicuous," "cryptic," and "contrasting" need to be defined in that different kinds of patterns and colors can be compared to their effectiveness in concealing or exposing an individual. One attempt to operationally define crypticity has been proposed by Endler (1978, 1984, 1990). This task is made more difficult by the need to consider the visual systems of the species involved, ambient light, and background coloration (Hailman, 1977, 1979; Endler, 1978, 1990, 1992). Although the perceptual world of birds is perhaps closer to humans than that of most other nonprimates, birds have four or five different cone types and are likely to perceive colors differently, and may be able to make more color distinctions and also perceive ultraviolet light (Goldsmith, 1990; Waldvogel, 1990).

In this chapter, I will first review all of the hypotheses for the functions of bird coloration, including a few that were not previously applied to birds. Although various aspects of avian coloration have been reviewed previously, but have tended to focus on particular subsets of hypotheses. Furthermore, I evaluate the evidence in support of each hypothesis with regard to birds, considering recent approaches that may be able to predict particular patterns or ornaments.

2. THE HYPOTHESES

I have divided the hypotheses into five categories (Table 1). The first includes hypotheses that deal with physiological benefits to individuals and do not involve interactions with other animals. Many of these are reviewed by Burtt (1981, 1986). The second group includes hypotheses that certain colors and patterns improve foraging, such as in aiding vision. The third involves some sort of interspecific interaction, where the coloration or ornament sends information (or misinformation) to members of another species (Cott, 1957; Edmunds, 1974; see also Hailman, 1977; Baker and Parker, 1979; Caro, 1986). Most of these hypotheses are concerned with avoiding predation. Fourth are the intraspecific signal hypotheses, where the color or ornament serves to communicate with conspecifics (Hailman, 1977; Butcher and Rohwer, 1989). Lastly I deal with some nonfunctional hypotheses, since the trait could have evolved through processes other than selection favoring that trait (Gould and Lewontin, 1979).

Altogether, I have identified 28 hypotheses proposing a specific function for some aspect of coloration and ornamentation in birds, plus two nonfunctional hypotheses summarized in Table 1. Some of these could be even further subdivided (for example, mimicry includes protective, aggressive, reproductive, and social mimicry; sexual selection includes both male-male competition and female choice). Testing all of these hypotheses in even a single species would be a daunting task indeed. Fortunately, many have limited applicability by only explain-

TABLE I
Hypotheses for the Evolution of Coloration and Ornamentation in Birds, the Types of Traits Each of these Hypotheses Explain, and the Type of Evidence Supporting Each Hypothesis[a]

Hypotheses	Types of traits explained	Supporting evidence[b]
Physiological Hypotheses		
Aerodynamic efficiency	Wing and tail elaborations	None[c]
Physical protection	Distribution of melanin	Experimental, comparative
Thermoregulation	Dark versus light coloration	Experimental,[d] comparative
UV protection	Dark coloration, especially on dorsal surface	Anecdotal
Facilitation of Foraging Hypotheses		
Glare reduction	Dark facial marks and dark bills	Experimental, comparative
Sighting lines	Facial stripes	Comparative
Visual clarity	Iris color	Anecdotal
Flushing or disrupting prey	Flash patterns in wings; conspicuous patterns on piscivores	Associative; experimental[e]
Interspecific Signaling Hypotheses		
Crypticity	Inconspicuous color patterns	Experimental,[e] comparative
Aposomatism	Conspicuous patterns	Anecdotal
Mimicry	Patterns that resemble other species	Anecdotal
Image Avoidance	Color polymorphisms (raptors, etc.)	Comparative
Signal of "unprofitable prey"	Conspicuous patterns	Experimental,[d] comparative
Predator deflection	Conspicuous patterns and flash patterns	None[c,e]
Startle/confusion effects	Flash patterns	None
Pursuit invitation	Conspicuous patterns	None
Pursuit deterrence	Flash patterns	Anecdotal
Intraspecific Signaling Hypotheses		
Indexic signal	Facial patterns	None
Feeding solicitation	Primarily bill or face markings	Experimental
Enhancement of displays	Usually conspicuous	Associative, comparative
Warning or alarm signal	Flash patterns	None
Flock attraction and cohesion	Colors conspicuous at a distance	Experimental (associative[c])
Species recognition	Conspicuous patterns?	Comparative, experimental[d]

TABLE I (*Continued*)

Hypotheses	Types of traits explained	Supporting evidence[b]
Individual recognition	Individually variable patterns	Experimental
Kin recognition	Individually variable patterns	Experimental
Status signal	Variable patterns	Experimental
Sexual selection	Usually conspicuous patterns (males)	Experimental, comparative
Female mimicry	Female-like (usually inconspicuous) young males	Experimental
Nonadaptive Hypotheses		
Genetic correlations	Pattern correlates with another trait	Experimental
Genetic drift	No specific predictions	Not usually testable

[a] See text for additional explanation and references.
[b] The type of evidence in support of each hypothesis. Experimental usually involves manipulations of the patterns in question or the use of models. Comparative evidence involves interspecific comparisons. Associative evidence indicates an association between the trait and its function, such as intraspecific comparisons using natural variability, and is not listed if experimental or comparative evidence is available.
[c] Studies failed to support the hypothesis.
[d] Studies have had mixed results.
[e] Only limited components of the hypothesis were tested.

ing a small set of patterns or ornaments; the number that would apply to any particular feature of any single species is far fewer.

2.1. Physiological Hypotheses

2.1.1. Hypothesis 1. Aerodynamic Efficiency

Some plumage features, ornaments could potentially have aerodynamic functions, would be likely only for elaborations of the flight feathers, such as elongated or atypically shaped retrices. Tails function during flight by providing lift, maneuverability and braking power (Pennycuick, 1972; Kerlinger, 1989; Balmford *et al.*, 1993). Both tail length and wing shape are related to habitat and the style of flying (Kerlinger, 1989). Balmford *et al.* (1993) investigated the aerodynamic properties of different tail shapes in birds. Their analyses suggest that graduated tails (where the feathers are progressively longer from the edge to the center) are the most aerodynamically costly (in terms of lift versus drag), pintails (only the central feathers are elongated) less so, and forked tails least costly or even aerodynamically advantageous.

Other studies investigated the effects of tail shape on maneuverability. The long, forked tails of swallows, for example, could serve to increase maneuverability, increase lift, and reduce stall speed (Pennycuick, 1972; Balmford *et al.*, 1993), but Møller (1989) has found that experimentally elongating the tail streamers of male Barn Swallows (*Hirundo rustica*) decreased their foraging efficiency. Similarly, manipulations of tail streamers in Scarlet-tufted Malachite Sunbirds (*Nectarinia johnstoni*) hindered the efficiency of hawking insects (Evans and Thomas, 1992). Long-tailed Widowbirds (*Euplectes progne*) are said to be incapable of flight when raining (Mackworth-Praed and Grant, 1960), and there is a trend for the longer-tailed widowbird species to occur in drier climates (Savalli, in press). The evidence available strongly suggests that greatly elongated tail feathers incur an aerodynamic cost rather than a benefit.

The dark coloration of primaries of soaring birds may help them absorb heat, thereby warming the feathers and increasing their flexibility. This flexibility would enable them to bend upwards at the tip, reducing drag due to reduced tip vortices (E. Burtt, personal communication). This possibility remains untested.

2.1.2. Hypothesis 2. Physical Protection

Dark feathers are more resistant to damage and abrasion than are light feathers due to the presence of melanin (Barrowclough and Sibley, 1980; Bergman, 1982; Burtt, 1979, 1986). Birds living in more abrasive environments should therefore have more melanin. Burtt (1986) compared desert birds from the Arabian peninsula with nondiving, nonburrowing oceanic birds and found that the desert birds had smaller proportions of nonmelanic areas in their plumage. In addition, the surfaces that are more vulnerable to wear and abrasion, such as the dorsal surfaces and the outer edges of the wings and tail, should be darker, and this, too, has been confirmed in a survey of wood warblers (Burtt, 1979, 1986). The physical protection hypothesis only applies to coloration based on melanin and not to other colors (such as carotene-based colors) and not to ornaments such as long tails. Conversely, patterns of wear may constrain the distribution of iridescent colors depending on multiple layers of melanin granules, since these would degrade if some of the layers are worn off (Greenewalt *et al.*, 1960).

2.1.3. Hypothesis 3. Thermoregulation

The coloration of a bird may have important thermoregulatory consequences, as it can for amphibians and reptiles (Hoppe, 1979). Dark colors absorb more radiant energy than light colors (Tyndall, 1897; Burtt, 1986; Hamilton and Heppner, 1967; Lustick, 1969; Lustick *et al.*, 1979; Ellis, 1980), and dark colors might be expected to be more frequent in colder climates. This is the case with exposed skin, in which there is no insulation and the absorbed radiation is transferred directly to the body as heat (Lustick *et al.*, 1979; Hill *et al.*, 1980; Burtt, 1986). Burtt (1986) found that the first species of wood warblers to arrive on their breeding grounds and the last species to leave, and those with the northernmost wintering distribution (i.e., those species exposed to the coldest conditions overall) tended to have darker colored legs than the species from warmer conditions. The situation is more complex for plumage color. Because of its poor conductive properties, plumage can function in a reverse manner (depending on the degree of feather erection and wind speed): dark plumage prevents radiation from passing through to the skin and keeps it near the surface where it may be lost through convection or reradiation, whereas light-colored plumage allows much of the radiant energy to pass through the feathers to reach the skin where it is absorbed and heats the animal (Walsberg *et al.*, 1978). Other data on heat absorption suggest that dark plumage does enable birds to absorb heat, though this was done without considering the effects of feather erection or wind (Hamilton and Heppner, 1967; Lustick *et al.*, 1978; Ellis, 1980). In Phainopeplas (*Phainopepla nitens*), a sexually dichromatic desert birds (males are black, females gray), Walsberg (1982) found that change in posture and choice of microclimate had a much greater effect on heat gain than did color. Similarly, there was no significant difference in daily energy expenditures between blue phase and white phase Snow Goose (*Chen caerulescens*) goslings (Beasley and Ankney, 1988). On the other hand, procellariform seabirds in tropical climates are darker than those in higher latitudes (Bretagnolle, 1993), a result also consistent with Hypothesis 4. Testing Hypothesis 3 will prove difficult because the predicted outcomes depend on the climactic conditions present. For example, in windy environments birds should be darker in warm climates, but if no wind is present, the reverse will be true. Since dark plumage absorbs heat, the dark coloration may also facilitate drying (Burtt, 1981) and thus could explain Gloger's rule, in which birds in humid climates tend to be darker.

2.1.4. Hypothesis 4. Protection from Ultraviolet Light

Ultraviolet-absorbing pigments, such as melanin and to a lesser extent, carotenoids, provide a shield from the potentially harmful effects of ultraviolet light (Burtt, 1979; Wake, 1979), and are thus expected to be heaviest in diurnal species of equatorial, of open areas, at high altitudes, and on dorsal surfaces. Dark dorsal surfaces are widespread in birds and most other animals, but this pattern is also consistent with abrasion resistance (Hypothesis 2), thermoregulation (Hypothesis 3), and countershading (a form of cryptic coloration; Hypothesis 10). *Ramphocelus* tanagers have greater concentrations of carotenoid pigments at higher altitudes (Brush, 1970), and this is consistent with ultraviolet protection (Burtt, 1979). Procellariform seabirds are darker in tropical latitudes than temperate latitudes (Bretagnolle, 1993), a result consistent with this hypothesis and Hypothesis 3. Albinos (or paler variants) are predicted to have higher rates of cancer and other indications of UV damage: domesticated forms that lack most pigment are likely to provide the best systems for testing this.

2.2 Facilitation of Foraging Hypotheses

2.2.1. Hypothesis 5. Glare Reduction

The upper mandible, lores, and areas around the eye can reflect light into the bird's eye, causing glare that can interfere with vision (Ficken *et al.*, 1971; Burtt, 1986). Dark bills, lores and eye-stripes (see Fig. 1) may reduce glare in species foraging in bright sunlight. Wood warblers with paler bills tended to forage in the shade more often than did dark-billed species (Burtt, 1986). An even more convincing test of this hypothesis involved experimentally whitening the upper mandible of Willow Flycatchers (*Empidonax traillii*): experimental birds foraged more often from shaded perches and appeared to have lower foraging success than controls (Burtt, 1984). Facial masks should be most prevalent in species foraging in bright, open environments. This hypothesis applies only to dark colors near the eyes.

2.2.2. Hypothesis 6. Sighting Lines

Ficken and Wilmot (1968) and Ficken *et al.* (1971) have suggested that eye-stripes (Fig. 1) may serve as sighting lines to aid in tracking and striking active prey. They showed that, among North American songbirds, such facial markings are more likely and more complex in species with a diet of active prey, such as insects, although phylogenetic effects were not excluded. Experimental alteration of such a line in the lizard *Anolis carolinensis* had no effect on the frequency or success of strikes at crickets (Burtt, 1981) but no experimental tests have been done with birds. This hypothesis applies only to contrasting markings in front of the eye.

FIGURE 1. Examples of bird color patterns. (a–d) Eye-stripes, contrasting facial patterns and masks may function as sighting lines, to reduce glare, or as indexing signals (respectively, a plover, a chickadee, a warbler, and a laughing-thrush). (e–g) Examples of flash patterns, such as contrasting patches in the wings of Black Turnstone (e) and Northern Mockingbird (g) and contrasting outer tail feathers of a Grey Fantail (f). (h) The breast stripe of *Spheniscus* penguins may aid in capturing prey by flushing fish schools. (i–k) Distinctive patterns, such as the width of the breast stripe of Great Tits (i), the amount of black on the chest of Harris Sparrows (j), and the red shoulders of Red-winged Blackbirds (k), can signal status and function in territory defense.

2.2.3. Hypothesis 7. Visual Clarity

Worthy (1974) suggested that light-colored irises may allow some light to pass through, especially at the edges, resulting in less sharp images on the retina than from dark irises which spill little light through the edges of the iris. On the other hand, light colored irises might allow more total light to reach the retina. Consequently, Worthy (1974) has suggested that more active species, such as those that pursue their prey, should have darker eyes than less active species such as sit-and-wait predators, though it is not clear why sit-and-wait predators would not also need good visual acuity. A more appropriate prediction is that species in bright, open environments should have dark irises while those in dark environments should have pale irises. This hypothesis is predicated on the untested assumption that the amount of light reflected is an indication of the light transmitted: if irises tend to have an opaque layer beneath the outer pigment, iris color would not affect vision.

2.2.4. Hypothesis 8. Flash Colors to Flush Prey

Some birds have highly contrasting (often white) patches in their wings and tail (Fig. 1). These patches are often hidden except when the bird is in flight or "flashes" its wings by rapidly raising them. Hailman (1960) has suggested that these behaviors and patterns serve to flush out insects on which the bird is foraging. In Northern Mockingbirds (*Mimus polyglottus*) and Red-backed Scrub-robins (*Erythropygia zambesiana*), two species with white wing patches, wing-flashing was associated with foraging (Hailman, 1960; Monroe, 1964) but Selander and Hunter (1960) suggested that wing-flashing is used by mockingbirds in social displays (Hypothesis 20).

Species of penguins that typically feed on schooling fish generally have conspicuous lateral markings (Fig. 1) that may enhance prey capture by causing the prey species to break up the schools, a phenomena proposed for certain dolphins as well (Wilson *et al.*, 1987). Captive anchovies were more likely to use a "flash expansion" escape response that depolarized the school when presented with a conspicuous *Spheniscus*-patterned model, while generally remaining polarized in a coordinated response when presented with a less conspicuous typical penguin model (Wilson *et al.*, 1987). However, no explanation for the anchovies' response is provided (hence, these markings are only tentatively treated as a pattern to flush prey), and the possibility that flash expansion may be more effective against *Spheniscus* penguins than other species cannot be ruled out (in which case the pattern would actually be maladaptive with regard to foraging since it allows anchovies to recognize and appropriately respond to *Spheniscus* penguins). Ultimately, tests need to show that the removal of flash patterns actually decreases foraging success. This hypothesis apparently applies only to flash patterns in the wings and perhaps to conspicuous patterns on birds that prey on schooling fish (i.e., *Spheniscus* penguins and perhaps some alcids).

2.3. Interspecific Signaling Hypotheses

2.3.1. Hypothesis 9. Crypticity

Crypticity (also referred to as concealing coloration or camouflage) is the matching of an animal's color pattern to the background so that the animal is rendered less conspicuous (G. Thayer, 1909; Cott, 1957; Edmunds, 1974; Endler, 1978; see also Pasteur, 1982) and includes such phenomena as countershading and disruptive coloration (see G. Thayer, 1909; Cott, 1957). This hypothesis was first given serious attention by Poulton (1890), Wallace (1895, reprinted in Beebe, 1944), and A. Thayer (1896). Crypticity is clearly a well established and widespread phenomena in the animal kingdom and numerous examples are known, particularly among insects (G. Thayer, 1909; Cott, 1957, 1985a; Edmunds, 1974; Endler, 1978). The adaptive significance of crypticity is obvious and there is little doubt that it is important in the evolution of bird coloration, as evidenced by the many birds that match the general color of their habitats, from white birds in the arctic to green forest birds (Cott, 1985a). Nonetheless, it is well advised to not rely solely on such anecdotal observations of an association of color with habitat. The Thayers (G. Thayer, 1909), for example, considered nearly all coloration to be cryptic, including the peacock's tail (artfully illustrated so its blue neck was against the sky and its train among flowers) and pink flamingoes matching the sunset! Crypticity has been best documented in comparisons of desert birds that match the local soil color (Niethammer, 1940; Koepcke, 1963). The only experimental study of crypticity in birds showed that Black-headed Gulls (*Larus ridibundus*) with their normally white underparts painted black had less success at capturing fish from a pool than did controls (Götmark, 1987). Experimental evidence of protective crypticity, including measurements of predation, is not available for birds (but see Hypothesis 13, below).

2.3.2. Hypothesis 10. Aposomatism

Animals may signal their distastefulness or hazardousness to potential attackers with conspicuous coloration known as aposomatism or warning coloration (Poulton, 1890; Cott, 1957; Edmunds, 1974; see also

Guilford, 1986; Guilford and Dawkins, 1993). Unfortunately, the palatability of birds to their potential predators is not well known. Cott (1947, 1985b; Cott and Benson, 1969) studied the palatability of a large number of bird species, using domestic cats (*Felis catus*), hornets (*Vespa orientalis*), and humans. The results indicated a general relationship between conspicuousness and unpalatability, but a recent re-analysis of the data only partially confirmed this relationship (Götmark, 1994). Even if the differences in palatability are real, they may not be sufficiently large to deter attacks by predators. Birds of the New Guinean genus *Pitohui* contain a toxin sufficiently strong to repel potential predators: their tissues and feathers contain homobatrachotoxin—a powerful neurotoxin otherwise known only from the South American poison dart frogs (Dendrobatidae)—in sufficient quantity to kill mice and cause numbing and burning on contact with human mucus tissues (Dumbacher *et al.*, 1992).

Domestic cats appeared to capture fewer colorful species of birds in Britain (Baker and Hounsome, 1983), but Merlin (*Falco columbarius*) had no aversion to colorful birds (Baker and Bibby, 1987). Problems with these data are discussed in Reid (1984), Lyon and Montgomerie (1985), and Hypothesis 13, below. Among studies using invertebrate or artificial prey, the evidence is most consistent with the hypothesis that warning colors function to increase recognition by predators, rather than unpalatability simply allowing for brighter colors (Guilford and Dawkins, 1993).

2.3.3. Hypothesis 11. Mimicry

In mimicry, the mimicking animal (or plant) resembles some other, unrelated species (the model) and gains some advantage (usually defensive) due to another species' (the dupe) resulting misidentification (Pasteur, 1982). In protective mimicry (including Batesian, Müllerian, and Mertensian mimicry), the mimic is similar to a dangerous or distasteful, aposomatically colored species (Cott, 1957; Wickler, 1968). Examples are rare in birds (Pough, 1988), though black flycatchers (*Melaenornis* spp.) in Africa may mimic aggressive and distasteful drongos (*Dicrurus* spp.); African forest flycatchers (*Stizorhina* spp.) may mimic the formic acid-smelling ant-thrushes (*Neocossyphus* spp.); Moluccan-New Guinean orioles (*Oriolus* spp.) may mimic the pugnacious friarbirds (*Philemon* spp.); *Malaconotus* and *Chlorophoneus* bush shrikes may mimic each other; some races of the Variable Pitohui (*P. kirhocephalus*) may mimic the more highly toxic Hooded Pitohui (*P. dichrous*); and wrynecks (*Jynx* spp.) are thought to mimic snakes

(Swynnerton, 1916; Hall *et al.*, 1966; Diamond, 1982; Edelstam, 1985; Dumbacher *et al.*, 1992). The ear tufts of owls may be used to intimidate conspecifics by mimicking the ears of mammalian predators on owls (Mysterud and Dunker, 1979). None of these supposed mimicry systems have been studied in any detail; the often cited mimicry of drongos by black flycatchers, for example, is based on the observations of a single cat presented with one dead bird of each of several species (Swynnerton, 1916:548). Kermadec and Herald Petrels (*Pterodroma neglecta* and *P. arminjoniana*) are similar in appearance to the aggressively kleptoparasitic jaegers and skuas (Stercorariinae), and do not get harassed by these birds (which do not parasitize each other) as do other dissimilar petrels (Spear and Ainley, 1993). Kermadec Petrels even kleptoparasitize other procellarids, presumably enabled to do so by their skua-like plumage and behavior.

In the tropics, many members of interspecific flocks have a strikingly similar appearance (Moynihan, 1960; Cody, 1973; Diamond, 1982). Moynihan (1960, 1968) suggested that this similarity, which he called "social mimicry," functions in reducing interspecific aggression and facilitates flocking. It may also result in interspecific territoriality, depending on the circumstances (Cody, 1973). Barnard (1979) has suggested that this may instead be an antipredator tactic: if predators preferentially pursue rarer phenotypes, then selection would favor similar coloration in order to blend into the "background" of the flock (but see Moynihan, 1981). The vulnerability of odd-colored prey is well documented, frequently with birds as predators (Mueller, 1974; Landeau and Terborgh, 1986; Wilson *et al.*, 1990), and has been demonstrated for Northern Goshawk (*Accipiter gentilis*) predation on color morphs of Rock Doves (*Columba livia*) (Pielowski, cited in Mueller 1974). (For stationary, difficult-to-find prey the reverse is true: foraging animals tend to form search images and concentrate on the more common type [Greenwood, 1985].) Interspecific flocking could lead to "character convergence" of signals that are used in interspecific communication among flock members, such as alarm signals (Moynihan, 1968; Barnard, 1982; see also Hypotheses 21 and 22 below). The similarities of members of mixed-species flocks may be due to convergence, however, and not truly mimetic (Burtt and Gatz, 1982; Pasteur, 1982). Besides anecdotal observations of similarities among flocking species, there is no comparative or experimental evidence to support social mimicry in birds. Perhaps the most practical way to test this hypothesis is to present mounts of the mimic, the model, and a related nonmimic to a potential predator (or other dupe). The predator is predicted to attack the nonmimic but should avoid both the model and mimic.

In aggressive (or Peckhammian) mimicry the mimetic resemblance allows close approach to, or even attraction of, potential prey (Cott, 1957; Wickler, 1968). Bird examples are rare and untested here as well. Zone-tailed Hawks (*Buteo albonotatus*) are very similar in coloration and manner of soaring to harmless Turkey Vultures (*Cathartes aura*), and this may allow them to approach their prey more easily (Willis, 1963). The best documented cases of avian aggressive mimicry are among the brood parasites (referred to as reproductive mimicry by Pasteur, 1982): the eggs of many cuckoos mimic their host's eggs (Payne, 1977; Davies and Brooke, 1989) and the mouth markings of nestling whydahs (*Vidua* spp.) correspond very closely to the mouth markings of their waxbill hosts' nestlings (Neunzig, 1929; Friedmann, 1960; Nicolai, 1964; Payne, 1967).

2.3.4. Hypothesis 12. Image Avoidance

Color polymorphisms not associated with age, sex, season or geographic location are known from a diverse variety of birds from the families Procellariidae, Anatidae (one species), Ardeidae, Accipitridae, Falconidae, Laridae, Scolopacidae (one species), Cuculidae, Strigidae, Emberizidae, and Malaconotidae (Paulson, 1973). The function of each polymorphism, if any, is often unclear, but among the predatory (Accipitridae, Falconidae and Strigidae) and parasitic (Cuculidae, Laridae) species, it may reduce the ability of the prey or host species to form recognition images of the predator or parasite and thus impede their ability to learn to recognize them (Payne, 1967; Paulson, 1973). Since migratory species are more likely than residents to have overlapping hunting ranges, especially during winter, the advantages derived from having two or more morphs should be greater for migratory than resident species (Rohwer and Paulson, 1987). A comparison of North American *Buteo* hawks supports this prediction: migratory species were more likely to be polymorphic and had more morphs than nonmigratory species. Another, as yet untested, prediction of this hypothesis is that the prey species should take longer to learn and respond to predators that are polymorphic. This hypothesis applies only to polymorphisms (which are generally uncommon) and perhaps only to predatory and parasitic species. It is possible that predators of birds may also form search images, as do predators of cryptic, immobile food (Greenwood, 1985), in which case polymorphic species may gain protection. If this were the case, however, polymorphisms should be more widespread, especially among the small, vulnerable species.

2.3.5. Hypothesis 13. Signal of "Unprofitable Prey"

The unprofitable prey hypothesis states that birds use bright and conspicuous plumage to signal their "unprofitability"—that is, how difficult they are to catch relative to their nutritional value—to potential predators and is thus similar to aposomatic coloration (Baker and Parker, 1979). Thus, distinctive colors should enable predators to learn which species are too difficult to catch to be worthwhile pursuing. Other species may mimic the unprofitable species (see above). Baker and Parker (1979) argued that the patterns of sexual, seasonal, and age-related dimorphism seen in European birds all support their hypothesis: since adult males are behaviorally more conspicuous (particularly in lekking species), they should incur less cost in increased conspicuousness than would females, which, since they must spend much time on the nest, are better off not being detected in the first place (especially if this could lead to the nest being discovered). Young birds who are inexperienced and poorer fliers are more profitable and so should not signal unprofitability (unless they were to mimic their fathers). Furthermore, Baker and Parker argued that seasonal patterns in plumage brightness of European birds are explicable by the unprofitable prey hypothesis, since during the summer there are many highly profitable young birds around, so it pays older birds to signal their relative unprofitability, but during the winter these young birds will be sufficiently experienced that there is little difference in profitability, so all birds should be cryptic. However, these patterns are also explicable by several other hypotheses, particularly sexual selection (below) (Krebs, 1979; M. Andersson, 1983). Andersson (1983) has further pointed out that molt into bright plumage does not correspond precisely to when young are present, but instead precedes courtship and pair formation; this is most apparent in ducks (Anatinae) where pair formation is considerably earlier (in fall and winter) than nesting (in summer); males are in their conspicuous plumage during the time of pair formation and not during the nesting season.

Data from cat predation of British songbirds suggests that the conspicuous species are less at risk than the cryptic species (Baker and Hounsome, 1983), but these results and interpretations have been criticized, because effects of body size and habitat could easily bias the results (Reid, 1984; Lyon and Montgomerie, 1985; but see Baker, 1985). In particular, it is possible, even likely, that those species more at risk from cats are the ones that are cryptic, while those safe from cats, by being larger or more arboreal perhaps, can afford the brighter colors. Merlins (*Falco columbarius*), on the other hand, captured prey independent of coloration, with a trend toward favoring more conspicuous prey (Baker and Bibby, 1987). This result contradicts the unprofitable prey and aposomatism hypotheses. The only experimental tests of this hypothesis showed that European Sparrowhawks (*Accipiter nisus*) and other raptors preferentially attacked mounts of dull (female) colored Pied Flycatchers (*Ficedulla hypoleuca*) over mounts of more conspicuously colored males (Götmark, 1992) and preferred dull-colored female Blackbirds (*Turdus merula*) over more conspicuous Great Spotted Woodpecker (*Dendrocopos major*) models (Götmark and Unger, 1994). Captive Peregrines (*Falco peregrinus*) did not show any preference when both sexes of Pied Flycatchers were fed to them, nor between Blackbirds and woodpeckers, suggesting that the sparrowhawks' aversions are not a result of aposomatism (Götmark, 1992; Götmark and Unger, 1994). On the other hand, no difference in predator attack rates was detected between mounts of cryptic Meadow Pipits (*Motacilla alba*) and conspicuous White Wagtails (*Motacilla alba*) (Götmark and Unger, 1994).

2.3.6. Hypothesis 14. Predator Deflection

Distinctive markings on an animal's periphery, far from vital parts, might attract a predator's attention and deflect its attack to these less vulnerable parts, allowing escape (Cott, 1957). Many lizards, for example, are able to autotomize their often brightly-colored tail, which remains wriggling to attract and delay predators (Cott, 1957; Dial and Fitzpatrick, 1983). This hypothesis could explain wing and tail marks (see examples in Fig. 1), as predators attacking these areas would get little more than a mouthful of feathers, but there is no evidence from birds.

Bright coloration could also serve to deflect predators away from the nest (Baker and Parker, 1979). The most obvious behavioral example is the "injury feigning" display of a number of birds (Cott, 1985a). Predator deflection is likely to lead to sexual dimorphism in species in which the male guards the nest or territory, but the female does all of the incubating and feeding. The male would serve to distract predators away from the nest (Baker and Parker, 1979). This hypothesis has many of the same difficulties discussed above, i.e. that the nesting season does not correspond precisely to the breeding plumage season (M. Andersson, 1983), and it also does not fit well with the general patterns of male parental care (in most species, regardless of whether they are dichromatic, males feed the young; Verner and Willson, 1969). An attempt to test this hypothesis in two species of New World orioles (*Icterus* spp.) (Flood, 1989) failed to support it: predation rates were similar for the nests of dull colored (first year) versus bright colored males and mobbing rates were similar for the bright males and duller females.

2.3.7. Hypothesis 15. Startle/Confusion Effects

Protean displays are behavior patterns of individuals or a group that serve to startle or confuse potential predators (Humphries and Driver, 1970; Edmunds, 1974; Caro, 1986). These include erratic movements, sudden bursts of motion, and sudden vocalizations that can be used by individual birds or groups, and group behaviors such as sudden scattering, maintaining very tight cohesion, and mobbing. Color patterns can function to enhance protean displays. Suddenly changing colors can confuse or startle predators, as can the sudden appearance of different colors (see Fig. 1) such as contrasting patches in wings and tail revealed as the bird takes flight. In a flock, flashing patterns on the wing can make it difficult for predators to isolate one individual. Flocking species should be more likely therefore to have wing patches than solitary species. This hypothesis applies to flash patterns: conspicuous markings that contrast with the rest of the plumage and can be kept hidden until suddenly revealed. Although wing and tail patches are common in birds, this hypothesis has not yet been tested.

2.3.8. Hypothesis 16. Pursuit Invitation

Smythe (1970) has proposed that the white rump and tail of many cursorial ungulates (and, by extension, similar patterns such as rump patches in birds) functions to trick a predator into pursuing it too early, and thus tiring before it can make the capture. This has been criticized by Coblentz (1980). He pointed out that two assumptions of the pursuit invitation hypothesis were unlikely to be met. First, it assumes that the predator can actually be tricked into pursuing too early. This is unlikely since predators would be expected to learn very rapidly how close they need to approach for a pursuit to be successful. Second, it assumes that it is energetically better for the prey individual to flee from the predator than just watch it and to remain just outside the critical flight distance. This assumption is also unlikely to be met. Caro (1986) suggests that it may be advantageous for mothers to distract predators from their young, in which case it becomes a distraction display (Hypothesis 14, above). This hypothesis also applies to flash patterns, particularly those that are revealed when a bird takes flight. This hypothesis has not been tested in birds, but seems unlikely on the grounds discussed by Coblentz (1980).

2.3.9. Hypothesis 17. Pursuit Deterrence

Individuals may signal to predators that they have been spotted. This is advantageous to both predators and prey since the predator will not continue wasting its efforts trying to stalk the prey and the prey will benefit by being left alone (Woodland *et al.*, 1980; Bildstein, 1983; Caro, 1986; reviewed by Hasson, 1991a). This hypothesis only applies to signals that can be turned on and off, such as behavior or coverable patches. It predicts that the signal should be used before the predator begins its pursuit. Woodland *et al.* (1980) have proposed that tail-flicking in Swamphens (*Porphyrio porphyrio*), which exposes the white undertail coverts, is such a signal since it appeared to be directed at the approaching human observers. However, Craig (1982) has argued instead that tail-flicking functions in status signalling among conspecifics.

2.4. Intraspecific Signaling Hypotheses

2.4.1. Hypothesis 18. Indexing Signals

Facial stripes and markings may help to indicate the direction that an animal is looking; in particular, they may help to unambiguously indicate at whom the gazer is staring. They may also be used to draw attention to a nest site (Hailman, 1977; Zahavi, 1981). There are no data that pertain to this hypothesis, and it is not clear how it could be tested. Manipulation of such markings would probably affect social interactions in some way, and such markings would be most likely to occur in social species.

2.4.2. Hypothesis 19. Feeding Solicitation

Bright grape colors and markings may stimulate a parent to feed the young and aid in locating the mouth in poorly lit nests (Swynnerton, 1916; Cott, 1985a). More relevant to adult birds, distinctive marks—such as the bill spot on some gulls (*Larus* spp.)—indicate to the young where to direct the pecking for food (Cott, 1985a). This has been demonstrated experimentally for bill color in Herring Gulls (*Larus argentatus*) (Tinbergen and Perdeck, 1950). It has been suggested, based on anecdotal evidence, that the black chest crescent on Northern Flickers (*Colaptes auratus*) may serve a similar purpose (Brackbill, 1955). Although this hypothesis could apply to any contrasting markings, it is likely to be limited to markings on the bill and around the head.

2.4.3. Hypothesis 20. Enhancement of Displays

Distinctive plumage features may serve to accentuate display movements that are used for intraspecific communication (Tinbergen, 1952). Such markings would not have a function without the display, but instead would enhance the behavioral signal, perhaps by making the signal more obvious and less ambiguous (see also Hasson, 1991b). Potential examples include the wing specula of ducks (*Anas* spp.), which are involved in a number of displays (Tinbergen, 1952), and the elongated crown feathers of the Steller's Jay (*Cyanocitta stelleri*), which make it obvious when the feathers are raised in an aggressive signal (Brown, 1964). In Old World warblers (*Phylloscopus* and *Regulus*), the presence of colored patches (wing bars, crown stripes, or rump patches) is associated with particular behavioral displays (Marchetti, 1993). Alternatively, associations between plumage patterns and behavioral displays could come about because the display functions in drawing attention to the plumage pattern, which is the actual signal. In wood warblers (Parulinae), the presence of contrasting wing and tail patches is associated with behavioral displays that reveal those patches, but the presence of displays in species that do not have patches suggests that the color pattern evolved to enhance the display (Burtt, 1986). Similarly, wing flashing displays are found in mimid species that lack wing patches as well as those that have them, suggesting again that the display evolved first and was subsequently enhanced by the contrasting color (Burtt *et al.*, 1994). In either case, an additional hypothesis is needed to explain the original function or presence of the behavioral display. This hypothesis could apply to a wide variety of colors, patterns, and ornaments, though in all cases they should be associated with display movements.

2.4.4. Hypothesis 21. Warning or Alarm Signal

Birds and mammals often signal to conspecifics that there is a predator present. Such signals are typically acoustic (e.g., Marler, 1955; Sherman, 1977), but can also be visual, such as the white tail flagging of deer or stotting of gazelles (Estes and Goddard, 1967; Bildstein, 1983; Caro, 1986). Distinctive colors that can be selectively displayed can serve as alarm signals. Potential alarm colors in birds include "flash patterns" such as wing patches, rump patches, and tail markings that are typically visible only in flight (such as when fleeing from a predator) or in preflight intention movements (Moynihan, 1960). Tail-flagging in deer appears to function as an alarm signal (Bildstein, 1983), but this hypothesis has not been tested in birds. Removal of the flash patterns should result in a flock's increased response time asynchrony to predators. This hypothesis only applies to signals that can be turned on and off or are only visible at certain times, such as when fleeing.

2.4.5. Hypothesis 22. Flock Attraction and Cohesion

Conspicuous coloration may enable birds to locate each other and thus facilitate flocking, particularly if trying to locate flocks or feeding aggregations from a large distance (Moynihan, 1960; see also Selander and Hunter, 1960; Simmons, 1972; Cott, 1985a). This may explain why many gregarious species, such as many corvids, icterine blackbirds, starlings, vultures, egrets, swans, gulls and terns, are either black, white or pied. Flash patterns can also facilitate flocking by enhancing intention movements that indicate a bird is about to take flight, thus allowing its flock-mates to prepare to join it (Moynihan, 1960; see also Hypothesis 21, above). The flock cohesion hypothesis can also apply to interspecific communication for species that form multispecies flocks. Egret models painted white had more waders land nearby than did blue-painted models or control areas without models (Kushlan, 1977). Male Red-winged Blackbirds (*Agelaius phoeniceus*) perched closer to all-black mounts than to control areas without mounts or to mounts with red epaulettes, suggesting that the black plumage, but not the epaulettes, functions to attract conspecifics (Røskaft and Rohwer, 1987). In the similar-looking widowbirds (*Euplectes* spp.), however, the seasonal pattern of flocking (mostly during the nonbreeding season) does not match the seasonal pattern of the males' black plumage (only during the breeding season); a result inconsistent with the hypothesis (Savalli, 1991).

2.4.6. Hypothesis 23. Species Recognition

Differences in color or ornamentation between species may serve as premating isolating mechanisms to prevent hybridization (Mayr, 1963) or to prevent mistaken aggression between heterospecifics (Lorenz, 1966). Highly distinctive appearances may be especially likely to evolve in species where there is no male parental care and thus no opportunity for imprinting (Sibley, 1957; Immelmann, 1975; but see Weary *et al.*, 1993). The expression of species recognition signals should be greatest for males during the breeding season, especially for polygynous species (Trivers, 1972).

Evidence for coloration being used for species recognition comes in several forms. Island-dwelling species, which often do not have any sympatric close relatives, are typically drab compared to their mainland counterparts (Sibley, 1957; Grant, 1965; Moreau, 1966). Stronger evidence comes from character divergence in areas of sympatry among close relatives (e.g., *Ficedula* flycatchers, Røskaft *et al.*, 1986; *Larus* gulls, N. Smith, 1966; *Agelaius* blackbirds, Orians and Christman, 1968; and *Sitta* nuthatches, Grant, 1975), although such examples are often controversial (see Alatalo *et al.*, 1990; Pierotti, 1987; and Hansen and Rohwer, 1986 for contradictory results for each of the first three examples), typically involve rather subtle characteristics, and are overall quite rare in birds (Rohwer *et al.*, 1980; Butcher and Rohwer, 1989). There was no clear evidence for character displacement in tail length for two pairs of *Euplectes* widowbird sibling-species (Savalli, 1995).

Some authors have suggested that the association of distinctive coloration with a high rate of interspecific hybridization in some birds, such as *Anas* ducks and hummingbirds, indicates that their coloration serves in species recognition (Sibley, 1957; Johnsgard, 1963) because the high rate of hybridization necessitates the very distinctive male plumages. This argument seems counterintuitive, however, since the high rate of hybridization remains *despite* the ease by which males can be recognized. Pierotti (1987), on the other hand, used the high rate of hybridization among sea birds with similarly colored bills and legs and lack of hybridization among forms that differ in bill and foot color as evidence that these traits are used in species recognition in many species. In short, associations of hybridization rates and coloration are open to contradictory interpretations and do not provide any clear evidence for or against species recognition.

Experimental manipulation of distinctive male plumage of six species did not appear to affect species recognition abilities (Mallard ducks, Nekipelov, 1969 cited in Kelso, 1970; Red-winged Blackbirds, D, Smith, 1972; Hansen and Rohwer, 1986; Yellow-headed Blackbirds, Rohwer and Røskaft, 1989; Common Yellowthroats, Lewis, 1972; Northern Orioles, Butcher, 1991; Yellow-shouldered Widowbirds, Savalli, 1995). In Zebra and Double-bar finches (*Poephila guttata* and *P. bichenovii*), on the other hand, individuals of both sexes show preferences for mates with leg bands of colors that match a portion of their own plumage and avoid colors found on the other species (Burley, 1986), although these findings do not exclude the possibility of differences in mate preferences that evolved independently of a need for species recognition. The manipulation of yellow head feathers in three species of penguins (*Eudyptes chrysocome*, *E. chrysolophus*, and *Aptenodytes patagonicus*) resulted in reduced pairing success compared to controls, which was interpreted as evidence that these plumes function in species recognition (Jouventin, 1982), but is also consistent with sexual selection (Hypothesis 27). Overall, the evidence for species recognition playing a role in bird color evolution is problematical and does not seem sufficient to account for the extreme development or bright colors and ornaments seen in many bird groups.

2.4.7. Hypothesis 24. Individual Recognition

Plumage features that are highly variable among individuals may serve to enhance the recognition of individuals, particularly with respect to the outcome of aggressive encounters, and have been called "arbitrary identity badges" (Shields, 1977; Whitfield, 1987; Rohwer and Røskaft, 1989). In Ruddy Turnstones (*Arenaria interpres*) the highly variable head and neck patterns were not correlated with dominance or territory quality, but did serve to aid in individual recognition: males discriminated between neighbor-like and stranger-like models (Whitfield, 1986). In Ring-billed Gulls (*Larus delawarensis*) adults can recognize their own chicks by facial markings (Miller and Emlen, 1975), and Cliff Swallow (*Hirundo pyrrhonota*) parents may be able to recognize their chicks by their highly variable face patterns (Stoddard and Beecher, 1983).

2.4.8. Hypothesis 25. Kin Recognition

Animals may use variation in plumage to recognize close kin or assess the degree of relatedness (Waldman, 1987). Benefits include being able to behave nepotistically and increase inclusive fitness (Hamilton, 1964; Alexander, 1974; Dawkins, 1982) and to achieve an optimal amount of outbreeding (Bateson, 1978, 1982). Kin recognition has been studied in a wide variety of animals. The best example of kin recognition based on appearance in birds is Bateson's (1980, 1982) visual choice tests with captive Japanese Quail (*Coturnix coturnix*) in which birds preferred novel members of the opposite sex that were of intermediate relatedness and avoided extremely different plumage types. As with individual recognition, this hypothesis requires individually variable plumage.

2.4.9. Hypothesis 26. Status Signal

A number of recent studies have demonstrated that distinctive plumages (e.g., see Fig. 1) can signal dominance status between age and sex classes of wintering birds (Rohwer, 1977, 1985; Parsons and Baptista, 1980; Fugle *et al.*, 1984; Watt, 1986; Holberton *et al.*, 1990; Møller, 1987; P. Ryan *et al.*, 1987). In only a few species, notably Least Auklets (*Aethia pusilla*), Great Tits (*Parus major*), and Chaffinches (*Fringilla coelebs*), is plumage variability associated with dominance among individuals within the same age and sex class (Järvi and Bakken, 1984; Whitfield, 1987; Jones, 1990). The presence of bright, iridescent plumage in female hummingbirds (Trochilidae) is associated with female defense of nectar sources (Wolf, 1969, 1975; Stiles and Wolf, 1970; Bleiweiss, 1985, 1992), and reduced sexual dimorphism in a Cuban population of Red-winged Blackbird (females are black like males) may result from both sexes defending the territory (Whittingham *et al.*, 1992). Status signals can also function during the breeding season (e.g., Bleiweiss, 1985; Møller, 1990), and if status signaling plays a role in competition for resources that attract mates, then the status signaling trait is sexually selected (see Hypothesis 27, below). If the signal is to indicate that an individual is territorial, it would be advantageous if it could be turned on or off (covered) in order to avoid harassment when not defending a territory, as is the case with Red-winged Blackbird epaulettes (Hansen and Rohwer, 1986; Metz and Weatherhead, 1992).

2.4.10. Hypothesis 27. Sexual Selection

Darwin (1871) suggested that bright colors and elaborate ornaments that appear to have no survival value function to assist members of one sex to acquire mates. This form of selection is known as sexual selection and it can operate by two methods: male–male competition (or "intrasexual selection") and female choice (or "intersexual selection"). Male–male competition is usually used to explain features of males, such as horns or large size, that seem to enhance their fighting capabilities (e.g., Hingston, 1933; Alexander *et al.*, 1979; Packer, 1983), while female choice has been presumed to select for gaudy, showy markings and displays (Darwin, 1871; see below)

Much theoretical work has been done on the evolution of direct female choice of males (see, e.g., Andersson, 1994; Bradbury and Andersson, 1987; Maynard Smith, 1991). Models for the evolution of female choice for male epigamic displays fall into two broad groups: "Fisherian" models (e.g., Fisher, 1930; Lande, 1981), and "good genes" models (e.g., Zahavi, 1975; Hamilton and Zuk, 1982; M. Andersson, 1986; Grafen, 1990). Distinguishing between these two types of models is difficult if not impossible, since many of their predictions are identical and they can operate simultaneously. However, there seems to be more evidence in support of the good genes models, particularly some recent evidence that suggests that the expression of male epigamic traits are positively correlated with parasite resistance (Loye and Zuk, 1991). Alternatively, elaborate male traits could evolve through the passive attraction of females to the most conspicuous males, with no active choice necessary (Parker, 1983). The kinds of traits favored under such a scenario could depend on biases in the female's sensory system (that evolved for other functions) (M. Ryan *et al.*, 1990; Endler, 1992) or be a result of generalization errors in the female's mate recognition system or during sexual imprinting that favors supernormal stimuli (Enquist and Arak, 1993; Weary *et al.*, 1993).

Traits that function in direct combat, such as large size, spurs, horns, or canines, are relatively easy to attribute to intrasexual selection. It is also possible, however, that some traits serve a more subtle role in mediating male–male competition by signaling the bearer's dominance or fighting abilities (Hingston, 1933; Noble, 1936; Zahavi, 1977, 1981; Rohwer, 1982; Hansen and Rohwer, 1986; see also Hypothesis 25, above) or by just indicating that a male is present and defending a particular site (Slagsvold and Lifjeld, 1988; Marchetti, 1993). In addition to helping acquire resources, intrasexual selection can take the form of courtship interference, which can have a significant effect on mate choice (Foster, 1983; Trail and Koutnik, 1986).

Sexual selection is widely regarded as one of the most important factors in the evolution of bright colors and dimorphic plumage in birds (M. Andersson, 1983; Butcher and Rohwer, 1989; but see Baker and Parker, 1979) and has also been implicated in the evolution of monomorphic plumage (Trail, 1990; Jones and Hunter, 1993; Møller, 1993). One important goal of sexual selection studies is to determine the relative importance of female choice versus male–male competition in the evolution of conspicuous plumage. A number of studies have demonstrated female choice in captive situations (e.g., Collias *et al.*, 1979; Burley, 1986; Hill, 1990; Zuk *et al.*, 1990; Johnson *et al.*, 1993), but since male–male competition is eliminated, we know nothing about its relative importance from these studies. A number of field studies of sexual selection have been done involving the experimental manipulation of traits (Table II). Eight of these demonstrated evidence consistent with female choice of male ornaments (though this interpretation is sometimes controversial; e.g., Savalli, 1989), while five studies provide evi-

TABLE II
Summary of Field Studies of Sexual Selection Using Experimental Manipulation of Plumage

Species	Trait	Evidence presented for			References
		Female choice	Male–male competition	Neither	
Indian Peafowl, *Pavo cristatus*	Eyespots on train	X			Petrie and Halliday, 1994
Great Snipe, *Gallinago media*	Tail spots	X			Höglund *et al.* 1990
Crested Auklet, *Aethia cristatella*	Crest	X			Jones and Hunter, 1993
Barn Swallow, *Hirundo rustica*	Tail streamers	X			Møller, 1988, 1992
Yellow-browed Warbler, *Phylloscopus inornatus*	Wing bars		X		Marchetti, 1993
Common Yellowthroat, *Geothlypis trichas*	Black face mask			X	Lewis, 1972
Northern Oriole, *Icterus galbula*	Orange color			X	Butcher, 1991
Yellow-headed Blackbird, *Xanthocephalus xanthocephalus*	Yellow head color			X	Rohwer and Røskaft, 1989
Red-winged Blackbird, *Agelaius phoenecius*	Red epaulette		X		Peek, 1972; D. Smith, 1972
Scarlet-tufted Malachite Sunbird, *Nectarinia johnstoni*	Pectoral tuft		X		Evans and Hatchwell, 1992a
	Tail streamer	X			Evans and Hatchwell, 1992b
Long-tailed Widowbird, *Euplectes progne*	Very long tail	X			M. Andersson, 1982
Yellow-shouldered Widowbird, *E. macrourus*	Long tail		X		Savalli, 1994
Jackson's Widowbird, *E. jacksoni*	Long tail	X			S. Andersson, 1992
House Sparrow, *Passer domesticus*	Black bib		X		Møller, 1990; Veiga, 1993
House Finch, *Carpodacus mexicanus*	Red coloration	X			Hill, 1991

dence for male traits functioning in intrasexual competition. Several other studies failed to demonstrate any role for sexual selection. The primary goal of most of these studies, however, was to provide evidence for female choice, so male–male competition may be even more prevalent than these studies indicate.

There is some reason to suspect that, in birds at least, females should choose mates primarily on the resources they provide. Biparental care is widespread and primitive in birds (McKitrick, 1992), while the diversity of morphological, vocal, and behavioral displays seen in male birds implies numerous more recent origins. This suggests that the primitive condition in birds is to select mates on the basis or resources, with alternative forms of mate choice evolving subsequently. In order to maximize their reproductive success, females would probably do best selecting quality resources or territories rather than good genes. This would result in males competing for those resources, with the consequence that females selecting quality resources will also mate with genetically superior males. Thus, there is no inherent advantage to females switching to mate choice based on plumage, even if the mating system changes to a lek system: the "best" territories will still contain the best males. Plus, there is the added advantage that fighting among males will eliminate cheaters from entering the system. Why, then, has female choice of male ornaments evolved at all? There are two possible advantages to selecting males directly rather than by their territories. First, it may be much easier to locate and evaluate a conspicuously colored male than to evaluate territory quality: this would be especially beneficial for migrant species with short breeding season, when there may not be enough time to find quality territories (since females typically arrive later than males). Second, females selecting for quality territories may have their choices constrained by the presence of other females: they may then use plumage characters to evaluate males for potential extrapair copulations in order to get higher quality genes. The importance of these benefits could influence the relative importance of female choice versus male–male competition in the evolution of bright colors and ornaments, but this theory has yet to be addressed.

Various components of sexual selection theory have been tested with comparative studies, such as the prediction that sexual dimorphism should be greatest in polygynous, especially lek-breeding, species (e.g., Höglund, 1989; Björklund, 1990, 1991; Oakes, 1992) and that brighter species should be subject to higher parasite loads (e.g., Hamilton and Zuk, 1982; Pruett-Jones *et al.*, 1990; Weatherhead *et al.*, 1991; Zuk, 1991 [and references therein]). In general, the results have been mixed: some are consistent with the predictions and others are not; in some cases even when using the same data set (e.g. Höglund, 1989; Oakes, 1992).

Some authors (studying frog vocalizations) have suggested recently that species preferences and mate choice (Hypothesis 27) may not be different phenomena (Ryan and Rand, 1993; Backwell and Jennions, 1993). While this may be true in some cases, it cannot be universal, since mate choice is sometimes directional (such as in a preference for longer tails), and species recognition cannot always be so. For example, although females of at least two widowbird species exhibit some preference for long-tailed males (M. Andersson, 1982; S. Andersson, 1992), for all but the longest-tailed species this would lead to selecting the wrong mate if this cue were also used to select species (Savalli, 1995).

Lastly, it should be noted that preferences for conspicuous individuals may extend to other social interactions besides mate choice: female American Coots (*Fulica americana*) preferentially feed young that have orange plumes over those on which the ornaments have been removed (Lyon *et al.*, 1994).

2.4.11. Hypothesis 28. Female Mimicry

The immatures of most species of sexually dimorphic birds resemble adult females, and in many species it takes males substantially longer to reach adult plumage than it does for females to reach sexual maturity (Rohwer *et al.*, 1980). The delayed plumage maturation of male birds may enable young males to mimic females and thus fool territorial males. This would enable them to enter territories without being challenged and either gain access to resources on that territory or perhaps acquire sneaky copulations with females on the territory (Rohwer, 1978; Rohwer *et al.*, 1980). This hypothesis, however, does not explain the occurrence of sexual dimorphism in the first place. In Pied Flycatchers, territorial males responded less aggressively to caged males that were femalelike (brown) than to those that were black and white, while females responded more aggressively to brown males (Slagsvold and Sætre, 1991). Consequently, brown males were able to nest closer to other males than were black and white males.

2.5. Nonadaptive Hypotheses

I have so far only considered functional explanations for the evolution of animal coloration. Some authors (e.g., Tinbergen, 1963; Sherman, 1988) would consider this sufficient to understand animal coloration at the functional level of analysis and that it would be invalid to compare these hypotheses with hypotheses from other levels. Gould and his coworkers (Gould and Lewontin, 1979; Gould and Vrba, 1982) have criticized this "adaptationist program," arguing that the term "adaptation" implies that the trait evolved through natural selection to fulfill its current role. They argue that adaptation (*sensu* Gould and Vrba, 1982) is not the only mode of evolution and that developmental, mechanical, allometric, and genetic constraints, and genetic drift must also be considered, regardless of the trait's current function. Such hypotheses are difficult to test, however, since they are historical and make no prediction of current utility (Coddington, 1988), although phylogenetic approaches can be used to identify adaptations (Coddington, 1988). Phylogenetic approaches have only been undertaken for studies of sexual selection (see Hypothesis 27). Most studies of coloration or ornamentation demonstrate only the current utility of specific colors or patterns, with no evidence that the trait is an adaptation sensu Gould and Vrba (1982).

2.5.1. Hypothesis 29. Genetic Correlations and Other Constraints

Genetic correlations, such as those due to pleiotropy or linkage, can result in one trait evolving due to selection on another correlated trait. Some difficulty can be avoided, however, by properly identifying the trait so as to consider all of its potential functions, rather than just its most obvious feature. For example, although the pigment melanin imparts certain colors to birds, it can also protect feathers from wear, a function independent of its color (see Hypothesis 3). By identifying the trait of interest as the presence of melanin rather than as the actual color, the possibility of a protective function is less likely to be overlooked.

Two potential candidates for nonadaptive colors will be briefly considered. The first is the carotenoid-based plumage color of a number of birds, such as flamingos, ibises, and House Finches. Since birds cannot synthesize carotenoids and must obtain them from their food (Brush, 1978), their coloration is at least partly dependant on diet (Hill, 1992). It is possible, therefore, that flamingos are pink not because it is advantageous to be so, but because high carotenoid levels in their diet ends up being shunted into the feathers, possibly as a consequence of some physiological mechanism, the function of which has nothing to do with coloration. Of course, the ability to deposit carotenoids in the plumage may be adaptive: male House Finches with redder plumage indicate their superior quality at providing for young (presumably because the amount of pigment relates to their own foraging success) and are preferred by females (Hill, 1991).

The second candidate is the expression of typically male traits in some females, such as the partially red epaulettes found in some female Red-winged Blackbirds. Mounts of dull- and brighter-plumaged females did not elicit differing responses from either males or females, suggesting that the more malelike plumage of some females has no function and instead evolved as a correlated response to selection on male plumage (Muma and Weatherhead, 1989). Similarly, the white forehead of male Pied Flycatchers is genetically correlated to the expression of that trait in females: the fathers of females with white patches had larger patches than did the fathers of females lacking the white patches (Potti, 1993).

2.5.2. Hypothesis 30. Genetic Drift

Particular color patterns and ornaments may have arisen not due to selection on it or another character, but instead due to random genetic drift (Kimura, 1983). Genetic drift is undoubtedly possible and can be demonstrated in living populations, but it is difficult to test as a historical explanation. Recent models have been developed that can demonstrate the past action of strong directional or stabilizing selection from the amount of morphological evolution, thereby rejecting genetic drift (Lande, 1976; Lynch and Hill, 1986; Turelli *et al.*, 1988). Unfortunately, most rates of morphological evolution are not sufficiently large to allow them to be distinguished from drift (Turelli *et al.*, 1988; Savalli, 1993). If a functional advantage to a trait can be found, then this hypothesis should be preferred over genetic drift (Mayr, 1983). Comparative studies demonstrating nonrandom distributions of particular colors or patterns indicate that some process besides drift must be at work.

3. FUTURE DIRECTIONS FOR THE STUDY OF ANIMAL COLORATION

3.1. Testing Hypotheses

Interest in and speculation about animal coloration has a long history, and, as a result, a great number of hypotheses have been generated to explain the great diversity of coloration and ornamentation seen in birds. Unfortunately, empirical testing of these hypotheses has not always kept pace: twelve (40%) of these hypotheses have no supporting evidence or only anecdotal evidence (Table I). At the other extreme, a few hypotheses, particularly status signaling and sexual selection, have received a great deal of attention in recent years. Overall, only half of the hypotheses are supported by experimental evidence, and in most of these, by just one or two tests and often only one component of the hypothesis. One hypothesis, aerodynamic efficiency, has been discounted by experimental evidence in two systems, while several other hypotheses have had mixed results.

Ten hypotheses (33%) have been tested using interspecific comparisons. Although interspecific comparisons can be a powerful tool for studying adaptation, the results must be interpreted with caution. With the exception of a number of comparative studies of sexual selection (see above), most have not included phylogenetic considerations. More importantly, most comparative studies investigated only a single hypothesis (a notable exception being Burtt's (1986) study of wood warblers), and consequently did not consider other hypotheses that were also consistent with the results. For example, the data supporting the unprofitable prey hypothesis (13) (Baker and Parker, 1979; Baker and Hounsome, 1983) are also consistent with either sexual selection (M. Andersson, 1983; Hypothesis 27) or aposomatism (Hypothesis 10). Without considering all relevant hypotheses, critical observations and experiments that can distinguish among them may never be done. By considering more than one hypothesis at a time difficulties in getting negative results published can also be avoided; difficulties which could explain the relative paucity of studies that failed to support hypotheses. A sampling of testable predictions both for within-species experimental studies and cross-species comparative studies, are presented in Table 3 for each of the hypotheses. Only by evaluating all possible hypotheses and rejecting those that do not explain the phenomena of interest can we expect to make progress (Platt, 1964).

Nonetheless, it is probably safe to say that most of the hypotheses discussed play at least some role in avian coloration. For any particular population, probably a number of selection pressures, some conflicting and some working in concert, are affecting its coloration. The viability of these hypotheses must be tested individually and their relative importance in accounting for the diversity of color patterns must be investigated. This is not to say that all 30 hypotheses should always be tested; this would be an impossible task. Fortunately, many hypotheses have rather limited applicability and make specific predictions about their occurrence, which allows most of them to be eliminated *a priori* for any specific species or color pattern. What hypotheses remain

TABLE III

Predictions of the Hypotheses for the Evolution of Coloration and Ornamentation in birds

Hypotheses	Intraspecific predictions	Interspecific predictions
1. Aerodynamic efficiency	Manipulations decrease flight performance	Tail shape correlates with foraging or foraging efficiency
2. Physical protection	Albinos experience greater wear	Species in abrasive environments are darker
3. Thermoregulation	Difficult to test accurately	Darkness of plumage should relate to climate
4. UV protection	Albinos should incur cancers and other damage	More pigments in low latitudes, high altitudes, open habitats
5. Glare reduction	Lightening masks reduces foraging efficiency	Masks prevalent in species of bright, open places
6. Sighting lines	Elimination of lines reduces foraging efficiency	More likely in species foraging on active prey
7. Visual clarity	Light irises must be shown to transmit light	Light eyes more common in dark habitats
8. Flushing or disrupting prey	Eliminating patterns reduces foraging efficiency	More likely in species foraging on active prey
9. Crypticity	Altering color patterns increases predator attacks	Color pattern varies with environment
10. Aposomatism	Predators attack duller individuals	Correlation between conspicuousness and distastefulness
11. Mimicry	Mimics and models responded to in same way	None?
12. Image Avoidance	Prey should take longer to learn and respond to	Polymorphisms most frequent in predators
13. Signal of "unprofitable prey"	Predators attack duller individuals	Correlation between conspicuousness and unprofitableness
14. Predator deflection	Removal of marks increases predation rate	Occurs in species more vulnerable to predators
15. Startle/confusion effects	Removal of marks increases predation rate	Occurs in species more vulnerable to predators
16. Pursuit invitation	Removal of marks increases predation rate	Occurs in species more vulnerable to predators
17. Pursuit deterrence	Removal of marks increases predator attack rate	Occurs in species more vulnerable to predators
18. Indexic signal	No clear predictions	More likely in social species
19. Feeding solicitation	Removal of mark interferes with begging/feeding young should respond to simplified models	None?
20. Enhancement of displays	Difficult to separate effect of pattern from display	Association between pattern and display

(*continued*)

TABLE III (*Continued*)

Hypotheses	Intraspecific predictions	Interspecific predictions
21. Warning or alarm signal	Removal of pattern should reduce response time to predators;	Occur in social species
22. Flock attraction and cohesion	Colors/patterns sufficient to attract conspecifics	Occur in social species
23. Species recognition	Alteration should lead to misdirected courtship/aggression	Character divergence in sympatry; patterns more complex where more species are sympatric
24. Individual recognition	Familiar, altered individuals treated as strangers strangers altered to look like familiars accepted	None?
25. Kin recognition	Pattern must reflect kinship; response to new birds dependant on own experience	None?
26. Status signal	Alteration of pattern affects dominance, competitive outcomes	Occurs in social and territorial species?
27. Sexual selection	Alteration of pattern affects mate choice	More elaborate in polygynous species?
28. Female mimicry	Males with female-like pattern treated like females	None?
29. Genetic correlations	Pattern correlates with another trait	None?
30. Genetic drift	No specific predictions	Pattern shows random distributions/associations

[a] See text for additional explanation and references.

should then be tested in detail. One particularly widespread pattern, for example, is color (or ornamentation) dimorphism: typically in dimorphic species of birds the males are brighter than the females, the breeding season plumage is brighter than the nonbreeding plumage, and adult birds are brighter than juveniles (M. Andersson, 1983; Butcher and Rohwer, 1989). Only a few of the hypotheses predict such a pattern: sexual selection (Hypothesis 27), status signaling (particularly in conjunction with sexual selection; Hypothesis 26), species recognition (Hypothesis 23), and possibly unprofitable prey (Hypothesis 13). A few other hypotheses could generate such dimorphisms only if there are strong, and unlikely, dimorphisms in behavior (e.g., a greater tendency for breeding adult males to form flocks could lead to dimorphic flock cohesion signals; Hypothesis 22), while genetic drift (Hypothesis 30) makes no predictions about the forms of traits at all. Similarly, flash patterns that can be revealed or hidden are predicted by only seven hypotheses (8, 14, 15, 17, 21, and possibly 20 and 26); to date no studies of birds have considered more than two of these at once.

3.2. The Relative Importance of the Hypotheses: Comparative Studies

In addition to knowing which hypotheses are possible, we also want to know how widely applicable the hypotheses are; that is, is the phenomenon they describe common? Although this could be done by testing the relevant hypotheses in a wide variety of bird species, such an approach is highly impractical given the effort involved in obtaining quality data. Even for such well-studied hypotheses as sexual selection it is not clear what the relative importance of female choice and male–male competition are. Instead, broad comparative approaches can indicate if the hypotheses have widespread applicability. There has been a recent resurgence of interest in comparative studies with development of new statistical techniques to control for phylogenetic effects (see Brooks and McLennan, 1991; Harvey and Pagel, 1991) and new phylogenetic data (e.g., Sibley and Ahlquist, 1990). The use of phylogenetic controls is essential to avoid spurious results due to lack of independence. As noted above, care must be take to carefully consider all hypotheses that are applicable. Although few phylogenetically controlled comparative studies of avian coloration and ornamentation have been done, this approach is likely to be productive in increasing or understanding and providing new insights, as it has for other taxonomic groups (e.g., Sillen-Tullberg, 1988; M. Ryan *et al.*, 1990; Basolo, 1990).

3.3. Toward Predictive Theories of Coloration and Ornamentation

Thus far, I have taken an explanatory approach to understanding existing patterns. Many of the hypotheses here explain bright and conspicuous colors, for example, but do not suggest why in some cases birds might be bright red, and in others have long tails, even though both function to make the bird conspicuous to potential mates or rivals. Consequently, we need an additional body of theory that can explain the specific characteristics of a particular color, pattern, or ornament, and thus lead to the ability to predict what an animal will look like. In addition to asking what function a particular trait might have, we need to ask how that trait's characteristics make it best suited for its function. There are two components of a signal's design that need to be investigated: the "strategic design," or how good a particular signal is at transmitting some piece of information, and the "tactical design" or "efficacy," that is, how easy it is to receive the message (Guilford and Dawkins, 1991).

3.3.1. Strategic Design: Signal Selection

The strategic design of a signal is those features that make the signal a good transmitter of information (or misinformation), that is, the semantic properties of a signal (Hailman, 1977). Zahavi (1975, 1977) has suggested that by being costly or posing a risk, elaborate male ornaments signal to females that the male is of superior quality despite such a handicap. Zahavi (1981, 1987, 1991; see also Hasson, 1991b) has expanded this idea to incorporate all kinds of signals in his theory of "signal selection," which he contrasts with natural selection (sexual selection being a special case of signal selection). This hypothesis postulates that virtually all signals are selected to be honest and reliable sources of information, and this is accomplished by the signal being costly so that cheaters cannot afford the signal. For example, species might signal their invulnerability to predators (because they are distasteful, for example) by increasing their vulnerability to predators with conspicuous colors, something a tasty species could not afford to do. Furthermore, signals should evolve that provide unambiguous information. Thus, to signal status and fighting ability, the color patterns should emphasize the true size of an animal rather than obscure or mislead the impression of size. Consequently, the form of a particular pattern should be closely linked with the information that is being signaled. For example, a line along an animal's body (the signal) may make it easier to judge its overall length (the message) (Zahavi, 1987). Contrasting edges and bars on feathers may enable feather wear and feather growth rates, respectively, to be more easily assessed (Hasson, 1991b). This approach has only yielded post hoc explanations for how particular patterns relate to specific functions, but this may be due to inadequate information theory rather than a limitation of the hypothesis.

Distinguishing handicap signals from alternative hypotheses is difficult. Guilford and Dawkins (1993) reviewed studies of aposomatism and concluded that conventional signaling (i.e., aposomatic colors make recognition easier) is more likely than handicap signaling or the null hypothesis (i.e., distastefulness merely allows conspicuous colors for other functions), though they conceded that these hypotheses are difficult to refute at present. The signal selection theory predicts that signals should be costly, and could be disproved if signals evolve towards being less costly. Borgia (1993) provides evidence that producing bower displays by the Satin Bowerbird (*Ptilonorhynchus violaceus*) has very little cost and suggests that bower building evolved to reduce the cost of conventional plumage displays, a finding that contradicts the signal selection theory.

3.3.2. Tactical Design: Background Light and Sensory Systems

The tactical design of a signal is those features that make the message easy to detect (Guilford and Dawkins, 1991), that is, the features that make a signal conspicuous for easy detection or, conversely, cryptic coloration for difficulty of detection. In order to study the tactical design of signals, we need information on environmental noise (in the case of visual communication, ambient light, and background coloration and patterns) and the sensory capabilities of the receiver (Hailman, 1977, 1979). The design features necessary for conspicuous and cryptic coloration, and the influence of background coloration and ambient light, have been emphasized by Hailman (1977, 1979) and Endler (1978, 1984, 1990).

3.3.2a. Sensory Exploitation and Drive. M. Ryan *et al.* (1990; Ryan and Keddy-Hector, 1992) suggest that the kinds of traits favored by sexual selection may be influenced by the sensory capabilities of the choosing individuals; hence males with the favored traits would be exploiting the females sensory system. Sensory exploitation is a subset of the more general sensory drive model, in which the environment influences (or drives) the evolution of sensory systems, which in turn influences signal evolution, which of course feeds back to sensory system evolution (Endler, 1992, 1993; Endler and McLellan, 1988). Thus, a sensitivity to red (because of a need to locate red food sources such as berries or flowers), for example, would render red a more conspicuous signal and would be more effective than other colors. Thus, the sensory system would bias, or drive, the evolution of certain kinds of signals over others. Unfortunately, our present knowledge of the sensory capabilities of most animals, including birds, is very limited.

The sensory exploitation model has been explicitly applied only to mate choice: in particular, that mate preferences are simply a by-product of the sensory system and sexual ornaments function to exploit the sensory system. This hypothesis makes a number of predictions: there should be a correspondence between the features of the male trait and the sensitivity of the female sensory system; there should be a correspondence between ecological factors, such as the kind of food eaten, and the male characteristics of the male trait; and—perhaps most intriguing and what distinguishes this hypothesis from conventional views of the evolution of female choice (see, e.g., Bradbury and Andersson, 1987; Kirkpatrick and Ryan, 1991)—specific mate "preferences" should sometimes exist for nonexistent male traits because of the female's sensitivity to such traits (M. Ryan *et al.*, 1990; Ryan and Keddy-Hector, 1992; Endler, 1992).

Female preference for a trait in species in which males do not have the trait has been demonstrated in one frog (M. Ryan *et al.*, 1990) and one fish (Basolo, 1990), but has received only partial support from studies of birds. In Zebra Finches, members of both sexes have preferences for leg bands of certain colors (that are different from actual leg color) and avoid birds with leg bands of unattractive colors (Burley *et al.*, 1982; Burley, 1986). The color preferences differ among this and another closely related species, however, and match colors already on other parts of the bird (such as the bill). Thus, it seems that preferences for particular colors coevolved with the color pattern, possibly as a species recognition mechanism, but those preferences led to a bias that was extrapolated to other parts of the bird. A similar preference for bands that match natural colors was found in American Goldfinches (Johnson *et al.*, 1993). Thus these studies provide only weak support for this hypothesis.

Although the sensory exploitation and recognition system models have so far been explicitly applied only to female choice, they could apply to all kinds of signals, particularly those requiring long distance transmission, rapid recognition and conspicuousness, such as those used in status signaling (agonistic displays) or aposomatism (Endler,

1992, 1993; Endler and McLellan, 1988). For example, a trait that makes a male more conspicuous to others might enhance his ability to advertise his presence on his territory and thereby improve his ability to defend it. Support for this is found in Marchetti's (1993) study of Old World warblers (*Phylloscopus* spp.), in which species in darker habitats tend to have more and brighter patches of color. When she added crown stripes to the Yellow-browed Warbler, *P. inornatus*, a species that lacks such stripes, Marchetti (1993) found that enhanced males obtained larger territories than controls. Sensory drive is not as likely to have dramatic effects for signals used in interspecific communication, since the signal would likely need to be appropriate for a variety of species, such as potential predators. Interspecific effects, such as predation, can influence the expression of intraspecific signals, such as by selecting for a signal that is conspicuous to conspecifics but difficult to detect with the predator's sensory system (Endler, 1992).

3.3.2b. Neural Networks. Another factor influencing the evolution of female preferences has been proposed by Enquist and Arak (1993). They used computer simulations of simple neural networks, which were subject to mutation and selection for those mutations that responded more to a conspecific shape than to a heterospecific shape with a shorter "tail" (Fig. 2). Once a certain reliability was reached, the networks were tested with a variety of shapes, with the result that supernormal shapes with long wings or long tails were preferred even over the appropriate ones. An evolutionary simulation confirmed that this could lead to increasingly exaggerated preferences, the final extent of which depended on the cost of the tail to survival (Enquist and Arak,

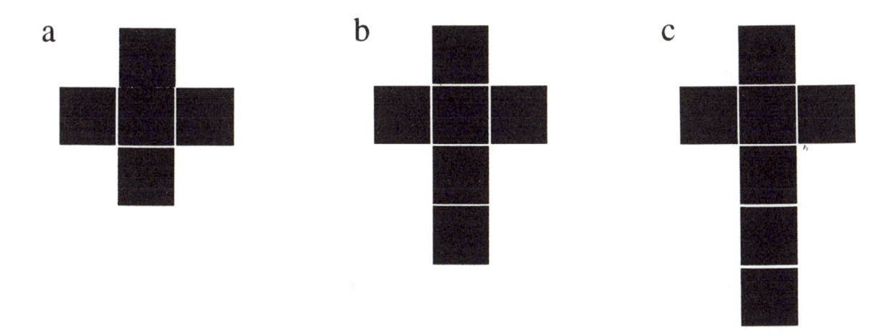

FIGURE 2. Patterns used in computer simulations of neural network evolution. The randomly mutated network was selected to respond to "conspecific" pattern (b) over heterospecific pattern (a). When tested against various shapes, supernormal stimuli (c) were preferred over the trained pattern (after Enquist and Arak, 1993).

1993). This suggests that species (and other) recognition systems must make generalizations that can lead to the evolution of supernormal traits and that mate choice is simply a by-product of species recognition, a sentiment that has been echoed by other recent authors (Ryan and Rand, 1993; Backwell and Jennions, 1993). One compelling test of this hypothesis would be to measure mate preferences in a species with an intermediate trait: females of species with tail lengths that fall between those of sympatric near relatives should exhibit stabilizing preferences since choosing a supernormal tail would lead to a species recognition error.

Although Enquist and Arak's (1993) model applied explicitly to tail length and to mate choice/recognition, this hypothesis could of course apply to other ornaments or coloration and other contexts as well: if a particular amount of contrast is needed to locate or recognize a conspecific rival, individuals showing greater contrast might elicit an even more effective response. Similarly, this could lead to the exaggeration of aposomatic traits.

4. CONCLUSIONS

Birds exhibit a diverse array of colors, patterns, and ornaments. This has spawned a great many hypotheses to explain plumage coloration and elaboration. Work to test hypotheses has not kept pace with the theoretical developments, however: 12 have received no or only anecdotal support, and most of the remainder are supported by only one or two studies. Furthermore, even when experimental or comparative studies were done, few considered more than one or two hypotheses at a time. Consequently, evidence in support of a number of hypotheses is also consistent with several others and the critical tests to distinguish among them have not been done. A lack of comprehensive comparative studies means that we have little indication of the relative importance of each of the hypotheses in explaining bird patterns and colors.

Crypticity may be the most widespread function for bird coloration, but the adaptive advantages to cryptic coloration are so obvious that it has received little empirical study by ornithologists. Instead, researchers have focused on the conspicuous and extravagant traits that are more difficult to explain (e.g., Butcher and Rohwer, 1989). For exaggerated colors and ornaments, sexual selection, especially mate choice, is by far the most popular explanation. Although briefly popular following Baker and Parker's (1979) introduction of the unprofitable prey hypothesis, the role of conspicuousness in signalling to predators is relatively neglected. This is unfortunate, as those issues have not been resolved. Indeed, ornithologists (unlike invertebrate biologists and ichthyologists) have largely neglected the role of predation in the evolution of coloration. Thus there has been few systematic studies to identify what factors allow some species to be brightly colored while others are constrained to be cryptic.

The most intriguing and exciting advances in avian coloration and ornamentation may well come from studies of signal selection, sensory drive, and neural network models. Although these theories are long way from providing a general understanding of visual communication —we still have much to learn about the visual systems of birds and the relationship between signals and their meaning—they could potentially lead us to predict the optimal color, pattern or ornament for a particular function. Developing these theories and testing them will keep students of avian coloration busy for some time to come.

ACKNOWLEDGMENTS. Edward Burtt, John Endler, Walter Koenig, and Thelma Rowell provided valuable comments on various versions of this manuscript. Edward Burtt kindly provided an unpublished manuscript.

REFERENCES

Alatalo, R. V., Eriksson, D., Gustafsson, L., and Lundberg, A., 1990, Hybridization between Pied and Collared Flycatchers: Sexual selection and speciation theory, *J. Evol. Biol.* **3**:375–389.

Alexander, R. D., 1974, The evolution of social behavior, *Ann. Rev. Ecol. Syst.* **5**:325–383.

Alexander, R. D., Hoogland, J. L., Howard, R., Noonan, K. M., and Sherman, P. W., 1979, Sexual dimorphism and breeding systems in pinnipeds, ungulates, primates and humans, in: *Evolutionary Biology and Human Social Behaviour* (N. A. Chagnon and W. D. Irons, eds.), Duxbury Press, North Scituate, Mass., pp. 402–435.

Andersson, M., 1982, Female choice selects for extreme tail length in a widowbird, *Nature* **299**:818–820.

Andersson, M., 1983, On the function of conspicuous seasonal plumages in birds, *Anim. Behav.* **31**:1262–1264.

Andersson, M., 1986, Evolution of condition-dependant sex ornaments and mating preference: sexual selection based on viability differences, *Evolution* **40**:804–816.

Andersson, M., 1994, *Sexual Selection*, Princeton University Press, Princeton, New Jersey.

Andersson, S., 1992, Female preference for long tails in lekking Jackson's widowbirds: experimental evidence, *Anim. Behav.* **43**:379–388.

Backwell, P. R. Y., and Jennions, M. D., 1993, Mate choice in a Neotropical frog, *Hyla ebraccata:* Sexual selection, mate recognition, and signal selection, *Anim. Behav.* **45**:1248–1250.

Baker, R. B., 1985, Bird coloration: in defense of unprofitable prey, *Anim. Behav.* **33:**1387–1388.
Baker, R. B., and Bibby, C. J., 1987, Merlin *Falco columbarius* predation and theories of the evolution of bird coloration, *Ibis:* **129:**259–263.
Baker, R. B., and Hounsome, M. V., 1983, Bird coloration: unprofitable prey model supported by ringing data, *Anim. Behav.* **31:**614–615.
Baker, R. B., and Parker, G. A., 1979, The evolution of bird coloration, *Philos. Trans. Royal Soc. Lond. Series B.* **287:**63–130.
Balmford, A., Thomas, A. L. R., and Jones, I. L., 1993, Aerodynamics and the evolution of long tails in birds, *Nature* **361:**628–631.
Barnard, C. J., 1979, Predation and the evolution of social mimicry, *Am. Nat.* **113:**613–618.
Barnard, C. J., 1982, Social mimicry and interspecific exploitation, *Am. Nat.* **120:**411–415.
Barrowclough, G. F., and Sibley, F. C., 1980, Feather pigmentation and abrasion: test of a hypothesis, *Auk* **97:**881–883.
Basolo, A. L., 1990, Female preference predates the evolution of the sword in swordtail fish, *Science* **250:**808–810.
Bateson, P., 1978, Sexual imprinting and optimal outbreeding, *Nature* **273:**659–660.
Bateson, P., 1980, Optimal outbreeding and the development of sexual preferences in Japanese Quail, *Z. Tierpsychol.* **53:**231–244.
Bateson, P., 1982, Preferences for cousins in Japanese Quail, *Nature* **295:**236–237.
Bateson, P. (ed.), 1983, *Mate Choice*, Cambridge University Press, Cambridge.
Beasley, B. A., and Ankney, C. D., 1988, The effect of plumage color on the thermoregulatory abilities of Lesser Snow Goose goslings, *Can. J. Zool.* **66:**1352–1358.
Beebe, W. (ed.), 1944, *The Book of Naturalists*, Princeton University Press, Princeton, New Jersey.
Bergman, G., 1982, Why are the wings of *Larus f. fuscus* so dark?, *Ornis Fennica* **59:**77–83.
Bildstein, K. L., 1983, Why white-tailed deer flag their tails, *Am. Nat.* **121:**709–715.
Björklund, M., 1990, A phylogenetic interpretation of sexual dimorphism in body size and ornament in relation to mating sytem in birds, *J. Evol. Biol.* **3:**171–183.
Björklund, M., 1991, Sexual dimorphism and mating system in the grackles (*Quiscalus* spp.: Icterinae), *Evolution* **45:**608–621.
Bleiweiss, R., 1985, Iridescent polychromatism in a female hummingbird: is it related to feeding strategies?, *Auk* **102:**701–713.
Bleiweiss, R., 1992, Reversed plumage ontogeny in a female hummingbird; implications for the evolution of irridescent colours and sexual dichromatism, *Biol J. Linnean Soc.* **47:**183–195.
Borgia, G., 1993, The cost of display in the non-resource-based mating system of the Satin Bowerbird, *Amer. Nat.* **141:**729–743.
Brackbill, H., 1955, Possible function of the flicker's black breast crescent, *Auk* **72:**205.
Bradbury, J. W., and Andersson, M. B. (eds.), 1987, *Sexual Selection: Testing the Alternatives*, John Wiley and Sons, New York.
Bretagnolle, V., 1993, Adaptive significance of seabird coloration: the case of procellariiforms, *Amer. Nat.* **142:**141–173.
Brooks, D. R., and McLennan, D. A., 1991, *Phylogeny, Ecology, and Behavior*, University of Chicago Press, Chicago.
Brown, J. L., 1964, The integration of agonistic behavior in the Steller's Jay *Cyanocitta stelleri* (Gmelin), *Univ. Calif. Publ. Zool.* **60:**223–328.

Brush, A. H., 1970, Pigments in hybrid, variant and melanic tanagers (birds), *Comp. Biochem. Physiol.* **36:**785–793.

Brush, A. H., 1978, Avian pigmentation, in: *Chemical Zoology,* Volume 10 (A. H. Brush, ed.), Academic Press, New York, pp. 141–161.

Burley, N., 1986, Comparison of the band-colour preferences of two species of estrildid finches, *Anim. Behav.* **34:**1732–1741.

Burley, N., Krantzberg, G., and Radman, P., 1982, Influence of colour-banding on the conspecific preferences of Zebra Finches, *Anim. Behav.* **30:**444–455.

Burtt, E. H., Jr., 1979, Tips on wings and other things, in: *The Behavioral Significance of Color* (E. H. Burtt Jr., ed.), Garland STPM Press, New York, pp. 75–110.

Burtt, E. H., Jr., 1981, The adaptiveness of animal colors, *BioSci.* **31:**723–729.

Burtt, E. H., Jr., 1984, Colour of the upper mandible: an adaptation to reduce reflectance, *Anim. Behav.* **32:**652–658.

Burtt, E. H., Jr., 1986, An analysis of physical, and optical aspects of avian coloration with emphasis on wood-warblers, *Ornithol. Monogr.* **38:**1–126.

Burtt, E. H., Jr., and Gatz, A. J., Jr., 1982, Color convergence: is it only mimetic?, *Am. Nat.* **119:**738–740.

Burtt, E. H., Jr., Swanson, J. A., Porter, A. B., and Waterhouse, S. M., 1995, Wing-flashing in mockingbirds of the Galápagos Islands, *Wilson Bull.,* **106:**559–562.

Butcher, G. S., 1991, Mate choice in female Northern Orioles with a consideration of the role of the black male coloration in female choice, *Condor* **93:**82–88.

Butcher, G. S., and Rohwer, S., 1989, The evolution of conspicuous and distinctive coloration in birds, in: *Current Ornithology,* Volume 6, (D. M. Power, ed.), Plenum Press, New York, pp. 51–108.

Caro, T. M., 1986, The functions of stotting: a review of the hypotheses, *Anim. Behav.* **34:**649–662.

Coblentz, B. E., 1980, On the improbability of pursuit invitation signals in mammals, *Am. Nat.* **115:**438–442.

Coddington, J. A., 1988, Cladistic tests of adaptational hypotheses, *Cladistics* **4:**3–22.

Cody, M. L., 1973, Character Convergence, *Ann. Rev. Ecol. Syst.* **4:**189–211.

Collias, E. C., Collias, N. E., Jacobs, C. H., McAlary, F., and Fujimoto, J. T., 1979, Experimental evidence for facilitation of pair formation by bright color in weaverbirds, *Condor* **81:**91–93.

Cott, H. B., 1947, The edibility of birds, *Proc. Zool. Soc. Lond.* **116:**371–524.

Cott, H. B., 1957, *Adaptive Coloration in Animals,* Methuen and Co., London.

Cott, H. B., 1985a, Coloration, adaptive, in: *A Dictionary of Birds* (B. Campbell and E. Lack, eds.), Buteo Books, Vermillion, South Dakota, pp. 97–99.

Cott, H. B., 1985b, Palatability of birds and eggs, in: *A Dictionary of Birds* (B. Campbell and E. Lack, eds.), Buteo Books, Vermillion, South Dakota, p. 427.

Cott, H. B., and Benson, C. W., 1969, The palatability of birds, mainly based upon observations of a tasting panel in Zambia, *Ostrich Suppl.* **8:**357–384.

Craig, J. L., 1982, On the evidence for a "pursuit deterrent" function of alarm signals of swamphens, *Am. Nat.* **119:**753–755.

Darwin, C., 1871, *The Descent of Man and Selection in Relation to Sex,* Modern Library, New York.

Davies, N. B., and Brook, M. de L., 1989, An experimental study of co-evolution between the cuckoo, *Cuculus canorus,* and its hosts. I. Host egg discrimination, *J. Anim. Ecol.* **58:**207–224.

Dawkins, R., 1982, *The Extended Phenotype,* Oxford University Press, Oxford.

Dial, B. E., and Fitzpatrick, L. C., 1983, Lizard tail autotomy: function and energetics of postautotomy tail movements in *Scincella lateralis*, *Science* **219**:391–393.

Diamond, J. M., 1982, Mimicry of friarbirds by orioles, *Auk* **99**:187–196.

Dumbacher, J. P., Beehler, B. M., Spande, T. F., Garraffo, H. M., and Daly, J. W., 1992, Homobatrachotoxin in the genus *Pitohui*: Chemical defense in birds?, *Science* **258**:799–801.

Edelstam, C., 1985, Mimicry, in: *A Dictionary of Birds* (B. Campbell and E. Lack, eds.), Buteo Books, Vermillion, South Dakota, p. 353.

Edmunds, M., 1974, *Defence in Animals*, Longman, New York.

Ellis, H. I., 1980, Metabolism and solar radiation in dark and white herons in hot climates, *Physiol. Zool.* **53**:358–372.

Endler, J. A., 1978, A predator's view of animal color patterns, in: *Evolutionary Biology, Volume 11* (M. K. Hecht, W. C. Steere, and B. Wallace, eds.), Plenum Press, New York, pp. 319–364.

Endler, J. A., 1984, Progressive background matching in moths, and a quantitative measure of crypsis, *Biol. J. Linn. Soc.* **22**:187–231.

Endler, J. A., 1990, On the measurement and classification of colour in studies of animal colour patterns, *Biol. J. Linn. Soc.* **41**:315–352.

Endler, J. A., 1992, Signals, signal conditions, and the direction of evolution, *Amer. Nat.* **139** Suppl:S125–S153.

Endler, J. A., 1993, Some general comments on the evolution and design of animal communication systems, *Phil. Trans. Royal Soc. Lond. Series B* **340**:215–225.

Endler, J. A., and McLellan, T., 1988, The processes of evolution: towards a newer synthesis, *Ann. Rev. Ecol. Syst.* **19**:395–421.

Enquist, M., and Arak, A., 1993, Selection of exaggerated male traits by female aesthetic senses, *Nature* **361**:446–448.

Estes, R. D., and Goddard, J., 1967, Prey selection and hunting behavior of the African wild dog, *J. Wildl. Mgmt.* **31**:52–70.

Evans, M. R., and Hatchwell, B. J., 1992a, An experimental study of male adornment in the scarlet-tufted malachite sunbird: I. The role of pectoral tufts in territorial defence, *Behav. Ecol. Sociobiol* **29**:413–419.

Evans, M. R., and Hatchwell, B. J., 1992b, An experimental study of male adornment in the Scarlet-tufted Malachite Sunbird: II. The role of the elongated tail in mate choice and experimental evidence for a handicap, *Behav. Ecol. Sociobiol* **29**:421–427.

Evans, M. R., and Thomas, A. L. R., 1992, The aerodynamic and mechanical effects of elongated tails in the Scarlet-tufted Malachite Sunbird: measuring the cost of a handicap, *Anim. Behav.* **43**:337–348.

Ficken, R. W., and Wilmot, L. B., 1968, Do facial eye-stripes function in avian vision?, *Amer. Midl. Nat.* **79**:522–523.

Ficken, R. W., Matthiae, P. E., and Horwich, R., 1971, Eye marks in vertebrates: aids to vision, *Science* **173**:936–939.

Fisher, R. A., 1930, *The Genetical Theory of Natural Selection*, Oxford University Press, Oxford.

Flood, N. J., 1989, Coloration in New World orioles 1. Tests of predation-related hypotheses, *Behav. Ecol. Sociobiol.* **25**:49–56.

Foster, M. S., 1983, Disruption, dispersion, and dominance in lek-breeding birds, *Am. Nat.* **122**:53–72.

Friedmann, H., 1960, The Parasitic Weaverbirds, *Smith. Inst. U.S. Nat. Mus. Bull.* **223**:1–196.

Fugle, G. N., Rothstein, S. I., Osenberg, C. W., and McGinley, M. A., 1984, Signals of status in wintering White-crowned Sparrows, *Zonotrichia leucophrys gambelii*, *Anim. Behav.* **32:**86–93.

Goldsmith, T. H., 1990, Optimization, constraint, and history in the evolution of eyes, *Quart. Rev. Biol.* **65:**281–322.

Götmark, F., 1987, White underparts in gulls function in hunting camouflage, *Anim. Behav.* **35:**1786–1792.

Götmark, F., 1992, Anti-predator effect of conspicuous plumage in a male bird, *Anim. Behav.* **44:**51–56.

Götmark, F., 1994, Are bright birds distasteful? A re-analysis of H. B. Cott's data on the edibility of birds, *J. Avian Biol.* **25:**184–197.

Götmark, F., and Unger, F., 1994, Are conspicuous birds unprofitable prey? Field experiments with hawks and stuffed prey species, *Auk*, **111:**251–262.

Gould, S. J., and Lewontin, R. C., 1979, The spandrels of San Marco and the Panglossian paradigm: a critique of the adaptationist program, *Proc. Roy. Soc. Lond. Series B* **205:**581–598.

Gould, S. J., and Vrba, E. S., 1982, Exaptation—a missing term in the science of form, *Paleobiol.* **8:**4–15.

Grafen, A., 1990, Sexual selection unhandicapped by the Fisher process, *J. Theor. Biol.* **144:**473–516.

Grant, P. R., 1965, Plumage and the evolution of birds on islands, *Syst. Zool.* **14:**47–52.

Grant, P. R., 1975, The classical case of character displacement, in: *Evolutionary Biology,* Volume 8 (T. Dobzhansky, M. K. Hecht, and W. C. Steere, eds.), Plenum Press, New York, pp. 237–337.

Greenewalt, C. H., Brandt, W., and Friel, D. D., 1960, Iridescent colors of hummingbird feathers, *J. Opt. Soc. Amer.* **50:**1005–1013.

Greenwood, J. J. D., 1985, Frequency-dependent selection by seed-predators, *Oikos* **44:**195–210.

Guilford, T., 1986, How do warning colors work? Conspicuousness may reduce recognition errors in experienced predators, *Anim. Behav.* **34:**286–288.

Guilford, T., and Dawkins, M. S., 1991, Receiver psychology and the evolution of animal signals, *Anim. Behav.* **42:**1–14.

Guilford, T., and Dawkins, M. S., 1993, Are warning colors handicaps?, *Evolution* **47:**400–416.

Hailman, J. P., 1960, A field study of the mockingbird's wing-flashing behavior and its association with foraging, *Wilson Bull.* **72:**346–357.

Hailman, J. P., 1977, *Optical Signals: Animal Communication and Light,* Indiana University Press, Bloomington.

Hailman, J. P., 1979, Environmental light and conspicuous colors, in: *The Behavioral Significance of Color* (E. H. Burtt Jr., ed.), Garland STPM Press, New York, pp. 289–354.

Hall, B. P., Moreau, R. E., and Galbraith, I. C. J., 1966, Polymorphism and parallelism in the African bush-shrikes of the genus *Malaconotus* (including *Chlorophoneus*), *Ibis* **108:**161–182.

Hamilton, W. D., 1964, The genetical evolution of social behavior I. II., *J. Theor. Biol.* **7:**1–52.

Hamilton, W. D., and Zuk, M., 1982, Heritable true fitness and bright birds: a role for parasites, *Science* **218:**384–387.

Hamilton, W. J., III, and Heppner, F., 1967, Radiant solar energy and the function of black homeotherm pigmentation: an hypothesis, *Science* **155:**196–197.

Hansen, A. J., and Rohwer, S., 1986, Coverable badges and resource defence in birds, *Anim. Behav.* **34**:69–76.
Harvey, P. H., and Pagel, M. D., 1991, *The Comparative Method in Evolutionary Biology*, Oxford University Press, New York.
Hasson, O., 1991a, Pursuit deterrent signals: communication between prey and predator, *Trends Ecol. Evol.* **6**:325–329.
Hasson, O., 1991b, Sexual displays as amplifiers: practical examples with an emphasis on feather decorations, *Behav. Ecol.* **2**:189–197.
Hill, G. E., 1990, Female House Finches prefer colourful males: sexual selection for a condition dependant trait, *Anim. Behav.* **40**:563–572.
Hill, G. E., 1991, Plumage coloration is a sexually selected indicator of male quality, *Nature* **350**:337–339.
Hill, G. E., 1992, Proximate basis of variation in carotenoid pigmentation in male House Finches, *Auk* **109**:1–12.
Hill, R. W., Beaver, D. L., and Veghte, J. H., 1980, Body surface temperatures and thermoregulation in the Black-capped Chickadee (*Parus atricapillus*), *Physiol. Zool.* **53**:305–321.
Hingston, R. W. G., 1933, *The Meaning of Animal Colour and Adornment*, Edward Arnold & Co., London.
Höglund, J., 1989, Size and plumage dimorphism in lek-breeding birds: A comparative analysis. *Amer. Nat.* **134**:72–87.
Höglund, J., Eriksson, M., and Lindell, L. E., 1990, Females of the lek-breeding Great Snipe, *Gallinago media*, prefer males with white tails, *Anim. Behav.* **40**:23–32.
Holberton, R. L., Hanano, R., and Able, K. P., 1990, Age-related dominance in male Dark-eyed Juncos: effects of plumage and prior residence, *Anim. Behav.* **40**:573–579.
Hoppe, D. M., 1979, The influence of color on behavioral thermoregulation and hydroregulation, in: *The Behavioral Significance of Color* (E. H. Burtt, Jr., ed.), Garland STPM Press, New York, pp. 35–62.
Humphries, D. A., and Driver, P. M., 1970, Protean defence by prey animals, *Oecologia* **5**:285–302.
Immelmann, K., 1975, The evolutionary significance of early experience, in: *Function and Evolution in Behaviour* (G. Baerends, C. Beer, and A. Manning, eds.), Clarendon Press, Oxford, pp. 243–253.
Järvi, T., and Bakken, M., 1984, The function of the variation in the breast stripe of the Great Tit (*Parus major*), *Anim. Behav.* **32**:590–596.
Johnsgard, P. A., 1963, Behavioral isolating mechanisms in the family Anatidae, *Proc. Int. Ornith. Congr.* **13**:531–543.
Johnson, K., Dalton, R., and Burley, N., 1993, Preferences of female American Goldfinches (*Carduelis tristis*) for natural and artificial male traits, *Behav. Ecol.* **4**:138–143.
Jones, I. L., 1990, Plumage variability functions for status signalling in Least Auklets, *Anim. Behav.* **39**:967–975.
Jones, I. L., and Hunter, F. M., 1993, Mutual sexual selection in a monogamous seabird, *Nature* **362**:238–239.
Jouventin, P., 1982, Visual and vocal signals in penguins, their evolution and adaptive characters, *Advances in Etholology* **24**:1–149.
Kelso, L., 1970, Importance of bright male coloration to sexual selection in Mallard Ducks [translated abstract of Nekipelov, N. V., 1969, *Izvestiya vost-siber. otdel Geograph. Soc. USSR* **66**:93–97.], *Bird-banding* **41**:259.
Kerlinger, P., 1989, *Flight Strategies of Migrating Hawks*, University of Chicago Press, Chicago.

Kimura, M., 1983, *The Neutral Theory of Molecular Evolution*, Cambridge University Press, Cambridge.

Kirkpatrick, M., and Ryan, M. J., 1991, The evolution of mating preferences and the paradox of the lek, *Nature* **350:**33–38.

Koepcke, M., 1963, Anpassungen und geographische Isolation bei Vögeln der peruanischen Küstenlomas, *Proc. Int. Ornith. Congr.* **13:**1195–1213.

Krebs, J. R., 1979, Bird colours, *Nature* **282:**14–16.

Kushlan, J. K., 1977, The significance of plumage colour in the formation of feeding aggregations of Ciconiiformes. *Ibis* **119:**361–364.

Lande, R., 1976, Natural selection and random genetic drift in phenotypic evolution, *Evolution* **30:**314–334.

Lande, R., 1981, Models of speciation by sexual selection on polygenic traits, *Proc. Natl. Acad. Sci.* **78:**3721–3725.

Landeau, L., and Terborgh, J., 1986, Oddity and the 'confusion effect' in predation, *Anim. Behav.* **34:**1372–1380.

Lewis, D. M., 1972, Importance of face mask in sexual recognition and territorial behavior in the Yellowthroat, *Jack-Pine Warbler* **50:**98–109.

Lorenz, K., 1966, *On Aggression*, Harcourt, Brace & World, New York.

Loye, J. E., and Zuk, M. (eds.), 1991, *Bird-Parasite Interactions: Ecology, Evolution, and Behavior*, Oxford University Press, Oxford.

Lustick, S., 1969, Bird energetics: effects of artificial radiation, *Science* **163:**387–390.

Lustick, S., Battersby, B., and Kelty, M., 1978, Behavioral thermoregulation: orientation toward the sun in Herring Gulls, *Science* **200:**81–83.

Lustick, S., Battersby, B., and Kelty, M., 1979, Effects of insolation on juvenile Herring Gull energetics and behavior, *Ecology* **60:**673–678.

Lynch, M., and Hill, W. G., 1986, Phenotypic evolution by neutral mutation, *Evolution* **40:**915–935.

Lyon, B. E., Eadie, J. M., and Hamilton, L. D., 1994, Parental choice selects for ornamental plumage in American coot chicks, *Nature* **371:**240–243.

Lyon, B. E., and Montgomerie, R. D., 1985, Conspicuous plumage of birds: sexual selection or unprofitable prey?, *Anim. Behav.* **33:**1038–1040.

Mackworth-Praed, C. W., and Grant, C. H. B., 1960, *Birds of Eastern and North Eastern Africa*, Volume 2, 2nd ed., Longmans, London.

Marchetti, K., 1993, Dark habitats and bright birds illustrate the role of the environment in species divergence, *Nature* **362:**149–152.

Marler, P., 1955, Characteristics of some animal calls, *Nature* **176:**6–8.

Maynard Smith, J., 1991, Theories of sexual selection, *Trends Ecol. Evol.* **6:**146–151.

Mayr, E., 1963, *Animal Species and Evolution*, Belknap Press, Cambridge, Massachusetts.

Mayr, E., 1983, How to carry out the adaptationist program?, *Am. Nat.* **121:**324–334.

McKitrick, M. C., 1992, Phylogenetic analysis of avian parental care, *Auk* **109:**828–846.

Metz, K. J., and Weatherhead, P. J., 1992, Seeing red: uncovering coverable badges in Red-winged Blackbirds, *Anim. Behav.* **43:**223–230.

Miller, D. E., and Emlen, J. T. Jr., 1975, Individual chick recognition and family integrity in the Ring-billed Gull, *Behaviour* **52:**124–144.

Møller, A. P., 1987, Social control of deception among status signalling House Sparrows *Passer domesticus*, *Behav. Ecol. Sociobiol.* **20:**307–311.

Møller, A. P., 1988, Female choice selects for male sexual tail ornaments in the monogamous swallow, *Nature* **332:**640–642.

Møller, A. P., 1989, Viability costs of male tail ornaments in a swallow, *Nature* **339:**132–134.

Møller, A. P., 1990, Sexual behavior is related to badge size in the House Sparrow *Passer domesticus*, *Behav. Ecol. Sociobiol.* **27**:23–29.

Møller, A. P., 1992, Female swallow preference for symmetrical male sexual ornaments, *Nature* **357**:238–240.

Møller, A. P., 1993, Sexual selection in the Barn Swallow *Hirundo rustica*. III. Female tail ornaments, *Evolution* **47**:417–431.

Monroe, B. L., Jr., 1964, Wing-flashing in the Red-backed Scrub-robin, *Erythropygia zambesiana*, *Auk* **81**:91–92.

Moreau, R. E., 1966, *The Bird Faunas of Africa and its Islands*, Academic Press, New York.

Moynihan, M., 1960, Some adaptations which help promote gregariousness, *Proc. Int. Ornith. Congr.* **12**:523–541.

Moynihan, M., 1968, Social mimicry; character convergence versus character displacement, *Evolution* **22**:315–331.

Moynihan, M., 1981, The coincidence of mimicries and other misleading coincidences, *Am. Nat.* **117**:372–378.

Mueller, H. C., 1974, Factors influencing prey selection in the American Kestrel, *Auk* **91**:705–721.

Muma, K. E., and Weatherhead, P. J., 1989, Male traits expressed in females: direct or indirect sexual selection? *Behav. Ecol. Sociobiol.* **25**:23–31.

Mysterud, I., and Dunker, H., 1979, Mammal ear mimicry: a hypothesis on the behavioural function of owl 'horns,' *Anim. Behav.* **27**:315–316.

Neunzig, R., 1929, Zum Brutparasitismus der Viduinen, *J. Orn.* **77**:1–21.

Nicolai, J., 1964, Der Brutparasitismus der Viduinae als ethologisches Problem. Prägungsphänomene als Faktoren der Rassen- und Artbildung, *Z. Tierpsychol.* **21**:129–204.

Niethammer, G., 1940, Die Schutzanpassung der Lerchen, *J. Ornith.* **88(Suppl.)**:74–83.

Noble, G. K., 1936, Courtship and sexual selection of the flicker (*Colaptes auratus luteus*), *Auk* **53**:269–282.

Oakes, E. J., 1992, Lekking and the evolution of sexual dimorphism in birds: comparative approaches, *Amer. Nat.* **140**:665–684.

Orians, G. H., and Christman, G. M., 1968, A comparative study of the behavior of Red-winged, Tricolored, and Yellow-headed blackbirds, *Univ. Calif. Publ. Zool.* **84**:1–81.

Packer, C., 1983, Sexual dimorphism: the horns of African antelopes, *Science* **221**:1191–1193.

Parker, G. A., 1983, Mate quality and mating decisions, in: *Mate Choice* (P. Bateson, ed.), Cambridge University Press, Cambridge, pp. 141–164.

Parsons, J., and Baptista, L. F., 1980, Crown color and dominance in the White-crowned Sparrow, *Auk* **97**:807–815.

Pasteur, G., 1982, A classificatory review of mimicry systems, *Ann. Rev. Ecol. Syst.* **13**:169–199.

Paulson, D. R., 1973, Predator polymorphism and apostatic selection, *Evolution* **27**:269–277.

Payne, R. B., 1967, Interspecific communication signals in parasitic birds, *Am. Nat.* **101**:363–375.

Payne, R. B., 1977, The ecology of brood parasitism in birds, *Ann. Rev. Ecol. Syst.* **8**:1–28.

Peek, F. W., 1972, An experimental study of the territorial function of vocal and visual display in the male Red-winged Blackbird (*Agelaius phoeniceus*), *Anim. Behav.* **20**:112–118.

Pennycuick, C. J., 1972, *Animal Flight*, Edward Arnold, London.

Petrie, M., and Halliday, T., 1994, Experimental and natural changes in the peacock's (*Pavo cristatus*) train can affect mating success, *Behav. Ecol. Sociobiol.* **35:**213–217.

Pierotti, R., 1987, Isolating mechanisms in seabirds, *Evolution* **41:**559–570.

Platt, J. R., 1964, Strong inference, *Science* **146:**347–353.

Potti, J. A., 1993, A male trait expressed in female Pied Flycatchers, *Ficedula hypoleuca:* the white forehead patch, *Anim. Behav.* **45:**1245–1247.

Pough, F. H., 1988, Mimicry of vertebrates: are the rules different?, *Am. Nat.* **131(Suppl.):**S67–S102.

Poulton, E. B., 1890, *The Colours of Animals*, Kegan Paul, Trench, Trübner, & Co., London.

Pruett-Jones, S. G., Pruett-Jones, M. A., and Jones, H. I., 1990, Parasites and sexual selection in birds of paradise, *Amer. Zool.* **30:**287–298.

Reid, J. B., 1984, Bird coloration: predation, conspicuousness and the unprofitable prey model, *Anim. Behav.* **32:**294–295.

Rohwer, S., 1977, Status signaling in Harris Sparrows: some experiments in deception, *Behaviour* **61:**107–129.

Rohwer, S., 1978, Passerine subadult plumages and the deceptive acquisition of resources: a test of a critical assumption, *Condor* **80:**173–179.

Rohwer, S., 1982, The evolution of reliable and unreliable badges of fighting ability, *Amer. Zool.* **22:**531–546.

Rohwer, S., 1985, Dyed birds achieve higher social status than controls in Harris' Sparrows, *Anim. Behav.* **33:**1325–1331.

Rohwer, S., and Paulson, D. R., 1987, The avoidance-image hypothesis and color polymorphism in *Buteo* hawks, *Ornis Scand.* **18:**285–290.

Rohwer, S., and Røskaft, E., 1989, Results of dyeing male Yellow-headed Blackbirds solid black: implications for the arbitrary identity badge hypothesis, *Behav. Ecol. Sociobiol.* **25:**39–49.

Rohwer, S., Fretwell, S. D., and Niles, D. M., 1980, Delayed maturation in passerine plumages and the deceptive acquisition of resources, *Am. Nat.* **115:**400–437.

Røskaft, E., and Rohwer, S., 1987, An experimental study of the function of the epaulettes and the black body colour of male Red-winged Blackbirds, *Anim. Behav.* **35:**1070–1077.

Røskaft, E., Järvi, T., Nyholm, N. E. I., Virolainen, M., Winkel, W., and Zang, H., 1986, Geographic variation in secondary sexual plumage colour characteristics of the male Pied Flycatcher, *Ornis. Scand.* **17:**293–298.

Ryan, M. J., and Keddy-Hector, A., 1992, Directional patterns of female mate choice and the role of sensory biases, *Amer. Nat.* **139(Suppl.):**S4–S35.

Ryan, M. J., and Rand, A. S., 1993, Species recognition and sexual selection as a unitary problem in animal communication, *Evolution* **47:**647–657.

Ryan, M. J., Fox, J. H., Wilczynski, W., and Rand, A. S., 1990, Sexual selection for sensory exploitation in the frog *Physalaemus pustulosus*, *Nature* **343:**66–67.

Ryan, P. G., Wilson, R. P., and Cooper, J., 1987, Intraspecific mimicry and status signals in juvenile African penguins, *Behav. Ecol. Sociobiol.* **20:**69–76.

Savalli, U. M., 1989, Female choice, *Nature* **339:**432.

Savalli, U. M., 1991, Sexual selection and the evolution of tail length in the Yellow-shouldered Widowbird, Ph.D. thesis, University of California at Berkeley.

Savalli, U. M., 1993, An application of the neutral model for the evolution of tail length in the genus *Euplectes* (Aves, Ploceidae), *Evolution* **47:**696–699.

Savalli, U. M., 1994, Tail length affects territory ownership in the Yellow-shouldered Widowbird, *Anim. Behav.* **48:**105–111.

Savalli, U. M., 1995, The evolution of tail length in widowbirds: tests of alternatives to sexual selection, *Ibis* **137**:in press.
Savalli, U. M., in press, Does rainfall constrain the evolution of tail length in widowbirds? *Ethol. Ecol. Evol.*
Selander, R. K., and Hunter, D. K., 1960, On the function of wing-flashing in mockingbirds, *Wilson Bull.* **72**:341–345.
Sherman, P. W., 1977, Nepotism and the evolution of alarm calls, *Science* **197**:1246–1253.
Sherman, P. W., 1988, The levels of analysis, *Anim. Behav.* **36**:616–619.
Shields, W. M., 1977, The social significance of avian winter plumage variability: a comment, *Evolution* **31**:905–907.
Sibley, C. G., 1957, The evolutionary and taxonomic significance of sexual dimorphism and hybridization in birds, *Condor* **59**:166–191.
Sibley, C. G., and Ahlquist, J. E., 1990, *Phylogeny and Classification of Birds*, Yale University Press, New Haven.
Sillen-Tullberg, B., 1988, Evolution of gregariousness in aposematic butterfly larvae: a phylogenetic analysis, *Evolution* **42**:293–305.
Simmons, K. E. L., 1972, Some adaptive features of seabird plumage types, *British Birds* **65**:465–479.
Slagsvold, T., and Lifjeld, J. T., 1988, Plumage colour and sexual selection in the Pied Flycatcher *Ficedula hypoleuca*, *Anim. Behav.* **36**:395–407.
Slagsvold, T., and Sætre, G.-P., 1991, Evolution of plumage color in male Pied Flycatchers (*Ficedula hypoleuca*): evidence for female mimicry, *Evolution* **45**:910–917.
Smith, D. G., 1972, The role of the epaulets in the Red-winged Blackbird, (*Agelaius phoeniceus*) social system, *Behaviour* **41**:251–268.
Smith, N. G., 1966, Evolution of some arctic gulls (*Larus*): an experimental study of isolating mechanisms, *Ornith. Monogr.* **4**:1–99.
Smythe, N., 1970, On the existence of "pursuit invitation" signals in mammals, *Am. Nat.* **104**:491–494.
Spear, L., and Ainley, D. G., 1993, Kleptoparasitism by Kermadec Petrels, jaegers, and skuas in the eastern tropical Pacific: evidence of mimicry by two species of *Pterodroma*, *Auk* **110**:222–233.
Stiles, F. G., and Wolf, L. L., 1970, Hummingbird territoriality at a tropical flowering tree, *Auk* **87**:467–491.
Stoddard, P. K., and Beecher, M. D., 1983, Parental recognition of offspring in the Cliff Swallow, *Auk* **100**:795–799.
Swynnerton, C. F. M., 1916, On the coloration of the mouths and eggs of birds, *Ibis* Ser. 10, Vol. **4**:264–294, 529–606.
Thayer, A. H., 1896, The law which underlies protective coloration, *Auk* **13**:124–129.
Thayer, G. H., 1909, *Concealing-coloration in the Animal Kingdom*, Macmillan, New York.
Tinbergen, N., 1952, "Derived" activities; their causation, biological significance, origin, and emancipation during evolution, *Quart. Rev. Biol.* **27**:1–30.
Tinbergen, N., 1963, On aims and methods in ethology, *Z. Tierpsychol.* **20**:410–433.
Tinbergen, N., and Perdeck, A. C., 1950, On the stimulus situation releasing the begging response in the newly hatched Herring Gull chick (*Larus argentatus argentatus* Pont.), *Behaviour* **3**:1–39.
Trail, P. W., 1990, Why should lek breeders be monomorphic?, *Evolution* **44**:1837–1852.
Trail, P. W., and Koutnik, D. L., 1986, Courtship disruption at the lek in the Guianan Cock-of-the-rock, *Ethology* **73**:197–218.

Trivers, R. L., 1972, Parental investment and sexual selection, in: *Sexual Selection and the Descent of Man 1871–1971* (B. Campbell, ed.), Aldine, Chicago, pp. 136–179.

Turelli, M., Gillespie, J. H., and Lande, R., 1988, Rate tests for selection on quantitative characters during macroevolution and microevolution, *Evolution* **42:**1085–1089.

Veiga, J. P., 1993, Badge size, phenotypic quality, and reproductive success in the House Sparrow: a study of honest advertisement, *Evolution* **47:**1161–1170.

Verner, J., and Willson, M. F., 1969, Mating systems, sexual dimorphism, and the role of North American passerine birds in the nesting cycle, *Ornith. Monogr.* **9:**1–76.

Wake, M. H. (ed.), 1979, *Hyman's Comparative Vertebrate Anatomy,* 3rd ed., University of Chicago Press, Chicago.

Waldman, B., 1987, Mechanisms of kin recognition, *J. Theor. Biol.* **128:**159–185.

Waldvogel, J. A., 1990, The bird's eye view, *Amer. Scientist* **78:**342–353.

Walsberg, G. E., 1982, Coat color, solar heat gain, and conspicuousness in the Phainopepla, *Auk* **99:**495–502.

Walsberg, G. E., Campbell, G. S., and King, J. R., 1978, Animal coat color and radiative heat gain: a re-evaluation, *J. Comp. Physiol.* **126:**211–222.

Watt, D. J., 1986, A comparative study of status signalling in sparrows (genus *Zonotrichia*), *Anim. Behav.* **34:**1–15.

Weary, D. M., Guilford, T. C., and Weisman, R. G., 1993, A product of discriminative learning may lead to female preferences for elaborate males, *Evolution* **47:**333–336.

Weatherhead, P. J., Bennett, G. F., and Shutler, D., 1991, Sexual selection and parasites in wood-warblers, *Auk* **108:**147–152.

Whitfield, D. P., 1986, Plumage variability and territoriality in breeding turnstone *Arenaria interpres:* status signalling or individual recognition?, *Anim. Behav.* **34:**1471–1482.

Whitfield, D. P., 1987, Plumage variability, status signalling and individual recognition in avian flocks, *Trends Ecol. Evol.* **2:**13–18.

Whittingham, L. A., Kirkconnell, A., and Ratcliffe, L. M., 1992, Differences in song and sexual dimorphism between Cuban and North American Red-winged Blackbirds (*Agelaius phoeniceus*), *Auk* **109:**928–933.

Wickler, W., 1968, *Mimicry in Plants and Animals,* World University Library, New York.

Willis, E. O., 1963, Is the Zone-tailed Hawk a mimic of the Turkey Vulture?, *Condor* **65:**313–317.

Wilson, R. P., Ryan, P. G., James, A., and Wilson, M.-P. T., 1987, Conspicuous coloration may enhance prey capture in some piscivores, *Anim. Behav.* **35:**1558–1560.

Wilson, S. E., Allen, J. A., and Anderson, K. P., 1990, Fast movement of densely aggregated prey increases the strength of anti-apostatic selection by wild birds, *Biol. J. Linn. Soc.* **41:**375–380.

Wolf, L. L., 1969, Female territoriality in a tropical hummingbird, *Auk* **86:**490–504.

Wolf, L. L., 1975, Female territoriality in the Purple-throated Carib, *Auk* **92:**511–522.

Woodland, D. J., Jaafar, Z., and Knight, M., 1980, The "pursuit deterrent" function of alarm signals, *Am. Nat.* **115:**748–753.

Worthy, M., 1974, *Eye Color, Sex, and Race: Keys to Human and Animal Behavior,* Droke House/Hallux, Anderson, South Carolina.

Zahavi, A., 1975, Mate selection—a selection for a handicap, *J. Theor. Biol.* **53:**205–214.

Zahavi, A., 1977, Reliability in communication systems and the evolution of altruism, in: *Evolutionary Ecology* (B. Stonehouse and C. M. Perrins, eds.), Macmillan, London, pp. 253–259.

Zahavi, A., 1981, Natural selection, sexual selection, and the selection of signals, in: *Evolution Today: Proceedings of the Second International Congress of Systematic and Evolution Biology* (G. G. E. Scudder and J. L. Reveal, eds.), Hunt Institute for Botanical Documentation, Pittsburgh, pp. 133–138.

Zahavi, A., 1987, The theory of signal selection and some of its implications, in: *Proceedings of the International Symposium on Biology and Evolution* (V. P. Delfino, ed.), Adriatica Editrica, Bari, Italy, pp. 305–327.

Zahavi, A., 1991, On the definition of sexual selection, Fisher's model, and the evolution of waste and of signals in general, *Anim. Behav.* **42:**501–503.

Zuk, M., 1991, Parasites and bright birds: new data and a new prediction, in: *Bird-Parasite Interactions: Ecology, Evolution, and Behavior* (J. E. Loye and M. Zuk, eds.), Oxford University Press, Oxford, pp. 317–327.

Zuk, M., Thornhill, R., Ligon, J. D., Johnson, K., Austad, S., Ligon, S. H., Thornhill, N. W., and Costin, C., 1990, The role of male ornaments and courtship behavior in female mate choice of Red Jungle Fowl, *Amer. Nat.* **136:**459–473.

CHAPTER 6

HATCHING ASYNCHRONY AND THE ONSET OF INCUBATION IN BIRDS, REVISITED

When Is the Critical Period?

SCOTT H. STOLESON and
STEVEN R. BEISSINGER

1. INTRODUCTION

In most animals, offspring from a reproductive bout usually hatch, emerge, or are born within a relatively short time of each other compared to the time required for their development. Thus, hatching or birthing in most animals is synchronous. This is especially likely to be true for animals with internal fertilization and development, where the birth of all offspring occurs simultaneously (e.g., some fishes, snakes, and most mammals). Synchronous reproduction also occurs in animals with external fertilization or development when all zygotes are subject to the same environmental conditions (e.g., many insects, anurans, and fishes). Thus, in most animals the behavior of parents has little effect on the time between the emergence of their first and last young.

In contrast, birds can influence birthing (i.e., hatching) intervals

---

SCOTT H. STOLESON and STEVEN R. BEISSINGER • School of Forestry and Environmental Studies, Yale University, New Haven, Connecticut 06511.

Current Ornithology, Volume 12, edited by Dennis M. Power. Plenum Press, New York, 1995.

through parental care. Avian parents can influence the onset of development of eggs, and the resultant synchrony of birth, by determining when to begin incubation. This is possible because avian eggs develop externally and development generally does not begin until a parent warms the eggs by initiating incubation (White and Kinney, 1974; Drent, 1975; O'Connor, 1984). Birds can lay only one egg daily, so if incubation is initiated before the last egg is laid (hereafter "early incubation"), eggs may hatch over a period of a few days to several weeks (Lack, 1968; O'Connor, 1984; Beissinger and Waltman, 1991). In contrast, if incubation is delayed until all eggs of a clutch have been laid, development and hatching will be synchronous. Eggs of a variety of birds hatch asynchronously (e.g., raptors, herons, parrots, and many passerines), and asynchronous hatching occurs in avian families nearly as often as synchronous hatching (Clark and Wilson, 1981; Slagsvold, 1986b; Ricklefs, 1993).

1.1. The Paradox of Hatching Asynchrony

Early incubation in altricial birds results in nestlings of different sizes (Lack, 1947; Bryant, 1978; Stokland and Amundsen, 1988). Often this size hierarchy contributes directly to the death of the later-hatched young (O'Connor, 1978; Mock *et al.*, 1990). This illustrates the *Paradox of Hatching Asynchrony*: avian parents are unique in having some control over birthing intervals, but many opt for a strategy that seems to be a maladaptive waste of parental effort and resources.

Historically, asynchronous hatching was viewed as a mechanism to promote adaptive brood reduction (Lack, 1954; Ricklefs, 1965). Recent debate has questioned whether asynchronous hatching in birds confers survival advantages for parents or selected offspring after hatching, or is a result of physiological constraints, resource limitations or behavioral factors that affect the onset of incubation. In many ways, the *Paradox of Hatching Asynchrony* reflects current concerns about our understanding of the evolution of parental care patterns: it is not always clear to what extent differences in parental care among species reflect variation in the benefits to offspring and costs to parents (Ridley, 1978; Tallamy, 1984; Clutton-Brock, 1991), or to what extent parental care is constrained by environmental limitations, endogenous factors, and phylogenetic history (Emlen and Oring, 1977; Silver *et al.*, 1985; Brooks and McLennan, 1991). Studies of the onset of incubation provide good tests of the role of adaptation, constraint, and phylogeny in the evolution of parental care patterns.

The *Paradox of Hatching Asynchrony* was recognized by early ornithologists because they were concerned with the adaptive significance of the early onset of incubation. For example, Dunlop (1910) first suggested that early incubation served to protect eggs. The mortality of smallest chicks was viewed as enigmatic, or in the case of siblicidal brood reduction in raptors, as an example of the ferocious nature of the species (Bent, 1961). David Lack (1947) presented a resolution to the paradox by suggesting an adaptive function for the mortality that results from asynchronous hatching. He proposed that when a species' food supply was unpredictable at the time of egg laying, parents should lay an optimistic clutch. If resources are scarce, asynchronous hatching enables parents to reduce the size of their brood to fit available food resources (the *Brood Reduction Hypothesis*).

Lack's hypothesis was quickly accepted, in part because it was an elegant and intuitive idea, and became the dominant paradigm of the field. An extensive theoretical basis was developed for this hypothesis, based on the trade-off between offspring quality and quantity (e.g., O'Connor, 1978; Lloyd, 1987; Godfray and Harper, 1990; Haig, D., 1990), and the trade-off in reproductive success between good and bad years (Pijanowski, 1992). Because Lack's hypothesis has dominated the field, hatching asynchrony has generally become equated with adaptive brood reduction, even when the conditions or assumptions of the hypothesis (such as unpredictability of food supply) are not appropriate. Despite the fact that the onset of incubation generally determines hatching patterns, scientific focus has remained fixed on the adaptive significance of hatching patterns (Bortolotti and Wiebe, 1993).

The *Brood Reduction Hypothesis* was not experimentally tested until the 1970's (Howe, 1976, 1978; Werschkul, 1979). Recently, a number of experiments have tested Lack's hypothesis in a variety of taxa, but have produced results that have rarely supported the predictions (Magrath, 1990; Amundsen and Slagsvold, 1991b). Perhaps because of the lack of verification of the *Brood Reduction Hypothesis,* a plethora of alternative hypotheses have been proposed (Magrath, 1990). However, these have lacked a cohesive conceptual organization, and have received scant experimental or theoretical attention. Few studies have examined environmental conditions or parental behaviors during egg-laying to assess possible benefits of early incubation that might offset the inherent costs of asynchrony. With few exceptions (e.g., Clark and Wilson, 1981; Hébert and Sealy, 1992; Veiga, 1992), most research has remained focused on adaptive hatching patterns.

The lack of unambiguous support for any single hypothesis has resulted in relatively little progress toward understanding the causes of hatching patterns in birds. Work on this subject has probably generated more hypotheses than it has understanding, prompting one anonymous reviewer of our recent National Science Foundation proposal to dismiss the subject of hatching asynchrony as "a stagnant backwater of population biology." We, however, feel this situation is common among complex but interesting biological phenomena. The failure of any one hypothesis to fully explain the wide variety of hatching patterns is characteristic of processes with multiple causal factors (Hilborn and Stearns, 1982). Significant understanding can be achieved by broadening our focus to examine all factors that may affect hatching patterns, especially those that influence the onset of incubation directly, and by going beyond the testing of single factors to consider multiple causes.

1.2. Goals of this Review

In this review, we examine the causes and consequences of the onset of incubation and resulting hatching patterns. Although Magrath (1990) recently reviewed the hatching asynchrony literature, the many disparate hypotheses that characterize this field were not synthesized into a cohesive framework. We develop a conceptual framework for understanding the effects and interactions of different factors affecting the onset of incubation and hatching asynchrony. In addition, since Magrath's review, several new hypotheses of potentially broad relevance have been proposed, and numerous experimental studies have been conducted.

Whereas most previous work has focused almost exclusively on hatching patterns, we will emphasize factors affecting the onset of incubation. First, we discuss the bases and patterns of the onset of incubation, as well as other factors that may influence hatching patterns. We critically review 17 hypotheses for hatching asynchrony that have been proposed in the literature, and concentrate on the assumptions, requirements, and critical predictions of each. Finally, we propose a stochastic modeling approach to evaluate empirical data as a method to assess multiple hypotheses.

2. THE ONSET OF INCUBATION AS THE PRINCIPAL PROXIMATE CAUSE OF HATCHING PATTERNS

The initiation of incubation and pattern of attentiveness of incubating birds controls the temporal pattern of hatching within a clutch. Therefore, to better understand hatching asynchrony, it is necessary to examine the proximate factors that affect incubation behavior. At the physiological level, the different phases of reproduction are regulated by various hormones. Patterns of attentiveness represent a compromise between a range of exogenous factors: the tolerance of embryos to survive when neglected by their parents, the energy needs and balance of parents, and the avoidance of predation (Vleck, 1981; Magrath, 1988; Weathers and Sullivan, 1989).

2.1. The Timing of Periods in the Nesting Cycle

Avian embryos generally do not begin to develop until parents supply heat by incubating (Drent, 1975). If parents defer incubation until all the eggs of a clutch are laid, embryos will develop simultaneously, and hatching and fledging will be relatively synchronous (Fig. 1a). When parents begin incubation on the first egg, development begins immediately, and eggs will hatch asynchronously (Fig. 1c). Beginning incubation on an intermediate egg will produce an intermediate hatching pattern (Fig. 1b). Differential rates of embryonic development and parental incubation efficiency can also affect the timing of hatching (Clark and Wilson, 1981; Bortolotti and Wiebe, 1993). Variable growth rates of nestlings can cause patterns of fledging to differ from hatching patterns, either by decreasing the degree of asynchrony (M. L. Morton, personal communication) or by amplifying the degree of asynchrony (e.g., Bryant, 1978; Emlen *et al.*, 1991; Viñuela and Bustamente, 1993).

The lengths of some periods in the nesting cycle vary with incubation and hatching patterns. The period from the start of egg-laying until the onset of incubation is longest when hatching is synchronous (Fig. 1), and nonexistent with completely asynchronous hatching. During this period eggs are most vulnerable to environmental conditions and predators. The period from the onset of incubation to the first hatch is nearly constant. Therefore, the time that a nest contains only eggs is maximized with synchronous hatching, and minimized with asynchrony. The time from first hatching to first fledging is nearly constant. The fledging period, from the fledging of the first young to the fledging of the last young, is longest with completely asynchronous hatching, and least with synchronous hatching (Fig. 1). Thus, the time that a nest contains nestlings is maximized with asynchrony, and minimized with synchrony. Synchronous hatching allows for rapid transitions from incubation to brooding and from tending nestlings to tending fledglings.

A. Incubation begun on last egg: synchronous hatching

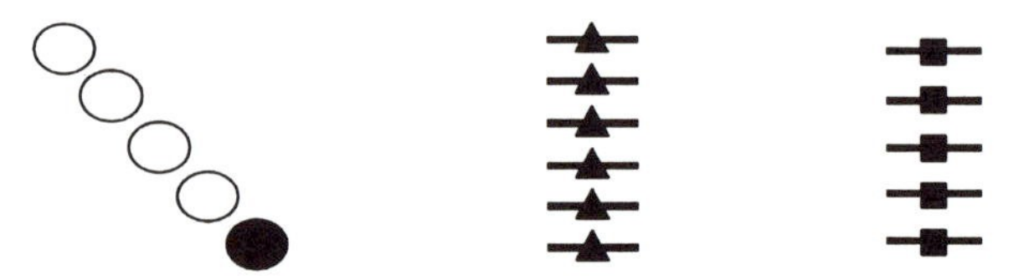

B. Incubation begun on intermediate egg: partially asynchronous hatching

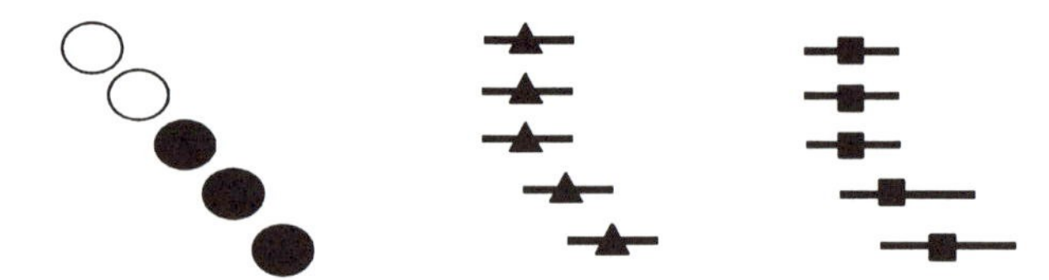

C. Incubation begun on first egg: asynchronous hatching

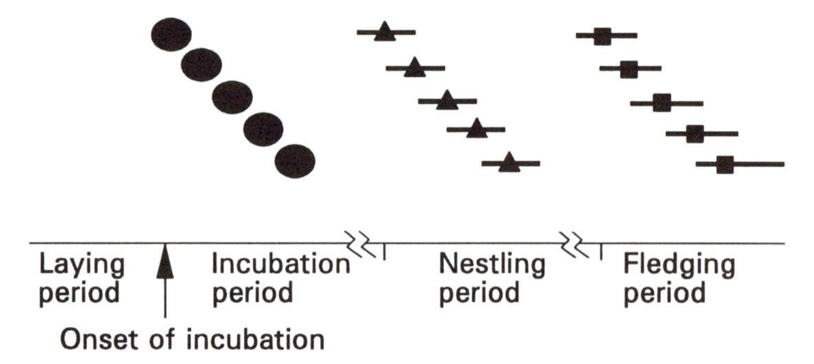

FIGURE 1. A graphic representation of the relationship between the onset of incubation, and hatching and fledging patterns. Ovals represent the laying of eggs, filled ovals indicate the onset of incubation. Triangles represent hatching, and squares represent fledging. Bars represent potential variation in hatching and fledging times due to differences in embryonic development or nestling growth. Initiating incubation on the last egg results in synchronous hatching (A), on an intermediate egg results in partially asynchronous hatching (B), and on the first egg produces completely asynchronous hatching (C).

2.2. Physiological Bases for the Onset of Incubation

The onset of incubation is controlled at the proximate level by hormones. Several hormones have been identified that control different aspects of egg-laying and incubation behaviors, based primarily on studies of domestic fowl and pigeons. Changes in photoperiod, rainfall, or other environmental cues trigger the release of FSH from the adenohypophysis, which stimulates the growth of ovarian follicles (Balthazart, 1983). An increase in plasma-luteinizing hormone causes the release of the ovum from its follicle. Ovulation occurs from 15 minutes to several hours after the laying of the preceding egg (Rol'nik, 1970; Sturkie and Mueller, 1976). The amount of time an egg spends within the oviduct, and therefore the laying interval, varies among taxa from 24 hours in chickens and most passerines to several days in larger nonpasserines (Rol'nik, 1970; Sturkie and Mueller, 1976).

The onset of incubation is associated with a decrease in levels of plasma-luteinizing hormone and steroidal hormones, frequently with a corresponding increase in levels of prolactin (Epple and Stetson, 1980; Hall and Goldsmith, 1983). However, the exact function of prolactin in incubation is unclear and controversial (Goldsmith and Williams, 1980; Goldsmith, 1983; Ball, 1991). Prolactin appears to initiate broodiness, and hence incubation in birds that delay incubation until their last or penultimate egg (Drent, 1975; Mead and Morton, 1985). Levels of prolactin increase through laying and peak during incubation in most species (Goldsmith and Williams, 1980; Balthazart, 1983; Goldsmith, 1983). However, prolactin levels of males also peak during the incubation period in species where males do not incubate (Goldsmith, 1983).

Other evidence suggests that prolactin may maintain incubation behavior, but does not trigger it (Drent, 1975; Goldsmith, 1983). In Ringed Doves (*Streptopelia risoria*), progesterone appears to induce incubation. Experimentally injecting doves with progesterone produced incubatory behavior (van Tienhoven, 1983). Tactile stimuli produced by the contact of the brood patch with eggs may induce both the onset of incubation and an increase in prolactin levels (Parsons, 1976; Goldsmith, 1983; Hall and Goldsmith, 1983). In general, ovulation and egg-laying are governed by various hormones, while prolactin is associated with incubation. However, the exact role of prolactin in incubation is not yet clear, and may vary among taxa.

2.3. External Factors Affecting the Onset of Incubation

Once laying has begun, eggs are exposed to various conditions that may affect their subsequent survival and hatchability. A parent's decision of when to begin to incubate involves a trade-off between ensuring its own survival, maximizing embryo survival, and maximizing expected survivorship of offspring after hatching.

Before embryological development begins, eggs can remain viable for a period of time without incubation, and can tolerate exposure to very low temperatures (Hussell, 1972; Drent, 1973; Webb, 1987). Embryos do not begin to develop until eggs reach a temperature of about 27–28°C, and this threshold is called "physiological zero" (Webb, 1987). Once development begins, eggs must remain within a narrow range of temperatures for development to be normal (Romanoff and Romanoff, 1972; Deeming and Ferguson, 1992). Prolonged exposure to temperatures above physiological zero, but below normal incubation temperatures, causes abnormal growth and mortality in embryos (Romanoff and Romanoff, 1972; White and Kinney, 1974). The embryological tolerance to chilling varies greatly among different taxa. Some species can neglect their eggs for long periods without effect (Drent, 1975; Roby and Ricklefs, 1984; Gaston and Powell, 1989; Ewert, 1992). Temperatures above 41–45°C are fatal to embryos regardless of the stage of development (Drent, 1973; White and Kinney, 1974; Webb, 1987). Therefore, birds in extremely hot climates must show constant attentiveness to their eggs, and exhibit mechanisms to keep eggs cooler than ambient temperatures, such as wetting the egg or using the brood patch to dissipate heat (Yom-Tov *et al.*, 1978; Grant, 1982; Walsberg and Voss-Roberts, 1983; Jehl and Mahoney, 1987). Similarly, high-altitude species probably must incubate when eggs are exposed to strong, direct solar radiation (Morton and Pereyra, 1985).

Although eggs can remain viable for an extended period without incubation if temperatures do not exceed physiological zero, this prolongs exposure to potential predators and brood parasites. Once incubation begins, the sitting adult may also be more vulnerable to predation, especially if its ability to escape is limited, as in cavity nesters (Drent, 1970; Nilsson, 1986; Martin, 1992).

The onset of incubation may also be constrained by the mating system of the species, and by which sex incubates. Nest attentiveness is usually greater and begins earlier when both sexes incubate than when only one sex incubates, or when the nonincubating sex provides the incubator with all of its food requirements (Skutch, 1957; Lyon and Montgomerie, 1985; Lifjeld and Slagsvold, 1986; Nilsson and Smith, 1988; Williams, 1991; Nilsson, 1993a). If females incubate alone and must forage for themselves, they may be constrained by the high nutrient requirements of egg production to delay incubation until after the clutch is nearly complete (Drent, 1975; Magrath, 1988; Slagsvold and

Lifjeld, 1989a). When laying females of some species were given supplemental food, they began incubating earlier than control females (Moreno, 1989a; Nilsson, 1993b; but see Wiebe and Bortolotti, 1994a). In species where males incubate, males may delay initiating incubation until a full clutch has been laid to guard their mates to prevent cuckoldry (Power *et al.*, 1981).

2.4. Patterns of the Onset of Incubation

Unfortunately, patterns of incubation have received scant attention, and what data exist are primarily circumstantial or qualitative. Few studies have published data on the actual timing of the onset of incubation, especially with respect to egg-laying (e.g., Haftorn, 1981; Zerba and Morton, 1983; Morton and Pereyra, 1985; Kennamer *et al.*, 1990; Meijer, 1990). Parent birds may be particularly sensitive to disturbance and prone to desert their nests during egg-laying (Lessells and Avery, 1989; Götmark, 1992). This often makes direct observation of the onset of incubation difficult. Sometimes incubation is inferred when a parent is seen leaving the nest area. However, attendance at the nest does not necessarily indicate incubation (Ligon, 1968; Haftorn, 1981).

As a result of these difficulties, incubation patterns are frequently inferred from hatching patterns, and it is assumed that eggs hatch in the order they are laid (e.g., Drent, 1975; Clark and Wilson, 1981; Slagsvold and Lifjeld, 1989a). This assumption has been documented in relatively few species (e.g., Cargill and Cooke, 1981; Beissinger and Waltman, 1991; Bowman, 1992). However, the hatching sequence of eggs may bear little relation to their laying order, because incubating birds may not be able to cover all eggs within a clutch (Bortolotti and Wiebe, 1993), or because last eggs may require less time to hatch than earlier-laid eggs (Viñuela, 1991). Thus, inferences about incubation patterns made from hatching patterns may not be valid. In addition, parents may not have as complete control over the spread of hatching of their young as has been assumed (Bortolotti and Wiebe, 1993). This is a potentially critical point, because hypotheses that posit an adaptive role to hatching patterns are based on parental control of those patterns (Magrath, 1990; Ricklefs, 1993).

Parents may initiate incubation in a variety of ways. Some species begin with full incubation (e.g., Wilson *et al.*, 1986; Ward, 1990; Beissinger and Waltman, 1991). Many bird species do not begin incubating fully, but instead gradually increase the amount of daylight hours spent incubating until complete incubation is achieved (e.g., Drent, 1970; Bengtsson and Rydén, 1981; Briskie and Sealy, 1989; Lessells and

Avery, 1989). In such cases the hatch spread is shorter than the laying spread. In other taxa, incomplete clutches are not incubated during the day but are brooded at night. In these species, full incubation begins after all eggs are laid and the resulting clutch hatches somewhat asynchronously (e.g., Gibb, 1950; Kavanau, 1987; Hébert and Sealy, 1992).

For both types of partial incubation, inferences made about the onset of incubation based on hatching patterns may differ from inferences made from occasional nest observations, because different incubation regimes can produce similar hatching patterns (Fig. 2). A clutch that hatches with an intermediate degree of asynchrony may be pro-

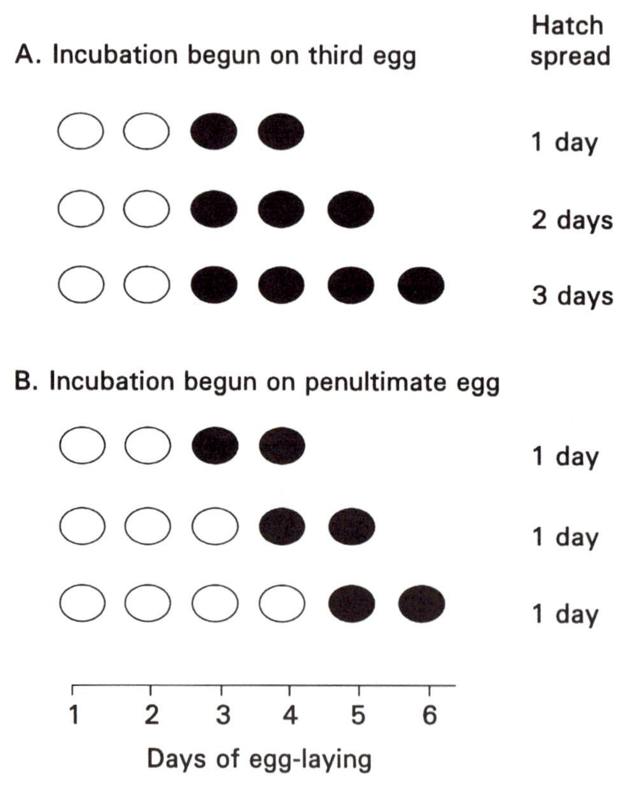

FIGURE 2. A graphic representation of the effects on hatch spread of different patterns of incubation for three clutch sizes. Filled ovals represent eggs incubated immediately after laying, and hollow ovals represent eggs laid before the onset of incubation. (A) When the onset of incubation is constant relative to the beginning of laying, regardless of clutch size, hatch spread varies with clutch size. (B) When the onset of incubation is constant relative to the end of laying, hatch spread is constant.

duced by beginning full incubation after laying half the clutch, or by beginning gradual nocturnal or diurnal incubation on the first egg.

The hatch spread of some bird species can vary with clutch size (e.g., Smith, 1988; Briskie and Sealy, 1989; Stouffer and Power, 1990; Hébert and Sealy, 1992). In species that begin incubation with a specific egg, the onset of incubation may be considered invariant with respect to laying order (Fig. 2a). Species that begin incubation on the first egg hatch their eggs completely asynchronously, and hatch spread is dependent on clutch size. For example, clutch size in the Green-rumped Parrotlet (*Forpus passerinus*) ranges from 4 to 10 eggs, and initiating incubation on the first egg produces hatch spreads from 6 to 15 days (Beissinger and Waltman, 1991). Species that have only partially asynchronous hatching can also exhibit an invariable onset of incubation. Incubation in the Dark-eyed Junco (*Junco hymenalis*) is usually begun on the third egg, but clutch size ranges from three to five eggs (Smith, 1988).

In other taxa, incubation may begin on a particular egg relative to the end of laying (e.g., the last or penultimate egg). If clutch size in these species varies, then the initiation of incubation will be variable with respect to laying order (Fig. 2b). The resulting hatch spread will be constant, and independent of clutch size (e.g., Zerba and Morton, 1983; Hébert and Sealy, 1992). This pattern is thought to characterize many passerine species, in which incubation is frequently begun on the penultimate egg (Clark and Wilson, 1981).

The lack of variation in a trait may be the result of strong selection, or may indicate ecological or phylogenetic constraints on the trait (Stearns, 1980). The lack of variation in the onset of incubation in some avian taxa may indicate selection for a particular pattern of incubation (Fig. 2a). Conversely, an invariable hatch spread, independent of clutch size, may indicate that the hatching patterns themselves are adaptive.

2.5. Other Proximate Factors Affecting Hatching Patterns

Nestling size hierarchies can be affected by intraclutch variation in egg size as well as by hatching asynchrony. Production of relatively small last eggs may be a mechanism to handicap last-hatched chicks to facilitate their elimination through brood reduction, either alone or in conjunction with asynchronous hatching. Penguins of the genus *Eudyptes* normally raise only a single chick from the larger second egg in two-egg clutches. These species exhibit the greatest degree of egg-size difference of any bird and the size disparity is sufficient to counter any effects of slight asynchrony in hatching (Lamey, 1990). Because chicks hatched from small eggs frequently have a lower probability of survival than their sibs (e.g., Parsons, 1975; Amundsen and Stokland, 1990), and because within-clutch variation in egg size can be greater in years of poor food availability (Pietiäinen *et al.*, 1986), it has been suggested that the combination of hatching asynchrony and small last eggs constitutes a "brood reduction strategy" (Slagsvold *et al.*, 1984).

Except in *Eudyptes* penguins, it is difficult to differentiate the effect of egg size from the effects of hatching asynchrony. Attempts to partition the variance in nestling size between these two factors have found a negligible role for egg size variation (Pierotti and Bellrose, 1986; Meathrel and Ryder, 1987; Stokland and Amundsen, 1988; Magrath, 1992; Sydeman and Emslie, 1992; Jover *et al.*, 1993). Reduced size of last eggs may not be an adaptation, but may simply indicate a decline in essential nutrients required for egg production in females (Murphy, 1986; Meathrel and Ryder, 1987; Bolton, 1991). Alternatively, egg size reduction may be mediated through the interactions of hormones governing egg production and incubation behavior (Parsons, 1976).

In many passerines egg size actually increases with laying order (e.g., Howe, 1976; Zach, 1982; Haftorn, 1986). Because hatching from a larger egg may improve the competitive ability of last-hatched young, this pattern has been suggested as a "brood survival strategy" (Howe, 1976, 1978; Slagsvold *et al.*, 1984), since it may promote the survival of chicks under conditions of hatching asynchrony. As yet, there is little evidence to support or refute the idea of an adaptive role for increasing egg size within a clutch. Therefore, designating an adaptive function to egg size variation must be considered premature (Pierotti and Bellrose, 1986).

In some precocial species, embryos can exert some control over the timing of hatching. Laboratory studies of quail and geese have shown that acoustic signals between embryos within a brood increases the synchronization of hatching of the brood (Vince, 1964, 1968; Drent, 1973; Davies and Cooke, 1983). Synchronization is achieved primarily through accelerating the development and hatching of later-laid eggs (Vince, 1964; Davies and Cooke, 1983). In Northern Bobwhites (*Colinus virginianus*), embryonic communication also causes more advanced embryos to retard their hatching as well (Vince, 1968). Thus, hatching patterns in precocial species may not reflect incubation behaviors. Many precocial species begin incubation early (e.g., Cannon *et al.*, 1986; Arnold *et al.*, 1987; Kennamer *et al.*, 1990) perhaps in response to some of the same selection pressures as altricial species. However, precocial species have generally been ignored in studies of asynchrony (see Arnold *et al.*, 1987 for a notable exception).

Laboratory evidence suggests that embryonic communication is possible in asynchronously-hatching altricial birds. Glaucous-winged Gull (*Larus glaucescens*) eggs incubated in contact with each other had shorter hatching spreads than eggs that were separated. However, this effect was not evident under field conditions (Schwagmeyer *et al.*, 1991). Embryos of the American White Pelican (*Pelecanus erythrorhynchos*) vocalize to parents to elicit greater attentiveness during pipping (Evans, 1988, 1990a). The general applicability and importance of embryonic control of hatching times in altricial and semiprecocial species is not yet known.

3. HATCHING PATTERNS IN BIRDS

All birds hatch their eggs asynchronously to some extent, since complete synchrony is virtually impossible due in part to within-clutch variation in hatching time (Clark and Wilson, 1981). Hatch spreads vary greatly, ranging from a few hours in many precocial species to two weeks or more in some owls and parrots with large clutches (Wilson *et al.*, 1986; Beissinger and Waltman, 1991). Within a species, hatch spreads may vary from early to late in the nesting season, or among broods which hatch at about the same time (Slagsvold and Lifjeld, 1989a; Stouffer and Power, 1990; Harper *et al.*, 1992).

3.1. Quantification of Asynchrony

Asynchrony has generally been classified by hatch spreads. Hatch spreads of less than 24 hours have been considered synchronous, and all spreads greater than 24 hours have been considered asynchronous (cf. Ricklefs, 1993). Unfortunately, this simple dichotomy neglects much of the variation that exists in hatching patterns. Because of their focus on the onset of incubation, Clark and Wilson (1981) chose to classify asynchrony according to which egg a bird begins to incubate fully. For example, incubation on the last egg is designated as n, on the penultimate as $n - 1$, and on the first egg as 1. This system has the advantages of being able to describe the whole range of possible incubation patterns, and being applicable to precocial species in which hatch spreads are reduced. However, it ignores partial incubation, which can influence hatching patterns (see section 2.4.). In addition, classifications can be ambiguous: the first egg of a two-egg clutch may be designated as 1 or $n - 1$ (Ricklefs, 1993). Finally, the correlation between incubation and hatching patterns can be poor (Bortolotti and Wiebe, 1993).

3.2. Phylogenetic Patterns of Incubation and Hatching Asynchrony

Much of the variation in patterns of incubation and hatching occurs at the taxonomic levels of family and order, as with many life history traits (Stearns, 1980; Harvey and Pagel, 1991). In this section we review the patterns of variation in the onset of incubation and hatching asynchrony in relation to phylogeny. Although it is controversial, we follow the taxonomy of Sibley and colleagues (Sibley and Ahlquist, 1990; Sibley and Monroe, 1990). Those taxa with single-egg clutches or for which we could find no information are not included. Details of incubation and hatching patterns are based on data from regional handbooks, monographs, and original papers. In many cases generalizations were made from relatively few examples.

3.2.1. The Eoaves and Lower Neoaves

The Eoaves comprise the orders Struthioniformes and Tinamiformes (Fig. 3). The most basal parvclass of the neoaves, the Galloanserae, comprises the Craciformes, Galliformes, and Anseriformes. All of these groups have precocial young, and generally hatch their young synchronously (e.g., Palmer, 1962; Marion and Fleetwood, 1978; Boag and Schroeder, 1992; Zwickel, 1992). There are a few exceptional species that exhibit hatch spreads of slightly over 24 hours (e.g., Cargill and Cooke, 1981). Several species begin incubation early, but embryonic communication facilitates the synchronous hatching of eggs (Vince, 1964; Arnold *et al.*, 1987). The taxonomic affinities of the button-quail (Turniciformes) remain uncertain (Sibley and Monroe 1990), but incubation, hatching, and developmental patterns are similar to the orders discussed above (Cramp, 1980).

3.2.2. The Higher Neoaves

The orders Piciformes and Galbuliformes have altricial young, yet hatch their young relatively synchronously (Skutch, 1969, 1983; but see Stanback, 1991). The Bucerotiformes and Upupiformes exhibit complete asynchrony, while the Trogoniformes are somewhat asynchronous (Kemp, 1978; Skutch, 1983; Cramp, 1985). The Coraciiformes are mixed with regard to hatching patterns. Within the Coraciiformes, the Cerylidae, Coraciidae, Dacelonidae, Meropidae, Momotidae, and Todidae tend to have moderate to extreme asynchrony of hatching (Parry, 1973; Kepler, 1977; Orejuela, 1977; Cramp, 1985; Scott and Martin, 1986; Bryant and Tatner, 1990; Wrege and Emlen, 1991), while the Al-

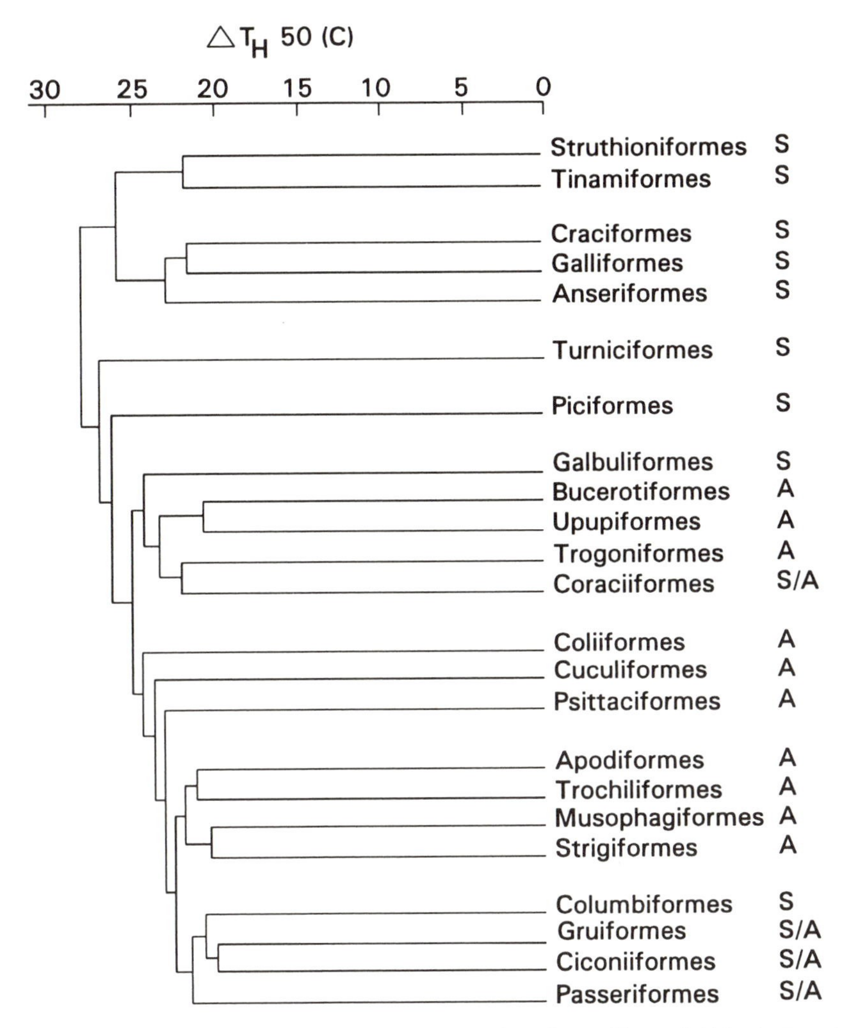

FIGURE 3. A phylogeny of orders in the Class Aves based on Sibley and Ahlquist (1990), indicating the predominant hatching pattern for each order. S = synchronous hatching (defined here as < 24 hrs), A = asynchronous hatching (≥ 24 hrs), S/A = both hatching patterns regularly occur within the order. Orders for which no information on hatching patterns could be found have been omitted.

cedinidae and Cerylidae appear to have relatively synchronous hatching patterns (Douthwaite, 1978; Boag, 1982; Skutch, 1983). However, the breeding biology of these families is poorly known, partly because they are predominantly tropical in distribution, and are all burrow or cavity-nesters.

Extreme asynchrony predominates in the Coliiformes, Cuculiformes, Psittaciformes, Apodiformes, and Trochiliformes (Rowan, 1967, 1983; Lack, 1973; Balph, 1975; Smith and Saunders, 1986; Strahl, 1988; Forshaw, 1989; Bryant and Tatner, 1990; Taplin and Beurteaux, 1992; Fig. 3). Birds in all of these groups usually begin incubation on the first or second egg. The hatch spreads of some psittacine species with large clutch sizes are among the most extreme recorded (Navarro and Bucher, 1990; Beissinger and Waltman, 1991). Extreme asynchrony results in offspring mortality in some psittacines (Beissinger and Stoleson, 1991), but not in others (Snyder *et al.*, 1987). All of the Trochilidae lay two eggs and begin incubation with the first (Skutch, 1964, 1969; Calder and Calder, 1992).

The Strigimorphae have somewhat more advanced development of hatchlings than the more basal Neoaves (semialtricial versus altricial; Ricklefs, 1983). The Strigidae and Tytonidae exhibit extreme asynchrony, and include another contender for the longest hatch spread recorded for any species: the Barn Owl (*Tyto alba;* Voous, 1975; Wilson *et al.*, 1986). All of the semiprecocial Caprimulgidae initiate incubation on the first of their two eggs (Skutch, 1972; Jackson, 1985; Csada and Brigham, 1992). The aberrant nocturnal, frugivorous Oilbird (*Steatornis caripensis*) also begins incubation on its first egg, and hatches its two to four eggs over a span of up to 12 days (Snow, 1961). Incubation and hatching patterns of the Musophagiformes are poorly known in the wild. Data from captive birds suggest incubation begins on the second egg, and hatch spread varies with clutch size (Rowan, 1983; Candy, 1984; Hewston, 1984).

The superorder Passerimorphae shows a wide variety of incubation and hatching patterns, both among and within orders (Fig. 3). The Columbiformes are limited to clutches or one of two eggs. Most species with two eggs appear to begin incubation gradually on the first egg, but generally hatch both eggs within a 24-hour period (Rowan, 1983; Skutch, 1983; Bowman, 1992; Mueller, 1992). The Gruiformes display a variety of incubation and hatching patterns (Fig. 4). The Eurypygidae, Gruidae, and at least some Otididae begin incubation with the first egg (Cramp, 1980; Thomas and Strahl, 1990; Tacha *et al.*, 1992). In contrast, the Heliornithidae may not begin to incubate until both eggs are laid (Alvarez del Toro, 1971). Hatching patterns within the Rallidae vary, with most species hatching large clutches over a 24- to 36-hour period, but some species are completely asynchronous (Cramp, 1980; Kaufmann, 1989; Horsfall, 1991; Meanley, 1992). Cranes (Gruidae) are highly asynchronous, show much sibling aggression, and siblicide is common (Tacha *et al.*, 1992).

The Ciconiiformes as delineated by Sibley and Ahlquist (1990)

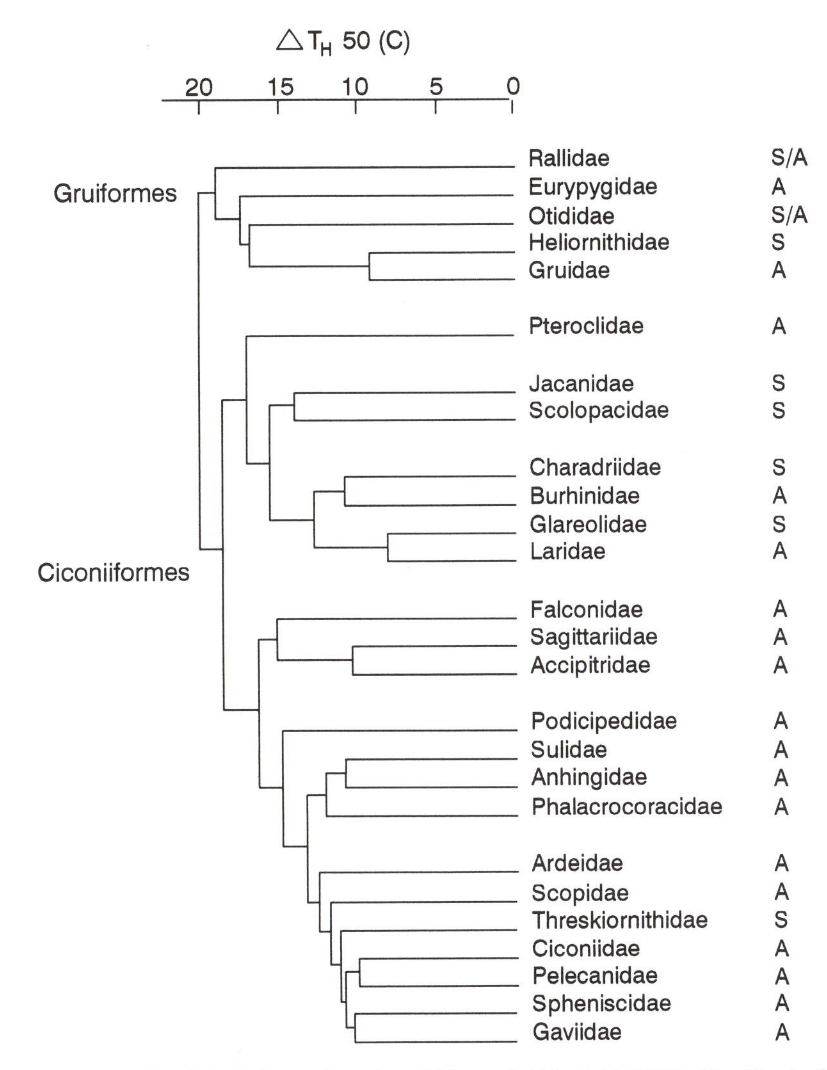

FIGURE 4. A detailed phylogeny based on Sibley and Ahlquist (19909) of families in the orders Gruiformes and Ciconiiformes, indicating the predominant hatching pattern for each family. Families with single-egg clutches or for which no information on hatching patterns could be found have been omitted. S, A, and S/A as in Fig. 3.

comprise numerous taxa that were previously grouped in several other orders, including the Gaviiformes, Podicipediformes, Pelecaniformes, Sphenisciformes, Charadriiformes, and Falconiformes, as well as the Ciconiiformes (Pettingill, 1985; Sibley and Monroe, 1990; Fig. 4). These differ markedly in morphology, ecology, and developmental mode (altricial to precocial-2), perhaps suggesting a dubious monophyly. Incubation and hatching patterns vary widely within this order as well, from complete synchrony to complete asynchrony (Fig. 4).

Within the Ciconiiformes (*sensu* Sibley and Monroe), most taxa with altricial or semialtricial development exhibit a high degree of hatching asynchrony (Fig. 4). Penguins (Spheniscidae) show a variety of patterns within the constraints of one or two egg clutches (Lamey, 1990). Most species hatch their eggs extremely asynchronously, and egg-size variation and brood reduction is common. However, Yellow-eyed Penguins (*Megadyptes antipodes*) hatch similar-sized eggs synchronously with little offspring mortality (van Heezik and Davis, 1990). The Ardeidae, Ciconiidae, and Scopidae tend to be completely asynchronous, and sibling aggression is frequent (Kahl, 1966; Anthony and Sherry, 1980; Thomas, 1984; Mock and Parker, 1986; Wilson *et al.*, 1987; Butler, 1992). In contrast, the Threskiornithidae tend to delay incubation until completion of the clutch and thus hatch synchronously (Palmer, 1962; Cramp, 1980). The Bald Ibis (*Geronticus eremita*) is an exception, being completely asynchronous, and the only ground-nesting species in this family for which information could be found (Cramp, 1980). The raptors (Accipitridae, Sagittariidae, and Falconidae) also begin incubation early and hatch very asynchronously (Brown and Amadon, 1968). Offspring mortality and sibling aggression is common, and siblicide is prevalent in the largest species (Brown and Amadon, 1968; Meyburg, 1978; Stinson, 1979; Edwards and Collopy, 1983). The altricial families characterized by having totipalmate feet (the Pelecanidae, Anhingidae, Phalacrocoracidae, and Sulidae) all begin incubation with their first or second egg. Hatching is extremely asynchronous, sibling aggression is common, and siblicide occurs in many species (Dorward, 1962; Olver, 1984; Shaw, 1985; Cash and Evans, 1986; Drummond, 1987; Stokland and Amundsen, 1988; Anderson, 1989).

Ciconiiform families with semiprecocial and precocial chicks are variable in hatching patterns (Fig. 4). The semiprecocial Laridae all begin at least partial incubation on the first egg, and the resulting asynchrony usually results in high mortality for last-hatched young (e.g., Nisbet and Cohen, 1975; Hahn, 1981; Hébert and Barclay, 1986; Bollinger *et al.*, 1990). Siblicidal aggression is common only in the skuas, subfamily Stercorariinae (Spellerberg, 1971; Williams, 1980), but has been reported in other species as well (e.g., Braun and Hunt, 1983). The Gaviidae and Podicipedidae, which have downy, mobile young that are not self-feeding at hatching (precocial-4; Ricklefs, 1983), begin incubation with the first egg, and so are completely asynchronous (Palmer,

1962; Ferguson and Sealy, 1983; Forbes and Ankney, 1987). Sibling aggression is common in some species, but siblicide is not known to occur.

The shorebirds (Charadriidae, Scolopacidae, Jacanidae, Glareolidae) generally do not begin incubation until the full clutch has been laid, and therefore hatch synchronously (Parmalee, 1970; Gibson, 1971; Jehl, 1973; Osborne and Bourne, 1977; Gratto-Trevor, 1992; Haig, S. M., 1992). There are exceptions, such as the Black-bellied Plover (*Pluvialis squaterola*) which partially incubates early eggs and has hatch spreads of over 24 hours (Hussell and Page, 1976). The predominantly tropical Thick-knees (Burhinidae) are also an exception to the general shorebird pattern. They begin incubation with the first egg, and chicks hatch over several days (Freese, 1975; Cramp, 1983). In these species the young, though mobile, are brooded in the nest until all eggs have hatched. Sandgrouse (Pteroclididae) are also completely asynchronous, but early-hatched young leave the nest with one parent while the other continues to incubate (Cramp, 1983).

There is much variation in incubation and hatching patterns among the Passeriformes (Fig. 3). The patterns in this group have been examined in some depth by Clark and Wilson (1981), Slagsvold (1986b), and Ricklefs (1993). Therefore we present only some broad generalizations. Passerines tend to begin full incubation on the penultimate or last egg, but frequently increase their attentiveness gradually prior to full incubation. Thus, most species are partially asynchronous with hatch spreads of 24 to 48 hours (Clark and Wilson, 1981). Few passerine species are known to hatch their eggs completely asynchronously. Among these are a tropical blackbird (*Agelaius icterocephalus;* Wiley and Wiley, 1980), a weaver (*Ploceus taeniopterus;* Jackson, 1993), tropical tityras (*Tityra;* Skutch, 1969), and the Australian butcherbirds (*Cracticus;* Courtney and Marchant, 1971).

3.2.3. Overview of Phylogenetic Patterns

Synchronous hatching appears to be the primitive condition in birds, and is widespread in the lower, most precocial orders (Fig. 3). A major transition in character state from synchronous to asynchronous hatching occurred in the Parvclass Coraciae (Galbuliformes through Coraciiformes in Fig. 3). In the Coraciiformes, Columbiformes, and some Gruiformes, Ciconiiformes, and Passeriformes, synchrony is secondarily derived and not necessarily associated with precocial development. Asynchronous hatching predominates in most higher orders, and is most often associated with altricial or semialtricial development.

However, very different hatching patterns can be found in closely related families, such as the Glareolidae and Laridae, or even within a family, as in the Rallidae (Fig. 4). In these taxa, asynchrony is more likely to be a true adaptation rather than a result of phylogenetic constraints.

3.3. Other Correlates of Asynchrony

Birds vary in the degree of nestling development at hatching, based on presence or absence of down, degree of mobility, and dependence on parents (Nice, 1962; Ricklefs, 1983). Hatching asynchrony is generally considered a phenomenon confined to altricial species, while precocial species hatch synchronously. Although most common in altricial and semialtricial species, hatching asynchrony occurs across a broad range of developmental modes. Semiprecocial chicks are mobile to a limited extent, but remain in the nest and are dependent on parental feeding. Complete asynchrony is the norm in several semiprecocial families including the Laridae (excluding the Alcidinae), Caprimulgidae, and Eurypygidae. Chicks that are categorized as precocial-four are downy and mobile at birth, but rely on parental feeding for a brief period after hatching (Ricklefs, 1983). In this category the Gaviidae, Podicipedidae, Burhinidae, Gruidae, and (some) Otididae are completely asynchronous, while the Rallidae range from completely asynchronous to synchronous hatching. Hatching asynchrony occurs in some families where chicks are self-feeding at hatching (Precocial-3,2,1) (Ricklefs, 1983), such as the Pteroclidae, which begin incubation on the first egg. Clearly asynchrony in these taxa cannot be explained by hypotheses based on limits to parental ability to feed young (e.g., *Brood Reduction, Peak Load Reduction, Hurry-up, Dietary Diversity,* and *Sibling Rivalry*).

Several researchers have identified correlates of hatching asynchrony other than developmental mode. Clark and Wilson (1981) found that hole-nesting species tended to hatch their eggs more asynchronously than species with open nests. They suggested that higher predation rates of open nests favor minimizing the length of the nestling period through the synchronization of hatching (see section 4.5.2.). They used the same argument of differential rates of nest predation to explain the predominance of asynchrony in tree-nesting species versus those that nest in bushes or on the ground, and in temperate species versus tropical species. Their analyses were reexamined by Slagsvold (1986b). He found asynchrony to be correlated with hole-nesting only in tropical areas when the effects of geographic distribution and clutch size were controlled. He found no other significant relation between asynchrony and geographic range.

Hatching patterns can vary independently of clutch size in species with partial asynchrony. Frequently the degree of hatching asynchrony increases during the course of the breeding season. For example, early clutches of the Common Moorhen (*Gallinula chloropus*) hatch completely synchronously, while late nests and second broods may require three or more days to hatch (Cramp, 1977). The usual explanation is that greater asynchrony reflects seasonal declines in food resources, and hence a greater utility of adaptive brood reduction (e.g., Bryant, 1978; Newton and Marquiss, 1984; Hussell and Quinney, 1987; Skagen, 1987; Perrins and McCleery, 1989). However, other hypotheses offer alternative interpretations of this phenomenon, and will be discussed in the following sections.

3.4. Summary of Hatching Patterns

Incubation patterns are poorly known, and difficult to quantify. Therefore, most studies have focused on hatching patterns. Hatching patterns range from complete synchrony to complete asynchrony, and vary greatly among and sometimes within taxa. Synchronous hatching is generally associated with precocial development, and is the primitive condition in birds. Asynchrony is associated with altricial development. Many species exhibit an increase in the degree of hatching asynchrony during the course of the nesting season. The onset of incubation may be affected by a number of factors, including the physiology of eggs and embryos, and the ecology of species. Other factors may influence hatching patterns directly. This multitude of factors has resulted in many hypotheses being proposed for the evolution of hatching asynchrony in birds.

4. HYPOTHESES FOR THE EVOLUTION OF HATCHING ASYNCHRONY

The plethora of hypotheses for the evolution of asynchronous hatching have been reviewed to varying extent by Slagsvold (1986b), Skagen (1988), Lessells and Avery (1989), and Magrath (1990). However, to date, these hypotheses have lacked a conceptual framework to relate hatching patterns with various selection pressures that potentially affect incubation behavior.

4.1. A Conceptual Framework

We categorize hypotheses for the evolution of hatching asynchrony into four groups (Table I), based on the factors that could potentially produce hatching asynchrony. Two hypotheses suggest that asynchrony is an epiphenomenon of constraints on incubation. According to these hypotheses, neither the onset of incubation nor the resulting hatching pattern necessarily have an adaptive function. Four hypotheses view asynchronous hatching as a consequence of factors selecting for the early onset of incubation, regardless of its effect on nestling survival. Most of these hypotheses ascribe some sort of protective function to early incubation. The subsequent asynchrony in hatching is a nonadaptive consequence which sometimes entails a cost in the form of mortality of later-hatched young. The majority of hypotheses (eight) consider the nestling size-hierarchy produced by asynchronous hatching to be adaptive for nestling survival or to decrease the costs of reproduction for parents. Finally there are three hypotheses that posit that selection acts on the timing of different phases of the nesting cycle through adult and offspring mortality and costs of reproduction.

The various hypotheses are based on the effects of different factors that are predominantly intrinsic to the organism (e.g., behavior or physiology) or predominantly extrinsic (e.g., environmental variation, predators, or nest site availability). For each hypothesis, there is a critical period in the nesting cycle which constrains reproductive success, and this period differs among the categories of hypotheses. The critical period for nonadaptive hypotheses and hypotheses based on adaptive hatching patterns is egg-laying, while the nestling period is critical for hypotheses invoking adaptive hatching patterns. Hypotheses based on both have two critical periods: egg-laying, and either brooding or fledging. It is the relative timing of these critical periods which is essential.

Critical periods in the nesting cycle should be considered when evaluating experimental tests of hatching asynchrony in birds. For example, manipulating the hatching pattern does not alter the pattern of incubation, and therefore can not address hypotheses for which egg-laying is the critical period. Thus, ambiguous or negative results of such manipulations suggest that hypotheses based on incubation patterns may be of greater importance than those based on hatching patterns for the species in question.

In the following sections we review the hypotheses proposed for the evolution of hatching asynchrony, and relate them to critical periods of the nesting cycle. For each hypothesis, we describe its theoretical

TABLE I

Hypotheses for the Evolution of Hatching Asynchrony.[a]

Evolutionary significance of asynchrony	Hatching asynchrony results from	Intrinsic or extrinsic	Constraint or selective force	Hypothesis	Critical period in the nesting cycle	Number of experimental tests
Nonadaptive	Constraint	Intrinsic	Physiology	Hormone	laying	1
				Energy constraints	laying	2
Adaptive	Adaptive incubation patterns	Extrinsic	Environment	Egg viability	laying	2
				Limited breeding opportunity	laying	0
			Predation	Egg protection	laying	0
			Social system	Brood parasitism	laying	0
	Adaptive hatching patterns	Intrinsic	Parental efficiency	Peak load reduction	nestling	3
				Dietary diversity	nestling	0
				Larder	nestling	0
				Sibling rivalry	nestling	4
			Sexual dimorphism	Sex ratio manipulation	nestling	0
		Extrinsic	Environment	Brood reduction	nestling	32
				Hurry-up	nestling	0
			Stochastic events	Insurance	nestling	2
	Incubation and hatching	Intrinsic	Mating system	Sexual conflict	laying/brooding	3
		Extrinsic	Predation	Nest failure	laying/fledging	4[b]
				Adult predation	laying/brooding	0

[a]For each hypothesis, the adaptive significance of asynchrony, the nature of the selective force and whether that force is predominately intrinsic or extrinsic to the organism, and the critical period in the nesting cycle are indicated. The number of experimental tests of each hypothesis is noted; references for these are annotated with "(T1)" in the References section.

[b]Not experimentally tested but vital rates measured and compared to results from a model.

basis and applicability, identify its important assumptions and predictions, review any correlational or experimental support or refutation, and suggest how it should be tested experimentally.

4.2. Factors Affecting the Onset of Incubation: Nonadaptive Constraints

Perhaps the simplest hypotheses consider hatching asynchrony to be the mechanistic result of physiological or environmental constraints on the initiation of incubation, so that neither asynchrony nor the resulting nestling size hierarchy are themselves necessarily adaptive.

4.2.1. Energy Constraints Hypothesis

Hatching patterns may reflect environmental constraints on egg-laying (the *Energy Constraints Hypothesis*). When females experience poor feeding conditions during laying, they may need to forage for a greater part of the day, allowing less time for incubation, and causing eggs to hatch more synchronously than normal (Greig-Smith, 1985; Slagsvold, 1986a; Enemar and Arheimer, 1989; Moreno, 1989a). If food availability increases through the course of a breeding season, females can spend less time foraging and more time incubating during laying, thereby hatching their eggs more asynchronously (Gibb, 1950; Nisbet and Cohen, 1975; Mead and Morton, 1985; Slagsvold, 1986b; Slagsvold and Lifjeld, 1989a; Slagsvold and Amundsen, 1992). For similar reasons, Pied Flycatcher (*Ficedula hypoleuca*) clutches hatched more asynchronously in high-quality habitats than in low-quality habitats (Slagsvold, 1986a). This hypothesis is not applicable to species in which males provide all the food requirements of incubating females. It implies that early incubation and asynchronous hatching are the preferred or default patterns, since synchrony is induced by environmental stress.

The *Energy Constraints Hypothesis* predicts that females provided with supplemental food prior to and during laying should hatch their eggs more asynchronously than controls (Wiebe and Bortolotti, 1994a). Marsh Tits (*Parus palustris*) provided with supplemental food did hatch their eggs more asynchronously than controls (Nilsson, 1993b). However, female American Kestrels (*Falco sparverius*) receiving food hatched their eggs more synchronously than control females, contrary to the predictions (Wiebe and Bortolotti, 1994a).

4.2.2. Hormone Hypothesis

Hatching asynchrony may result if the same hormone or hormones that cause the termination of ovulation also stimulate the initiation of incubation (the *Hormone Hypothesis;* Mead and Morton, 1985). In most birds, eggs are laid one day after ovulation (Woodard and Mather, 1964; Gilbert, 1971). If one hormonal mechanism controls both functions simultaneously, females should begin incubation on the penultimate egg. Initiating incubation on the penultimate egg may be the most common pattern in passerine birds (Clark and Wilson, 1981). However, many species begin incubation gradually, or on the first or second egg, and much intraspecific variation in the onset of incubation occurs (Lessells and Avery, 1989; Magrath, 1990; Hébert and Sealy, 1992; Ricklefs, 1993).

The hormonal mechanisms regulating avian reproduction are imperfectly known (Balthazart, 1983; Ball, 1991). In many species prolactin is associated with incubation but plasma-luteinizing hormone appears to control egg-laying (e.g., Myers *et al.*, 1989). It is unclear whether prolactin induces incubation or is itself a result of incubation behavior (Goldsmith and Williams, 1980; Balthazart, 1983). The roles of various hormones in controlling parental behaviors must be clarified for this hypothesis to be tenable.

When applicable, this hypothesis makes several testable predictions. Full incubation attentiveness should begin with the penultimate egg, regardless of clutch size. Hatch spreads produced by these patterns of incubation should not vary with clutch size, nor with the experimental addition of extra eggs during egg laying (Mead and Morton, 1985). This hypothesis has been tested only in the Yellow Warbler (*Dendroica petechia*). Hatch spreads in the warbler varied with clutch size, and experimentally adding eggs during early egg-laying advanced the onset of incubation and increased hatch spreads, contrary to predictions (Hébert and Sealy, 1992).

4.3. Factors Producing Adaptive Incubation Patterns

The focus on adaptive hatching patterns has drawn attention away from the behavior that actually causes hatching asynchrony—the onset of incubation—and the mechanisms that control it. The following four hypotheses are based on the potential benefits of early incubation, and consider the resulting hatching asynchrony and nestling size hierarchy as incidental. Possible functions of early incubation include maintaining the viability of eggs, and protecting the clutch from predators and conspecifics that might depredate or parasitize it, or usurp the nest site.

4.3.1. Egg Viability Hypothesis

Initiating incubation before the clutch is completed could serve to maintain the viability of first-laid eggs (the Egg *Viability Hypothesis;* Arnold *et al.*, 1987). Avian embryos do not begin to develop appreciably until egg temperatures reach about 34°C, and no development occurs below physiological zero (Drent, 1973; O'Connor, 1984; Webb, 1987). Once embryonic development begins, embryos are much more sensitive to exposure to temperatures above or below incubation temperature. Prolonged exposure to temperatures above physiological zero, yet below normal incubation levels (36–38°C), results in abnormal development and embryo mortality (Romanoff and Romanoff, 1972; White and Kinney, 1974; Webb, 1987). Survivorship of embryos depends upon the temperature, and the duration and frequency of exposure (Rol'nik, 1970; Drent, 1975; O'Connor, 1984), and varies among taxa (Roby and Ricklefs, 1984; Webb, 1987; Gaston and Powell, 1989; Astheimer, 1991). In all avian species studied, high temperatures have a much greater effect on embryo survival than lower temperatures (Yom-Tov *et al.*, 1978; Bennett and Dawson, 1979; Carey, 1980; Grant, 1982; Walsberg and Voss-Roberts, 1983; Webb, 1987).

Clutch size in waterfowl may be constrained by the decreasing viability of unincubated eggs (Arnold *et al.*, 1987). Waterfowl chicks are precocial and incubation is usually initiated before the termination of egg-laying. Yet eggs normally hatch synchronously. Arnold *et al.* (1987) demonstrated that egg hatchability declined during the laying period if eggs were unincubated. They suggested that early incubation may prevent a reduction in egg viability, but would constrain the number of eggs that could hatch synchronously (see also Cannon *et al.*, 1986; Kennamer *et al.*, 1990; Arnold, 1993).

Veiga (1992) extended this idea to asynchronously hatching altricial species. He demonstrated that House Sparrow (*Passer domesticus*) eggs that were left unincubated for three or more days had lower hatching success than those unincubated for shorter periods, and the degree of hatching asynchrony increased when ambient temperatures regularly exceeded physiological zero later in the season. Veiga concluded that hatching asynchrony may be a nonadaptive result of a behavioral mechanism to maintain the viability of first-hatched eggs.

The Egg *Viability Hypothesis* applies only to species whose laying period is long enough to affect egg viability, and whose laying takes place when ambient temperatures rise above physiological zero or below freezing for extended periods. Under these conditions, the hypothesis predicts that the hatchability of unincubated eggs should decrease with increasing length of exposure to ambient temperatures, and that development of embryos should occur in the absence of incubation. This can be tested experimentally by removing eggs immediately after laying, exposing them to ambient temperatures for specific time periods, and then allowing them to be incubated to complete development. Hatchability of experimental eggs should be compared to that of appropriate controls. Results from our preliminary experimental tests of this hypothesis with the Green-rumped Parrotlet indicated a significant drop in hatchability may occur after just one day of exposure, and a rapid decline with longer exposure times.

Three general predictions can be derived from this hypothesis. With an increasing average ambient temperature, such as with a latitudinal or seasonal gradient, species that hatch their eggs synchronously should show either a decrease in average clutch size, or a decrease in the average hatchability of eggs. Alternatively, there may be an increase in the prevalence and degree of hatching asynchrony as latitude decreases or the breeding season progresses. A latitudinal decline in clutch size has been well documented and is the subject of considerable debate (Klomp, 1970; Murray, 1985; Skutch, 1985; Godfray *et al.*, 1991). Koenig (1982) found a highly significant decline in hatchability with decreasing latitude from a sample consisting of 42 different avian families. Numerous studies have reported a seasonal decline in average clutch size (e.g., Gibb, 1950; Klomp, 1970; Hussell, 1972; Bryant, 1978; Slagsvold, 1982; Ferguson and Sealy, 1983). Although usually attributed to seasonal declines in food resources, declines in clutch size can be independent of food supply in some species (De Steven, 1980; Stutchbury and Robertson, 1988; Arnold, 1993).

Data on the relative occurrence of asynchronous hatching in the tropics versus the temperate zone are ambiguous or lacking. Clark and Wilson (1981) reported a tendency for greater synchrony in the tropics, based on a survey of 87 species. However, their sample of tropical species included a disproportionate number of passerines, and their analysis did not account for smaller clutches typical of tropical passerines (Slagsvold, 1986b).

Two caveats must be made concerning the *Egg Viability Hypothesis*. First, asynchronous hatching may lead to reduced hatchability of final eggs due to parental neglect (Slagsvold, 1985; Evans, 1990b, Nilsson, 1993b). Parents may spend less time incubating terminal eggs to provide care to first-hatched young. Reduced attentiveness has been shown to reduce hatchability (Evans, 1990c; Beissinger and Waltman,

1991). Experimentally increasing the hatch spreads of several temperate passerines produced a significant decline in the hatchability of final eggs compared to control nests (Slagsvold, 1985).

Second, a decline in the viability of unincubated eggs may be a consequence of early incubation rather than a cause. If a species has evolved complete asynchrony for some reason other than the maintenance of egg viability, eggs may no longer be subject to selection to remain viable when not incubated. Experiments that demonstrate a decline in hatchability when eggs are not incubated do not necessarily prove that egg viability was the ultimate selective force causing asynchrony. Comparative and experimental analyses will be needed to discriminate between cause and effect.

4.3.2. Egg Protection Hypothesis

Early incubation may be an adaptation to protect eggs from dangers other than unfavorable environmental conditions (Swanberg, 1950; Oring, 1982). Dunlop (1910) first proposed that parent birds could reduce the risk of predation on eggs by beginning incubation on the first egg (the *Egg Protection Hypothesis*). Incubation may be a more efficient defense against predation of eggs than other active behaviors (Thompson and Raveling, 1987). Parental brooding for the purpose of protecting eggs has been noted in some fish (Salfert and Moodie, 1984) and invertebrates (Milne and Calow, 1990). For egg protection to be a significant function of early incubation, a species must be subject to a substantial rate of egg loss prior to the start of incubation due to predation or to destruction by conspecifics. The latter case represents a form of reproductive interference. The net benefits obtained by protecting early eggs must be greater than the potential costs of nestling mortality resulting from asynchrony.

Colonial species do not defend all-purpose territories. They might be expected to begin incubation early to reduce losses of first-laid eggs due to intraspecific interference, and to hatch their eggs more asynchronously than territorial species. Egg protection has been proposed as the function of early incubation in colonial Common Terns (*Sterna hirundo;* Bollinger *et al.*, 1990) and Herring Gulls (*Larus argentatus;* Parsons, 1976). Embryonic synchronization of hatching (Vince, 1964, 1968) may enable precocial species to protect eggs through early incubation with little hatching asynchrony (Cannon *et al.*, 1986). Cavity-nesting species are frequently thought to experience lower rates of predation than open-nesting species, and so might be expected to hatch their eggs more synchronously. However, conclusions about rates of predation on cavity nests are based on studies of birds using nest boxes, and may not reflect rates of predation on natural cavity nests (Møller, 1989; Robertson and Rendell, 1990; Kuitunen and Aleknonis, 1992). Open nests and natural cavity nests may be subject to equal rates of predation (Nilsson, 1986). Therefore, cavity nesters may not necessarily hatch their eggs more synchronously than open nesting species.

Testing the *Egg Protection Hypothesis* experimentally is problematic because it is difficult to manipulate parents to lay eggs but not incubate them. One approach is to correlate the constancy of incubation during the laying period with the rate of egg loss. For example, incubation constancy was lower for Common Tern nests that were depredated than for those that were not in nests with two or more eggs (Bollinger *et al.*, 1990). However, such correlations may not take into account the ability of parents to defend nests by means other than incubation.

4.3.3. Limited Breeding Opportunities Hypothesis

Species that use nest sites that are limited in number may be forced to protect the site. For species that do not defend an all-purpose territory, this might be best accomplished by occupying the nest itself. In this way one bird would be free to forage, for itself and its mate, while the other bird defended the nest site by initiating incubation (the *Limited Breeding Opportunities Hypothesis;* Beissinger and Waltman, 1991). In the Green-rumped Parrotlet there are high rates of interactions between pairs at nest sites and a significant proportion of the population consists of nonbreeders (Beissinger and Bucher, 1992; Waltman and Beissinger, 1992). This pattern appears to be common in psittacines, which typically hatch their young very asynchronously (Saunders, 1986; Snyder *et al.*, 1987; Beissinger and Bucher, 1992). Female Elf Owls (*Micrathene whitneyi*) occupy their nest cavities for several weeks prior to laying, presumably to avoid losing them to other species, and also hatch their eggs very asynchronously (Ligon, 1968).

Because nest sites tend to be limited for secondary cavity-nesters in general (Alerstam and Högstedt, 1981; Nilsson, 1986; Brawn and Balda, 1988; Land *et al.*, 1989; Sedgwick and Knopf, 1990; Caine and Marion, 1991; Martin, 1993), this hypothesis predicts that asynchronous hatching should be more common in secondary cavity nesters than in primary cavity nesters or in open nesting species. This trend has been demonstrated, although with small sample sizes (Slagsvold, 1986b). The prevalence or degree of asynchrony should be correlated with the proportion of nonbreeders in the population, or with the abundance of nest-site competitors.

Because birds can sit on their nests without incubating (Swanberg, 1950; Vinuela, 1991), this hypothesis may be insufficient by itself to explain hatching asynchrony. However, nest site defense may function in conjunction with other factors, such as protecting eggs, minimizing the time females are in the nest, or by accelerating the hatching and fledging of first-laid eggs (Beissinger and Waltman, 1991).

4.3.4. Brood Parasitism Hypothesis

Early incubation may help to protect a clutch from brood parasitism (the *Brood Parasitism Hypothesis*) as well as predation. By initiating incubation early, females may reduce the opportunity for intraspecific (Kendra *et al.*, 1988; Lombardo *et al.*, 1989; Romagnano *et al.*, 1990) or interspecific parasitism (Wiley and Wiley, 1980). Interestingly, Jackson (1993) suggested that extreme asynchrony can promote intraspecific brood parasitism. Northern Masked Weavers (*Ploceus taeniopterus*) hatch their broods completely asynchronously, and third or fourth-hatched chicks rarely survive. Females may derive greater reproductive success from third or fourth eggs by laying them in another bird's nest than by laying in their own nest, if they hatch as the first or second chick (Jackson, 1993).

The early initiation of incubation could be advantageous if a clutch is parasitized, because the resulting brood will be larger than the original clutch and asynchronous hatching may facilitate the adjustment of the brood size through brood reduction (see section 4.4.1.; Wiley and Wiley, 1980; Magrath, 1990). In joint-nesting species (Brown, 1987), there may be a similar competition between laying females at a nest. Females may initiate incubation as soon as possible after the laying of their first egg to increase the probability that their own young will hatch first and thereby be competitively superior (Ricklefs, 1993). This may explain why Acorn Woodpeckers (*Melanerpes formicivorus*) and Pied Kingfishers (*Ceryle rudis*) exhibit significant asynchrony when relatively synchronous hatching is the norm in the Picidae and Alcedinidae (Stanback, 1991; Ricklefs, 1993).

This hypothesis can be tested experimentally by simulating nests in the process of laying without incubating parents, either by using artificial nests or by temporary removal of parents. The hypothesis predicts that nests without incubating birds should receive parasitic eggs with a greater frequency than normally incubated nests. Comparisons of different populations or species should show that the degree of asynchrony is correlated with the prevalence of brood parasitism.

4.4. Factors Selecting for Adaptive Hatching Patterns

The following hypotheses posit that asynchronous hatching patterns are adaptive in their own right, and that early incubation functions to produce those patterns. The first three hypotheses use nestling size-hierarchies to prioritize young within a brood, so that low priority young are eliminated through sibling competition or parental neglect. In the remaining hypotheses, asynchronous hatching serves to increase parental efficiency.

4.4.1. Brood Reduction Hypothesis

David Lack (1947, 1954) proposed that for bird species in which the food supply varies unpredictably, the optimal clutch size should reflect the average maximum number of young that can be raised under favorable conditions. In the event of food scarcity, the youngest nestlings are either actively neglected or are outcompeted by their larger sibs, and starve to death (Lack, 1954; Ricklefs, 1965). Brood size is adjusted to the parental ability to supply food by the elimination of the nestling(s) in which parents have invested the least or that will require the most future investment (Hébert and Barclay, 1986). If hatching was synchronous, all nestlings would be equally competitive, and all would suffer undernourishment or starvation (Lack, 1954). This hypothesis is best known as the *Brood Reduction Hypothesis*, but recently was called the *Resource Tracking Hypothesis* to differentiate it from the following two hypotheses for which asynchrony also functions to allow an adaptive reduction in brood size (Forbes, 1991; Ploger, 1992). Although the latter term describes the concept of the hypothesis more accurately, for the sake of clarity we will use the more familiar term.

Hatching asynchrony is not necessarily a prerequisite for a size hierarchy to develop among nestlings (Nelson, 1964; Bengtsson and Rydén, 1983; Groves, 1984; Amundsen and Slagsvold, 1991a). Also, brood reduction can occur in synchronously-hatched nests (Clark and Wilson, 1981; Bancroft, 1985). Therefore, the hypothesis has been modified to state that hatching asynchrony facilitates efficient brood reduction (Lack, 1966; Husby, 1986; Mock and Parker, 1986; Magrath, 1989). Plants also may adjust parental investment to track available resources by terminating flowering, ovary development, or fruit set. This process is most efficient when flowering is staggered, i.e., asynchronous (Lloyd, 1980; Haig, D., 1992).

Brood reduction in asynchronously-hatched nests can, but does not always, improve the condition or growth rate of surviving offspring (Gibb, 1950; Ricklefs, 1965; Stouffer and Power, 1991). For example, surviving nestlings in broods of common Starlings (*Sturnus vulgaris*) did not show an increase in growth rate or condition following the mortality of their youngest sibs (Stouffer and Power, 1991).

Numerous models have been constructed to provide a theoretical framework for the concept of adaptive brood reduction. O'Connor (1978) used Hamilton's (1964) concept of inclusive fitness to predict that there should be greater conflict between parents and young over when brood reduction should occur when parents have a relatively high investment in each chick (small broods). This should occur because surviving offspring benefit more from brood reduction than do parents. This parent-offspring conflict (Trivers, 1974) can result in the larger offspring initiating brood reduction through sibling aggression (Mock, 1984a; Mock and Parker, 1986; Drummond and García-Chavelas, 1989). Siblicidal behavior that seems to be contingent on inadequate food supplies is termed facultative siblicide (Mock, 1984a). For example, siblicidal brood reduction is much more frequent in Black-legged Kittiwakes (*Rissa tridactyla*) when food is scarce and nestling growth rates are low (Braun and Hunt, 1983). Drummond and García-Chavelas (1989) demonstrated that sibling aggression in the Blue-footed Booby (*Sula nebouxii*) was inversely proportional to food intake and weight increase. However, sibling aggression may not necessarily be correlated with food supply (Mock, 1984b, 1985b, 1987; Sullivan, 1988).

There is a growing body of evidence that some nestling mortality resulting from asynchrony is unrelated to food supply. In most species the feeding capacity of the parents is not exceeded in very young broods, the time when most chick mortality occurs (Bengtsson and Rydén, 1981; Graves *et al.*, 1984; Amundsen and Stokland, 1988). Steidl and Griffin (1991) used very high growth rates as evidence of abundant food in a colony of Ospreys (*Pandion haliaetus*), yet noted widespread brood reduction. The probability of chick mortality was found to be related with the degree of hatching asynchrony but not food abundance for numerous species with variable degrees of hatching asynchrony (e.g., Bryant, 1978; Strehl, 1978; Stouffer and Power, 1990; Seddon and van Heezik, 1991a,b; Stanback, 1991). In unmanipulated Common Starling nests, clutches that hatched synchronously had lower levels of nestling mortality than similarly-sized asynchronously-hatched nests (Stouffer and Power, 1990), indicating a clear cost of asynchrony. Penultimately hatched and last-hatched chicks in large broods of Green-rumped Parrotlets had very low fledging success (Beissinger and Waltman, 1991). Manual feeding with a commercial parrot nestling formula increased the probability of survival of last chicks but did not affect the survival of penultimate chicks (Stoleson and Beissinger, in prep.).

Other models have shown that parents derive the most benefit from brood reduction when it occurs while their investment in the youngest chick is small, during the early part of the nestling period (Lloyd, 1987; Forbes and Ydenberg, 1992). Pijanowski (1992) created a model that included a cost to asynchronous hatching in good food years. Asynchrony to facilitate brood reduction is still favored over synchronous hatching when good food years are infrequent, when the cost of asynchrony in good years is small, when the survival of chicks in synchronous broods in bad food years is uniformly low, or when bad food years are not severe (Pijanowski, 1992).

Several predictions can be derived from the *Brood Reduction Hypothesis*. The quantity of food available at the time of laying and the onset of incubation should be unrelated to the quantity of food available after hatching. Demonstrating variability in the food supply is insufficient to support this prediction. Also, the frequency and extent of partial brood losses should be directly related to food scarcity. This is the only hypothesis for which there is a well-established experimental methodology. Synchronously-hatching broods are created by swapping eggs or newly hatched chicks between nests. The hypothesis predicts that when food is scarce, asynchronous broods should produce more fledglings, fledglings of higher quality, or both, than synchronously-hatched broods. Asynchronous broods need not show greater reproductive success than synchronous broods when food is abundant (Pijanowski, 1992). However, if comparisons are based on broods manipulated to create a degree of synchrony not normally encountered in the species under study, parent birds may invest an "imprudent" degree of effort into the brood, ultimately with a negative effect on lifetime reproductive success (D. W. Mock, personal communication). Therefore parental effort should be monitored. Contrary to much published work, this hypothesis does *not* predict that last-hatched young in asynchronous broods are the most likely to die. Such mortality is the phenomenon the hypothesis was formulated to explain, and therefore is a premise of the hypothesis. Proving a premise false does not disprove a hypothesis, but rather indicates that the hypothesis is not relevant to the situation (Copi, 1972).

4.4.2. Insurance Hypothesis

In some bird species, the last egg appears to serve strictly as a replacement for earlier eggs or chicks that fail (the *Insurance Egg Hypothesis;* Stinson, 1979; Cash and Evans, 1986). For example, Hooded Grebe (*Podiceps gallardoi*) parents abandon the second of two eggs if the first hatches, but continue incubating the second egg if the first one fails (Neuchterlein and Johnson, 1981). Herring Gulls (*Larus argentatus*) may also abandon their third egg after successfully hatching their first two eggs (Graves *et al.*, 1984).

The *Insurance Hypothesis* pertains primarily to the evolution of clutch size in birds. Specifically, it is an explanation for why birds lay more eggs than they can normally fledge. Applied to asynchronous hatching, this hypothesis suggests that hatching asynchrony functions to facilitate the elimination of last-hatched chicks if and when they become redundant (Forbes, 1990). The *Insurance Egg Hypothesis* can be considered a food-independent analog of the *Brood Reduction Hypothesis*—asynchronous hatching is viewed as an adaptation to variability in egg hatchability or the mortality of first-hatched nestlings due to causes other than starvation (e.g., accident, predation, or congenital defect; Stinson, 1979; Anderson, 1990; Forbes, 1990). The benefits derived from insurance eggs are through reduced variance in reproductive success (Forbes, 1991). The *Insurance Hypothesis* suggests that parents initiate the onset of incubation to maximize benefits for the "base" clutch. Asynchronous hatching of insurance eggs would be a consequence of such behavior, and would not necessarily have an adaptive significance.

In species with regular brood reduction, an extra egg can have insurance value as well as extra reproductive value, depending on the fate of older chicks. Mock and Parker (1986) defined two different types of reproductive value for last-hatched nestlings in heron broods: extra reproductive value, when a chick survives in addition to older chicks, and insurance reproductive value, when a chick serves as a replacement for an older chick that died. For example, in Little Blue Herons (*Egretta caerulea*), last-hatched young normally die unless an older sibling dies first, in which case youngest chicks have a high probability of fledging (Werschkul, 1979). The dominant component of the total reproductive value of last-laid and penultimately laid Green-rumped Parrotlet eggs in small and medium sized clutches was insurance reproductive value (Beissinger and Waltman, 1991).

This hypothesis is most relevant for species with obligate brood reduction, where last-laid eggs are thought to serve strictly as insurance. In such species under normal conditions last-hatched chicks never survive due to sibling aggression. Such mortality appears to be independent of food supply. Anderson (1990) noted that the Masked Booby (*Sula dactylatra*) lays two eggs, but because of siblicide almost never fledges more than one young. Its eggs have a low probability of hatching, perhaps a consequence of its ground-nesting habits, where its eggs are vulnerable to ground predators and high surface heat. In contrast, the sympatric Red-footed Booby (*Sula sula*) builds tree nests and exhibits high hatchability, but lays a single egg.

The *Insurance Hypothesis* predicts that in the absence of synchrony, the elimination of excess chicks should occur less often, or at a greater age, than with asynchrony. In Masked Boobies, the probability and timing of siblicidal brood reduction was highly correlated with the degree of asynchrony (Anderson, 1989). However, some experiments with obligately siblicidal species found that one chick was eliminated even in experimentally synchronized broods (Dorward, 1962; Meyburg, 1978; Gargett, 1982), suggesting asynchrony may not be necessary for brood reduction to occur.

Testing the *Insurance Hypothesis* as a cause of hatching asynchrony is a different, and more difficult issue, than demonstrating that parents derive benefits from laying insurance eggs. Normally asynchronous clutches should produce more fledglings, on average, than either synchronized clutches of equal size, or asynchronous clutches that have had their last egg removed. The hypothesis requires that the survival of the smallest chicks be correlated with the mortality of earlier eggs or chicks, and that the mortality of the smallest chicks is independent of food supply.

4.4.3. Sex Ratio Manipulation Hypothesis

Selection may favor equal parental investment in offspring of each sex (Fisher, 1930). In sexually dimorphic species, the larger sex should have greater food demands than the smaller sex, and therefore be more expensive in terms of parental investment per chick (Slagsvold *et al.*, 1986; Teather and Weatherhead, 1988; Breitwisch, 1989, but see Stamps *et al.*, 1987). Parents should invest more in the smaller sex or manipulate sex ratios to maintain equal levels of investment (Trivers and Willard, 1973; Bednarz and Hayden, 1991). Alternatively, parents may facultatively manipulate the sex ratio of their offspring to capitalize on current ecological or social conditions (Myers, 1978; Gowaty,

1991). Because unequal sex ratios at hatching are uncommon in birds (Clutton-Brock, 1986; but see Gowaty and Lennartz, 1985; Ligon and Ligon, 1990; Gowaty, 1991), hatching asynchrony may provide a means for parents to manipulate the sex ratio of broods by selectively starving later-hatched young of the more expensive sex (the *Sex Ratio Manipulation Hypothesis;* Slagsvold, 1990).

Sex-biased mortality has been shown in some sexually dimorphic species when food is scarce (Howe, 1976; Cronmiller and Thompson, 1981; Roskaft and Slagsvold, 1985; Bortolotti, 1986; Teather and Weatherhead, 1989; Bednarz and Hayden, 1991; Slagsvold *et al.*, 1992). But such results appear to be a nonadaptive consequence of sex-specific susceptibility to starvation, rather than manipulation on the part of parents (Weatherhead and Teather, 1991). Slagsvold *et al.* (1986) noted that parent Rooks (*Corvus frugilegus*) did not preferentially feed offspring of one sex over the other, yet the larger male offspring fledged at a much lower rate. Harris' Hawks (*Parabuteo unicinctus*) fledge more young when the oldest nestling is the smaller sex (male) than when the first nestling is female, and the sex of first-hatched young is usually skewed towards males (Bednarz and Hayden, 1991). Furthermore, in the Blue-footed Booby, extreme size dimorphism does not seem to affect nestling dominance hierarchies based on hatching order, and there is no sex-biased mortality (Drummond *et al.*, 1991).

The strongest evidence of parental manipulation of nestling sex-ratios comes from two studies of captive birds. Color-banded Zebra Finches (*Poephila guttata*) with favored band colors preferentially fed offspring of their own sex, resulting in sex-biased mortality (Burley, 1986). Both male and female Budgerigars brought food to female-biased broods at a higher rate than other broods. However, females were not fed preferentially within broods, and males and and females fledged at similar ages and weights (Stamps *et al.*, 1987). Manipulation of offspring sex-ratio has been documented in rodents. When food was severely restricted, lactating wood rats (*Neotoma floridana*) actively discriminated against male young, resulting in female-biased litters (McClure, 1981).

This hypothesis is relevant only to species with a relatively large degree of sexual size dimorphism. It predicts that offspring of the larger sex suffer higher mortality during the period of prenatal care (Slagsvold, 1990). Therefore, the sex ratio of fledglings should be biased towards the smaller sex (Slagsvold *et al.*, 1992). In experimentally synchronized broods, parents should over-invest in the larger sex, and should result in a fledgling sex ratio that is skewed towards the larger sex, compared to asynchronous broods.

4.4.4. Peak Load Reduction Hypothesis

Asynchronous hatching may serve to maximize parental efficiency without promoting offspring mortality. Ingram (1959) suggested that the extreme nestling size hierarchy produced by asynchronous hatching in large broods of Short-eared Owls (*Asio flammeus*) might serve to spread out the total food demand of the brood. Hussell (1972) argued that such a mechanism may exist in species with sharp peaks in food demands of individual nestlings (the *Peak Load Reduction Hypothesis*). For example, House Martins appear to have a sharp peak in nestling energy demands, and the total brood demand is reduced by 7% to 8% through asynchronous hatching (Bryant and Gardiner, 1979).

Models by Mock and Schwagmeyer (1990) suggest that any energy savings due to asynchrony would be minimal, except in species with very large brood sizes (up to 10 chicks) and extreme hatching asynchrony (10–14 days). In addition, the food demands of nestlings must show a peak for any benefits to accrue. The hypothesis predicts that the peak energy expenditure of parents tending asynchronous broods would be less than the peak energy expenditure of parents tending synchronous broods of equal size. Although feeding rates may be used as an estimate of energy expenditure (Bryant and Tatner, 1991), it is preferable to measure energy expenditure directly using the doubly-labeled water method (Lifson and McClintock, 1966; Speakman and Racey, 1988; but see cautions in McNab, 1989).

4.4.5. Dietary Diversity Hypothesis

In some species asynchronous hatching may not stagger the peak energy demands, but instead staggers the demand for a special food resource needed only during a portion of the nestling period (the *Dietary Diversity Hypothesis;* Magrath, 1990). For example, hatchling Yellow Warblers were fed minute geometrid larvae almost exclusively, while older chicks received a wider variety of food. Asynchronous hatching may reduce the number of nestlings requiring the specific food items at any one time (Hébert, 1993a).

This hypothesis requires that nestlings are dependent on a specific food source at some age, and that the food itself is limited in quantity, or that parents are limited in their ability to procure the food. It predicts that parents at synchronous broods should be unable to raise their entire brood, and that nestling mortality should be due to starvation at a particular period in the nestling period. The *Dietary Diversity Hypothesis* can be tested by supplementing the supply of the special food resource at a subset of synchronized broods, and comparing fledging success to control asynchronous nests.

4.4.6. Hurry-Up Hypothesis

By beginning incubation during laying, first-hatched nestlings will hatch and fledge earlier than they would otherwise. This should increase the probability of nesting success for species which experience a dwindling food supply as the breeding season progresses (the *Hurry-up Hypothesis*), and would ensure that at least some young fledge (Hussell, 1972; Slagsvold, 1986a). The *Hurry-up Hypothesis* has been suggested to explain why many species hatch their eggs more asynchronously as the breeding season progresses (Hussell, 1972; Skagen, 1987).

This hypothesis requires a rapid decline in food availability with time. Few studies have documented such a decline, and most of these used parental provisioning rates as an index of food availability (e.g., Nisbet and Cohen, 1975; Skagen, 1987). Other measures, such as a decline in clutch size or fledging success, are also used, but may confound the effects of parental quality, territory quality, or thermoregulatory abilities of nestlings with food availability (van Balen and Cavé, 1970; Bryant, 1978; Newton and Marquiss, 1984; Skagen, 1987; Stutchbury and Robertson, 1988). Ideally, the degree of hatching asynchrony should be correlated with a direct measure of food availability. Bryant (1978) documented a seasonal decline in the food supply of House Martins (*Delichon urbica*), but found no correlation with hatch spread.

This hypothesis is probably irrelevant for species that lay multiple clutches during a breeding season without any variation in the degree of asynchrony. For example, Green-rumped Parrotlets are completely asynchronous and can raise up to three broods per season (Beissinger and Waltman, 1991; Waltman and Beissinger, 1992).

4.4.7. Sibling Rivalry Hypothesis

Asynchronous hatching may serve to impose a stable dominance hierarchy on a brood. This would avoid wasteful scramble competition (Hahn, 1981) and make more efficient use of parental resources regardless of the food supply (the *Sibling Rivalry Hypothesis*). Hahn found that natural, asynchronous broods of the Laughing Gull (*Larus atricilla*) had a higher fledging success than artificially synchronized broods. She suggested a reduction in sibling competition may have contributed to this result, although no measures of sibling rivalry were reported.

This hypothesis assumes that a stable dominance hierarchy among nestlings is facilitated by hatching asynchrony. However, stable dominance orders can be established in synchronously-hatched broods (Bancroft, 1984; Groves, 1984; Magrath, 1990), whereas the dominance hierarchies imposed by asynchronous hatching are not necessarily stable (Greig-Smith, 1985).

The hypothesis predicts that asynchrony reduces the level of sibling competition. The only direct support for this hypothesis comes from studies of Cattle Egrets (*Bubulcus ibis*), where experimentally synchronized broods fought more often than asynchronous broods (Mock and Ploger, 1987). Also, parent American Kestrels provisioned synchronized broods at higher rates than asynchronous broods, suggesting that asynchrony may reduce energy expenditures of nestlings (Wiebe and Bortolotti, 1994b). However, direct measures of energy expenditure of nestling bee-eaters (*Merops viridis*) determined that daily energy costs were lower in broods with same-sized chicks than in broods with a pronounced size hierarchy (Bryant and Tatner, 1990). This suggests that a greater degree of synchrony may be more energetically efficient.

Most sibling rivalry is nonaggressive, and is manifested through begging or maneuvering for preferred positions within the nest (Rydén and Bengtsson, 1980; Lamey and Mock, 1991; Redondo and Castro, 1992; McRae *et al.*, 1993). Often, the average begging level of a brood appears to determine the rates of parental feeding (Bengtsson and Rydén, 1983; Hussell, 1991; Smith and Montgomerie, 1991; Redondo and Castro, 1992). Parents generally tend to feed the largest or closest nestling (Rydén and Bengtsson, 1980; Bengtsson and Rydén, 1981, 1983; Nuechterlein, 1981; Smith and Montgomerie, 1991; but see Stamps *et al.*, 1985). Therefore, the smallest nestlings in asynchronous broods normally are often not fed, and expend more time and energy than their nestmates in begging and maneuvering, i.e., sibling rivalry (Bengtsson and Rydén, 1983; Redondo and Castro, 1992).

For the *Sibling Rivalry Hypothesis* to be viable, the costs inherent to maintain high levels of begging in smaller nestlings must be offset by benefits to older nestlings, probably through reduced begging costs. The hypothesis predicts that the total energy expenditure of synchronously-hatched broods will be greater than that of asynchronously-hatched broods because of the extra energy expended on scramble competition. Begging rates can be used as an index of energy expenditure in species without overt aggression, or energy expenditure can be measured directly using the doubly-labeled water method (Lifson and McClintock, 1966; Speakman and Racey, 1988).

Ricklefs (1993) proposed that complete asynchrony may eliminate the evolutionary consequences of sibling competition. If hatching completely asynchronously does in fact predetermine the dominance hierarchy among nestlings, then there should be no selection for accelerated embryo growth and earlier hatching. Prolonged incubation periods may result. There may be fitness advantages to prolonged incubation periods that are as yet unidentified. Ricklefs proposed a possible link between slow embryonic growth, longevity, and delayed senescence, mediated through extended maturation of the immune system (Ricklefs, 1992, 1993). Thus, complete asynchrony may be favored when fitness advantages can be gained through prolonged embryonic development and consequent effects on other life-history characters.

4.4.8. Larder Hypothesis

Later-laid eggs may represent a food larder for the oldest chicks (the *Larder Hypothesis;* also called the *Ice-box Hypothesis,* Alexander, 1974). By investing energy in extra eggs during the laying stage, parents may reduce their energetic demands later in the nestling stage because older chicks could eat their younger sibs (Murton and Westwood, 1977). However, laying, incubating, and feeding the last young would involve a significant waste of energy (Magrath, 1990). In addition, the amount of food energy contained in the extra nestlings is likely to be insignificant compared to both the total amount of food required by older nestlings through the entire nestling cycle, and to the amount of energy invested by parents in maintaining the extra nestlings until needed (Magrath, 1990). Although obligate siblicide is widespread in raptors (Stinson, 1979), skuas (Spellerberg, 1971), and pelicans (Drummond, 1987) cannibalism by nestlings is rare (Anderson, 1990; Magrath, 1990; Ploger, 1992; Stanback and Koenig, 1992; but see Bortolotti *et al.*, 1991; Viñuela, 1991). It is noteworthy that both a chrysomelid beetle (*Labidomera clivicollis*) and a land snail (*Arianta arbustorum*) lay large clutches that hatch asynchronously in which the first-hatched larvae cannibalize their younger sibs (Eickwort, 1973; Baur and Baur, 1986). In these species, there is no parental care beyond egg-laying, so laying extra eggs to act as a larder may be the only mechanism for females to increase investment in their first-hatched offspring.

This hypothesis is unlikely to be relevant for birds. It should be considered only for species that show regular cannibalization of youngest chicks by their older sibs. The hypothesis predicts that older chicks will eat younger ones, at least during periods of food shortage. The amount of food that parents can supply is likely to have a limit that is below the peak demand in the nestling period. Therefore the hypothesis predicts that parents should be unable to successfully raise broods if the extra nestlings are removed before being consumed.

4.5. Factors Influencing the Timing of the Entire Nesting Cycle

The onset of incubation determines the lengths of the laying and fledging periods in the nesting cycle (Fig. 1). Parents may manipulate the lengths of these periods to maximize benefits by reducing the probability of total nest failure or their own depredation. Alternatively, females may manipulate the length of these periods to minimize their own share of parental care at the expense of their mates.

4.5.1. Sexual Conflict Hypothesis

Slagsvold and Lifjeld (1989b) proposed a unique interpretation of asynchronous hatching as a result of parental behavior based on sexual conflict (the *Sexual Conflict Hypothesis*). Because a parent may attempt to minimize its share of parental care at the expense of its mate (Trivers, 1972), a conflict could result over the degree of parental investment made by either sex. Initiating incubation before the end of egg-laying may be a tactic that permits females to increase the parental effort of their mates. Early incubation would increase the time that females are on eggs, and asynchronous hatching would lengthen the time that females brood young nestlings. Thus, the total time that males must provision the female would be greater in asynchronously hatching broods than if eggs hatched synchronously. In addition, the early presence of nestlings may reduce a male's opportunity to attract a second female (Slagsvold and Lifjeld, 1989b).

The *Sexual Conflict Hypothesis* assumes that incubation is less energy demanding than alternative activities. Experimental evidence generally supports this supposition (Walsberg and King, 1978; Ricklefs and Hussell, 1984; Westerterp and Bryant, 1984), except for species at high latitudes that experience low ambient temperatures during incubation (Biebach, 1981; Moreno and Carlson, 1989). This hypothesis applies only to species with uniparental incubation where males provide food for incubating and brooding females (Slagsvold and Lifjeld, 1989b). However, incubation feeding by males may enhance the reproductive success of both sexes, and may indicate sexual cooperation rather than conflict (Lyon and Montgomerie, 1985; Lifjeld and Slagsvold, 1986; Alatalo *et al.*, 1988). Males of some raptors provide most of the food for females before and during laying (Poole, 1985;

Beissinger, 1987; Bortolotti and Wiebe, 1993). Yet incubation may not be initiated with the first egg, again suggesting cooperation between the sexes rather than female manipulation. Cooperation between the sexes is especially likely in species with long-term monogamous pair bonds, in which reproductive success is frequently correlated with the length of association of pairs (Mader, 1982; Mock, 1985a; Alatalo *et al.*, 1988; Bradley *et al.*, 1990; Emslie *et al.*, 1992).

The *Sexual Conflict Hypothesis* makes several testable predictions. Females may continue brooding beyond the point that the brood is capable of effective thermoregulation. Effective thermoregulatory ability can be determined for broods of different ages using the methodology of Dunn (1975) and Ricklefs (1987). Also, females at asynchronous broods should expend less energy than females at synchronous broods, and males at asynchronous broods should be manipulated into expending more energy than males at synchronous broods. Energy expenditures can be estimated using feeding rates as an index (Hébert, 1993b), or measured directly using the doubly-labeled water method (Lifson and McClintock, 1966; Speakman and Racey, 1988). Mass loss in females may be used as an index of parental effort (Bryant, 1988; Slagsvold and Lifjeld, 1989b), unless mass loss is part of an adaptive parental strategy (Norberg, 1981; Ricklefs and Hussell, 1984; Gaston and Jones, 1989; Moreno, 1989b).

Two tests of this hypothesis have been published. Females at asynchronous broods appeared to expend less energy than females at synchronous broods in both Pied Flycatchers and Yellow Warblers (Slagsvold and Lifjeld, 1989b; Hébert, 1993b). However, the patterns of effort by males of both species did not support the predictions. Thus, neither test fully supported the premise that there is conflict between the sexes. Recently, Slagsvold *et al.* (1994) showed that in Blue Tits (*Parus caeruleus*), female parents of broods with a reduced degree of asynchrony had a higher subsequent survival rate than those with more asynchronous broods. The opposite result was found for males. Because females appear to benefit from synchronous hatching, these results contradict the original hypothesis. However, they demonstrate that sexual conflict over the timing of hatching can exist, even if the reasons for the conflict are unclear.

4.5.2. Nest Failure Hypothesis

Relative to synchronous hatching, asynchronous hatching reduces the amount of time that a nest contains only eggs, and reduces the amount of time before the first chick fledges, but increases the total time a nest contains nestlings. Hussell (1972) suggested that asynchronous hatching is an adaptation to minimize total nest loss due to predation. The *Nest Failure Hypothesis* was formulated as a model to predict the optimal degree of asynchrony based on the probability of total nest failure during different phases of the nesting cycle (Clark and Wilson, 1981, 1985). In this model it is not the absolute rate of nest failure that is important, but rather the nest failure ratio: the ratio of the daily probability of nest failure during the nestling period to the daily probability of failure during the egg period (NFR, Clark and Wilson, 1981). Because the time from the onset of incubation to the first hatch and from the first hatch to the first fledge does not depend on when incubation is begun (Fig. 1), the relevant model parameters are the probability of nest loss during laying and during fledging (Hussell, 1985). A greater probability of nest failure during fledging favors increased synchrony, while an equal or greater probability of failure before incubation is begun favors increased asynchrony. The model predicts that even in the absence of brood reduction, most species should begin incubation before the last egg is laid.

This hypothesis cannot be tested experimentally, but nest failure rates can be measured for the different periods of the nesting cycle. The observed degree of asynchrony in Snow Buntings (*Plectrophenax nivalis*) was consistent with that predicted by the model (Hussell, 1985). Bancroft (1985) modified the hypothesis to include partial brood losses; his model predicted greater levels of asynchrony than he observed in Boat-tailed Grackles (*Quiscalus major*). Least Flycatchers (*Empidonax minimus*) and Yellow Warblers also hatched their eggs more synchronously than predicted by the model (Briskie and Sealy, 1989; Hébert and Sealy, 1993). It is unclear in these tests if the differences between the predictions of the model and observed hatching patterns are because the estimates of model parameters are wrong, or because the hypothesis has been falsified (Magrath, 1990). Recent work by Murray (in press) questions the validity of some of the assumptions of this model.

4.5.3. Adult Predation Hypothesis

The risk of predation on incubating or brooding adults may affect the optimal time to initiate incubation (Hussell, 1972; Weathers and Sullivan, 1989). Magrath (1988) included the effects of predation on adults in the Nest Failure model (Clark and Wilson, 1981, 1985; Hussell, 1985) to show that the rate of nest-content loss, the risk to incubating adults, and the probability of adults breeding again can interact to affect the timing of the onset of incubation (the *Adult Predation Hypothesis*). As in the *Nest Failure Hypothesis*, the relative rates of survival during different phases of the nesting cycle determine the optimal hatching pattern (Magrath, 1988). The two crucial values are the rate of survival of females during laying before and after incubation starts. A lower probability of survival while incubating should cause birds to delay the onset of incubation to minimize the time spent incubating, and would cause birds to hatch their eggs more synchronously than otherwise expected.

Like the *Nest Failure Hypothesis,* this hypothesis cannot be tested experimentally. The rates of nest failure and survival of females must be calculated and used in the model. To date, this hypothesis has not been tested, perhaps because it is difficult to determine survival rates of females.

4.6. Summary of Hypotheses

The categorization of hypotheses in Table I illustrates that a single species may be subject to multiple, potentially conflicting, selection pressures in different periods of the nesting cycle that affect patterns of incubation and hatching. The actual patterns observed in any species represent a compromise between the relevant intrinsic and extrinsic selective influences on both incubation and hatching patterns. The great variety of hatching patterns that exists in birds results from differences in the costs and benefits derived from these factors among taxa with different life history strategies.

Not all of the hypotheses should be considered of equal importance. Some will only apply to a subset of species with particular natural history traits. More importantly, however, different factors represented by various hypotheses affect reproductive success in a hierarchical manner. For example, for factors affecting nestling survival to be relevant, eggs must survive until hatching. Patterns of incubation that promote the survival of eggs may not necessarily produce hatching patterns that maximize fledging success. Therefore, factors that promote egg survival may constrain the influence of those that affect nestling survival.

The various hypotheses reviewed above differ in how amenable they are to experimental testing, to being compared, and to being incorporated into a model. The *Nest Failure* and *Adult Predation* hypotheses are not testable by experimental manipulation. However, they are the only hypotheses that yield quantitative predictions. The *Egg Protection, Limited Breeding Opportunites, Hurry-up,* and *Larder* hypotheses are difficult to test. Some hypotheses, such as the *Egg Viability Hypothesis,* can be tested singly, such that other factors are excluded. For many others, the proposed methodology cannot isolate single factors. For example, the differences in fledging success between synchronized and asynchronous broods may be due to any or all of the hypotheses based on adaptive hatching patterns (section 4.4.). Furthermore, the effects of different factors are measured using different currencies. Most can be assessed in terms of how they may affect the expected number of fledglings per brood, but the *Limited Breeding Opportunities* hypothesis deals with breeding versus not breeding, the *Sexual Conflict* hypothesis is based on benefits to just females in breeding pairs, and the *Adult Predation* hypothesis is based on parental survival. It may be difficult to compare or model the costs and benefits of factors if their effects are not expressed in the same units.

5. FINDINGS FROM EXPERIMENTAL TESTS

5.1. Experimental Designs

Although asynchronous hatching is a common phenomenon among altricial birds, there have been relatively few experimental tests of the hypotheses compared to the number of observational studies (Clark and Wilson, 1981). Magrath (1990) and Amundsen and Slagsvold (1991b) summarized results of 29 manipulative experiments designed to test hypotheses for the evolution of hatching asynchrony. Several more studies have been published or presented since these reviews (Beissinger and Stoleson, 1991; Seddon and van Heezik, 1991b; Stanback, 1991; Bowman, 1992; Harper *et al.*, 1992; Hébert and Sealy, 1992, 1993; Ploger, 1992; Bortolotti and Wiebe, 1993; Hébert, 1993a,b; Slagsvold *et al.*, 1994; Wiebe and Bortolotti, 1994a,b).

Nearly all experimental studies have focused on adaptive hatching patterns (Table I). These typically employed an experimental design that synchronized broods by moving either eggs or nestlings among nests (35 studies). Success at experimentally synchronous nests was compared to nests with a natural degree of hatching asynchrony. Six studies included results from nests manipulated to have an exaggerated degree of hatching asynchrony. Most experimental studies have explicitly or implicitly tested the *Brood Reduction Hypothesis* (Table I), although many did not state any hypotheses to be tested *a priori*. Many studies have simply described the outcome of synchronization of hatching without testing any specific predictions.

5.2. Summary of Results

All of the hypotheses based on adaptive hatching patterns (Table I) predict equal or greater fledging success for nests with a natural degree of hatching asynchrony than experimentally synchronized nests.

Fledging success of experimentally synchronous nests was frequently equal to or greater than that of asynchronous broods (Table II). Only one study reported consistently greater success in asynchronous broods (Hahn, 1981). In another, asynchronous broods fledged more young in one of two years (Hébert, 1993a). However, the *Brood Reduction Hypothesis* predicts greater success from asynchrony only when food is limited (Magrath, 1989; Pijanowski, 1992). Two studies manipulated food availability. Synchronous broods of Eurasian Blackbirds showed lower fledging success when food was limited (Magrath, 1989), but there was no similar trend in captive Zebra Finches (*Poephila guttata*; Skagen, 1988). Three studies monitored natural food supplies. Fledging success of synchronous broods of White-crowned Pigeons (*Columba leucocephala*) and House Wrens (*Troglodytes aedon*) did not differ from that of asynchronous broods during periods of food shortage (Bowman, 1992; Harper et al., 1992). Synchronized pigeon broods fledged more young than asynchronous broods when food was abundant (Bowman, 1992).

Hypotheses based on food limitations or parental efficiency (Table I) predict that synchronously-hatched nests should experience lower growth rates and lower fledging masses of nestlings, increased parental effort, and assume a greater age of mortality for those chicks that die

TABLE II
Summary of the Results of 35 Experimental Tests of Hatching Asynchrony that Compared Asynchronous with Synchronized Broods. Numbers indicate the number of studies in each category. S = Synchronous, A = Asynchronous. References for experimental tests are annotated with "(T2)" in the Literature Cited. Modified after Magrath (1990) and Amundsen and Slagsvold (1991b).

	Relative outcome of experiment			
Measure	S = A[a]	S > A	S < A	Not reported
Number of young fledging	15.5[b]	17.5[b]	2	0
Post-fledging survival	4	0.5[b]	0.5[b]	30
Age of nestling mortality	6	9	0	20
Fledging weight	18	2	6	9
Growth rate	11	3	0	21
Parental effort	4	2	0	29
Parental survival	1[c]	0	0	34

[a]Includes small differences that were not statistically significant.
[b]Opposite effects in good and bad years are counted as 0.5 for each.
[c]Overall survival equal, but sexes differed by treatment (see text).

compared to nestlings in asynchronous broods. None of the 14 experimental studies that measured growth rates found significant differences between average growth rates of young from synchronous and asynchronous broods of equal size (Table II). Of 26 studies that reported prefledging masses of nestlings, most (69%) found no significant differences between masses of nestlings from synchronous and asynchronous broods. One found that only male nestlings from synchronous broods were significantly lighter (Howe, 1976). Only five studies reported postfledging survival, and four found no differences based on synchrony (Hébert and Barclay, 1986; Beissinger and Stoleson, 1991; Harper *et al.*, 1992; Slagsvold *et al.*, 1994); the fourth found that young fledged from asynchronous broods had a higher survival rate than young fledged from synchronous broods in a food-poor year, and the opposite in a food-abundant year (Magrath, 1989). Only six studies have measured parental effort, using either feeding rates (Fujioka, 1985; Héberg and Barclay, 1986; Mock and Ploger, 1987; Beissinger and Stoleson, 1991; Bowman, 1992) or using female mass loss (Amundsen and Slagsvold, 1991a). Two of these studies found higher feeding rates for parents of synchronous broods of Cattle Egrets (Fujioka, 1985; Mock and Ploger, 1987), but the others found no differences (Table II). Only one study measured the effects of experimental synchrony on the subsequent survival of adults and found no differences when sexes were pooled, but significant differences between the sexes (Slagsvold *et al.*, 1994). In 60% of the 15 studies that included information on the timing of offspring mortality, chicks died in synchronous nests at a later age than did chicks in asynchronous nests; the remainder reported no significant differences. Overall only three experimental tests unambiguously support predictions of the *Brood Reduction Hypothesis* (Hahn, 1981; Magrath, 1989; Hébert, 1993a), although the methodology and analyses used in Hahn's study make her results difficult to interpret (Table II).

We conclude that most of these experiments provide little evidence that early incubation and asynchronous hatching confer advantages for offspring survival after hatching. Asynchronous and synchronous broods generally produced equal numbers of fledglings, of similar quality, with similar degrees of parental care. However, the absence of supporting evidence does not constitute rejection of the *Brood Reduction Hypothesis*, especially since the hypothesis may have been misapplied to species that do not experience unpredictable food supplies.

Few hypotheses other than *Brood Reduction* have been tested experimentally (Table I). The nest failure model produced predictions congruent with observed hatching patterns in only one of four studies

(Bancroft, 1985; Hussell, 1985; Briskie and Sealy, 1989; Hébert and Sealy, 1993). Various studies have supported predictions of the *Insurance Egg Hypothesis* (Anderson, 1990; Bollinger *et al.*, 1990), the *Egg Viability Hypothesis* (Viñuela, 1991; Veiga, 1992; Stoleson and Beissinger, in prep.); and the *Sibling Rivalry Hypothesis* (Wiebe and Bortolotti, 1994b). Predictions were not supported in tests of the *Sibling Rivalry Hypothesis* (Bollinger *et al.*, 1990), the *Sexual Conflict Hypothesis* (Slagsvold and Lifjeld, 1989b; Hébert and Sealy, 1993; Slagsvold *et al.*, 1994), or the *Peak Load Reduction Hypothesis* (Beissinger and Stoleson, 1991; Wiebe and Bortolotti, 1994b). Unfortunately, there are few predictions exclusive to a single hypothesis (Magrath, 1990), and in most studies alternative hypotheses could not be eliminated.

5.3. Problems with Experimental Tests of Hypotheses for Hatching Asynchrony

Of the 39 experimental tests of hatching asynchrony, 34 have been confined primarily to colonial water birds with small clutch sizes (12 studies), and passerines with slight degrees of hatching asynchrony (23 studies). Little experimental work has been done on groups in which asynchrony is the norm, such as raptors, grebes, gruiformes, coraciiformes, or psittacines (one study each of a bee-eater, a raptor, a parrot, a pigeon, and a swift). In addition, almost all studies have been conducted with temperate zone species. Tropical species are thought to have lower energetic requirements (Weathers, 1979; Bryant and Hails, 1983), slower growth (Ricklefs, 1976), lower prevalence of blood parasites (Ricklefs, 1992), and higher nest failure rates (Ricklefs, 1969), and consequently are probably subject to different selective pressures than temperate species.

Only 14 studies manipulated asynchronous broods to control for the possible effects of experimental manipulation. Negative effects of brood manipulations, such as decreased attentiveness, lower egg hatchability, and lower feeding rates, have been demonstrated (Bryant and Tatner, 1990; Götmark, 1992). Therefore caution should be used when interpreting results from studies that compared unmanipulated asynchronous broods to manipulated synchronous broods. Most studies have not quantified important measures, such as nestling growth rates, postfledging survival, and parental effort, that are crucial to assess the effects of asynchronous hatching on parental investment and reproductive success.

Finally, sample sizes in most experimental studies have been small, and therefore have lacked sufficient statistical power to detect any but gross differences between treatments. This problem is not confined to hatching asynchrony (Peterman, 1990; Graves, 1991; Taylor and Gerrodette, 1993). Asynchrony studies have typically employed sample sizes of 25 or less for synchronized broods (median sample size = 21). Testing for differences in the means of two treatments with equal variances with a student's *t*-test (two-tailed) using $\alpha = 0.05$ would have a statistical power of 0.11 assuming a "small" but real effect of treatment. A small effect as defined by Cohen (1988) signifies an approximate 15% nonoverlap of the distributions of synchronous and asynchronous scores. In other words, sample sizes of 25 would give an 11% probability of detecting a real difference between treatments. Therefore, failure to detect a difference between treatments would be ambiguous. Statistical power would increase slightly to 0.41 for a "medium" effect (approximately 33% nonoverlap of distributions; Cohen, 1988). Cohen (1988) proposed a statistical power of 0.80 as a convention for determining adequate sample sizes; this equates to a 20% probability of accepting a false null hypothesis (Type II error). In the example above, the sample sizes necessary to obtain a power of 0.80 are 310 for a small effect and 50 for a medium effect. Few experimental tests of hatching asynchrony have had sample sizes over 50 for both treatments (Slagsvold, 1982, 1986a; Bowman, 1992; Harper *et al.*, 1992).

6. SYNTHESIS: INTEGRATING MULTIPLE HYPOTHESES THROUGH MODELING

6.1. Prior Efforts to Integrate Multiple Hypotheses

Numerous authors have suggested that the broad array of hatching patterns exhibited by birds may be the result of trade-offs between multiple factors affecting the onset of incubation and hatching patterns. Clark and Wilson (1981, 1985) suggested that observed hatch spreads may represent a trade-off between nest predation and brood reduction. Mock and Parker (1986) postulated both brood reduction and insurance functions of smallest nestlings. Others have recognized the probability that multiple factors interact to produce observed hatching patterns (e.g., Bollinger *et al.*, 1990; Viñuela, 1991; Bowman, 1992; Veiga and Viñuela, 1993).

Few researchers have attempted to integrate the effects of multiple factors on hatching patterns. In a different context, Arnold *et al.* (1987) combined the effects of egg viability and risk of nest loss to determine constraints on clutch size in waterfowl. The risk of predation of adults was combined with the nest failure model by Magrath (1988). Stouffer

(1989) created a stochastic model based on empirical data to evaluate the effects of nest failure, brood parasitism, egg removal, and brood reduction on reproductive success in Starlings.

Recently, Konarzewski (1993) created a mathematical model that includes the influence of hatching asynchrony, hatching failure, nestling failure, brood reduction, and environmental variation on clutch size. This model is a useful heuristic tool to understand the joint effects and relative strengths of different factors on reproductive success. However, its general applicability is limited for several reasons. The model is based on incubation beginning after a base clutch is laid, and examining the effects of laying additional eggs that hatch progressively more asynchronously. This pattern may apply to some species, such as the Dark-eyed Junco (Smith, 1988), but may be inappropriate for others in which asynchrony is not a function of clutch size, or for those that begin incubation on the first egg. Egg hatchability was considered only in terms of how it may affect the insurance value of extra eggs. The model does not consider the fact that viability of eggs may be a function of the onset of incubation. Some of the required parameters of the model are difficult to measure empirically, such as the additional effort expended by parents prior to the death of extra chicks. Konarzewski included environmental variability in the model as the proportion of "good" years. He defined good food years as those years in which an extra nestling survives even when all of its older sibs hatch. Thus, the model ignores mortality of last-hatched nestlings due only to size differences, though empirical evidence suggests this is a widespread cost of hatching asynchrony (Stanback, 1991). Furthermore, Konarzewski suggested that his model shows environmental variation has a significant effect on the fitness benefits of extra, asynchronously-hatched eggs. This conclusion is hardly surprising since environmental variation was defined in terms of its effects on fitness. Thus, his model is of limited value in evaluating the relative effects of factors influencing the onset of incubation among different species, and is mostly irrelevant for the interpretation of results of experimental tests of hypotheses for hatching asynchrony.

6.2. Modeling Trade-offs in the Onset of Incubation

A clear methodology is needed to integrate the explicit trade-offs between factors affecting the onset of incubation. We advocate the use of stochastic models based on empirical data, such as that used by Stouffer (1989), to quantitatively assess the effects of multiple factors on fledging success. Because reproduction is a hierarchical process, transition probabilities of survivorship for each egg from one reproductive stage to the next (e.g., preincubation, incubation, brooding, postbrooding, and fledging) can be determined in relation to the effects on survivorship of the onset of incubation. Perhaps the most readily quantifiable factors that affect survivorship are: (1) the duration of exposure to eggs before incubation begins and its effects on nest predation and egg hatchability; (2) hatching spread and its effect on brood reduction; and (3) the time between the fledging of first and last chicks and the likelihood of nest predation (Fig. 5). Delaying incubation may cause a decline in the viability of first-laid eggs. But initiating incubation early results in an increased likelihood of brood reduction of last-hatched chicks (Fig. 6). Different aspects of nest failure are affected as incubation is delayed. Delaying incubation increases the time until fledging of the first young and so increases the chances of total brood failure. However, initiating incubation early causes nests to contain chicks for a longer period of time (Fig. 6). In many species nestlings may be more likely to be depredated than eggs, so this may increase the risk of nest failure.

The trade-offs that a parent makes when initiating incubation can be stochastically simulated by using transition probabilities to estimate the survivorship of individual eggs in a clutch and expected reproductive success under different scenarios for the onset of incubation. Survival probabilities from egg viability, brood reduction, and nest failure can be parameterized for each egg based on empirical data (Fig. 5).

In our model (Figure 7), each egg is given a probability of surviving from laying to hatching (P_h). This probability is the product of the probability of an egg remaining viable (P_v), derived from the relation between duration of exposure (*i*) and egg viability, and the probability of nest failure during the laying period (P_n). Each hatchling is given a

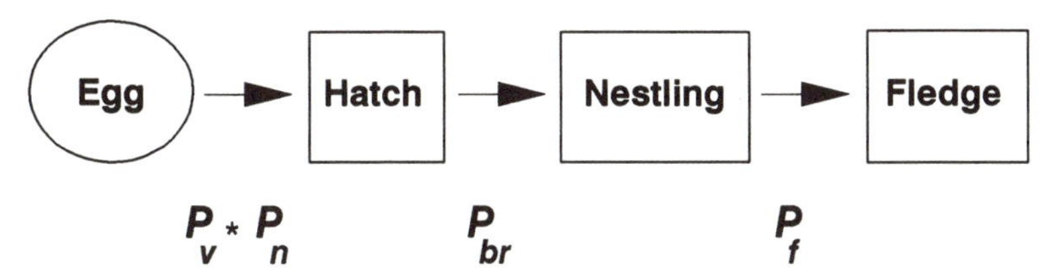

FIGURE 5. Transition probabilities used to construct a simple model of hatching asynchrony. P_v = the probability of an egg remaining viable before the onset of incubation. P_n = the probability of the nest surviving the period before the onset of incubation. P_{br} = the probability of a nestling surviving the period of brood reduction. P_f = the probability of a nestling surviving the period from the fledging of the first nestling until it fledges itself.

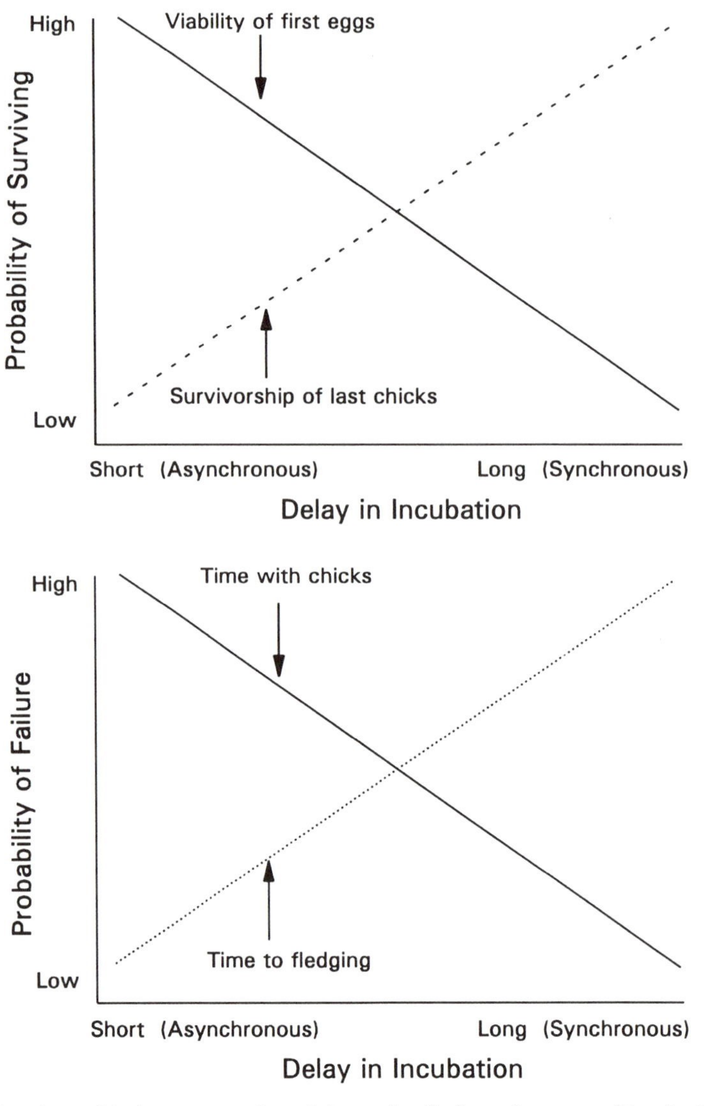

FIGURE 6. A graphical representation of the trade-offs from the onset of incubation on offspring survival. As synchrony increases, delaying incubation may cause a decline in the viability of first-laid eggs, but may improve the likelihood of survival for last-hatched chicks. Delaying incubation increases the time until fledging of the first young, thus increasing the chances for total brood failure. However, delaying incubation also decreases the amount of time the nest contains chicks, which may decrease the probability of nest failure.

probability of surviving from hatching through the period of brood reduction (P_{br}). This probability is a function of the age difference between the current nestling and the first-hatched nestling (*a*). However, other conditions may influence it as well. The probability of surviving the brood reduction stage may also depend on the number of surviving older siblings, the total brood size, the sex of the individual nestling, or the food supply. The transitional probability P_{br} must be determined for individual species based on empirical data. The likelihood of total nest failure from the onset of incubation until the first fledging is ignored in the model, because those stages of the nesting cycle are constant in duration and unaffected by changes in the onset of incubation (Fig. 1). Those nestlings that survive the brood reduction period are given a probability of surviving the period of fledging (P_f, or P_4 of Clark and Wilson, 1981). This probability is a function of the interval between the fledging of the first young and the current young (*f*), and may be considered an index of predation pressure exerted on nestlings as a result of asynchrony.

In the stochastic simulation, the survival of each egg or chick is determined by drawing a random number from a binomial distribution with parameters based on the transitional probabilities for that particular stage of the nesting cycle. The number of young surviving to fledging from the brood is then tallied for each iteration (Fig. 7). The mean number of chicks that fledge can then be compared between different incubation regimes.

Transition probabilities can be either absolute measures of survival during the nesting stages, or relative measures. Relative measures indicate the effect of individual factors on survivorship. For example, Veiga (1992) reported hatching success for experimental and unmanipulated control eggs in his viability experiments. The reported hatching success of experimental eggs is an absolute measure of P_v, and includes hatching failures for causes other than loss of viability due to exposure to ambient temperatures. The difference in hatching success between experimental eggs and their controls is a relative measure, and represents that fraction of hatchability that is due to the viability effect. The use of relative rather than absolute measures involves a loss of quantitative accuracy for predictions of fledging success, but provides a better assessment of the relative importance of individual factors. Also when using relative measures, the use of appropriate experimental controls becomes critical.

It may be possible to incorporate other hypotheses into this modeling framework if their effects on offspring survival can be quantified. For example, Stouffer (1989) included the effects of brood parasitism

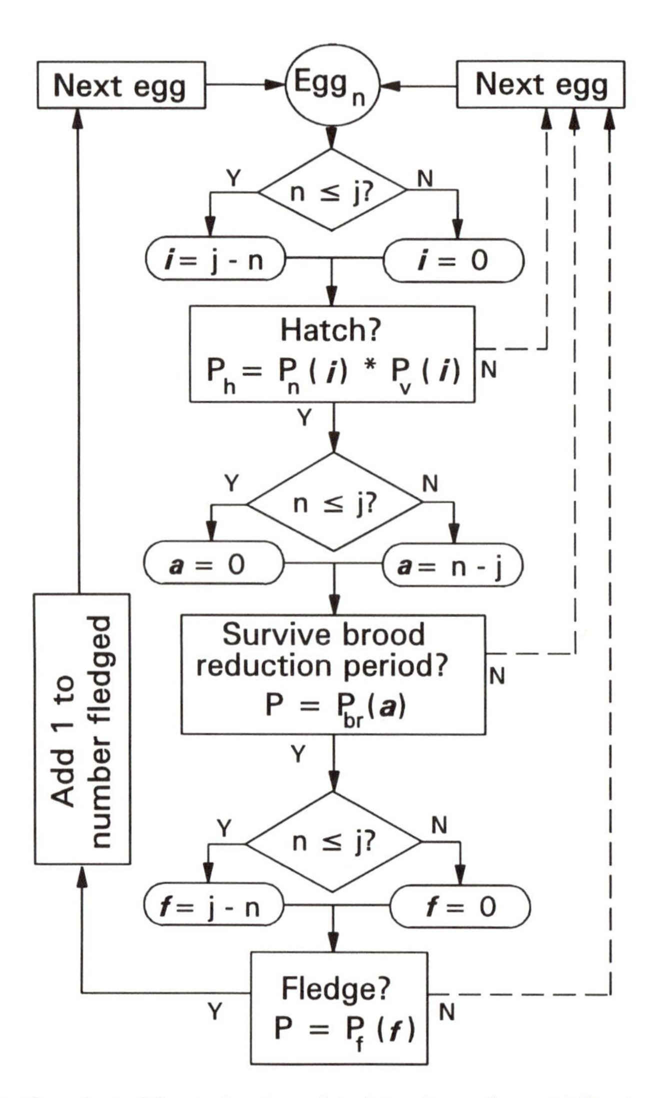

FIGURE 7. Flow chart of the stochastic model of the effects of egg viability, brood reduction, and total nest failure on breeding success in the House Sparrow. Each iteration determined the fate of all eggs in a clutch of five, based on initiating incubation on egg j. For each egg *n*, the interval *i* between the laying of the first egg and the *n*th egg was calculated as a function of j. The egg was assigned a probability $P_n(i)$ of escaping total nest failure during interval *i*, and a probability $P_v(i)$ of remaining viable as a function of the duration of *i*. These two probabilities were compared to two random numbers from binomial distributions. If both $P_n(i)$ and $P_v(i)$ were greater than their respective random numbers, then egg *n* hatched (Y). Otherwise, the egg did not hatch (N) and the model considered the next egg. If egg *n* hatched, the spread *a* between the hatch of egg *n* and the

on reproductive success. It may also be possible to incorporate environmental variation in the model if it can be defined in a noncircular manner. If successful, such a modeling procedure would allow the evaluation of the relative importance of various hypotheses for hatching asynchrony, and identify the critical period in the nesting cycle for the species in question. The model may also provide a framework for designing studies of and making predictions for other species.

6.3. Modeling Asynchrony: An Example Using the House Sparrow

We present an example that incorporates empirical data from the literature into the model to illustrate its parameterization and use. Unfortunately, adequate data do not exist for any single species. We chose the House Sparrow because it is the only altricial species for which the effects of incubation patterns on egg viability have been published (Veiga, 1992). These include separate values for the effects of exposure for nests initiated early and late in the season (Table III). We calculated the relative effect of exposure on hatchability as the difference in hatchability between control and experimental eggs. Nest failure rates are taken from Table I of Clark and Wilson's review (1981) for a different population of sparrows. There are no data on the daily probability of nest failure prior to the onset of incubation or after the first fledging, so we used failure rates for incubation and rates from first hatch to first fledge, respectively (P_2 and P_3 of Clark and Wilson). The length of these stages are from Lowther and Cink (1992). Survival probabilities for the brood reduction stage were estimated from Veiga (1990) and Veiga and Viñuela (1993). These values for P_{br} are likely to be overestimates because the original data present the probability of survival as a function of hatch order rather than as a function of a, the hatch spread, and cover the entire nesting period rather than a well-defined brood reduction period. Our model used only the modal clutch size of five eggs.

---

first egg was calculated. The nestling was assigned a probability $P_{br}(a)$ of surviving the brood reduction period as a function of a. If $P_{br}(a)$ was greater than a random number from a binomial distribution, then nestlings survived (Y). Otherwise, the nestling died (N), and the model considered the next egg. For surviving nestlings, the interval f between the fledging of the first chick and chick n was calculated as a function of j. Chicks were assigned a probability $P_f(f)$ of escaping total nest failure during the fledging period as a function of f. If $P_f(f)$ was greater than a random number from a binomial distribution, then the nestling fledged (Y). Otherwise, the nestling died before fledging, and the model considered the next egg (N). The number of eggs that survive to fledge was tallied for each clutch.

TABLE III
Parameter Values Used in Stochastic Simulations of Breeding in the House Sparrow[a]

Interval i, a, or f	P_v Early	P_v Late	P_v Average	P_{br}	P_n	P_f
0	0.988	0.951	0.965	1.000	1.000	1.000
1	0.986	0.960	0.972	0.950	0.983	0.984
2	0.930	0.955	0.940	0.930	0.967	0.967
3	0.874	0.785	0.834	0.870	0.950	0.951
4	0.829	0.723	0.774	0.550	0.934	0.936

[a] P_v = the probability of an egg being viable; P_{br} = the probability of a nestling surviving the brood reduction period; P_n = the probability of nest failure prior to the onset of incubation, P_f = the probability of nest failure after the first fledge; i = the interval in days between the laying of an egg and the onset of incubation; a = the interval in days between the hatching of the first nestling and the hatching of the *nth* nestling; f = the interval in days between the fledging of the first nestling and the *nth* nestling.

Separate simulations of the model were performed beginning with incubation on each of the five eggs, and for both early and late season nests. All simulations were run for 1000 iterations. Results of the simulations were in the form of integer values of the expected number of young fledged. These values were compared among simulations differing in the onset of incubation using Chi-square tests (Manly, 1991).

Results of all simulations showed that the maximum number of fledglings was produced by initiating incubation on the third egg (Fig. 8). Using early season values for egg viability, initiating incubation on the second, third, or fourth eggs did not produce significantly different fledging success (pairwise comparisons, all $\chi^2 < 5.36$, df = 2, $P >$ 0.137). Fledging success produced by beginning incubation on the first or fifth egg differed from each other and from yields for other incubation patterns (pairwise comparisons, all $\chi^2 > 28.0$, df = 2, $P < 0.001$). For late season viability values, fledging success differed significantly among all incubation strategies (all pairwise comparisons, $\chi^2 > 9.5$, df = 2, $P < 0.01$). Early and late season viability values produced similar fledging success when incubation was begun on the first, second, or third eggs (pairwise comparisons, all $\chi^2 < 1.78$, df = 3, $P > 0.60$; Fig. 8). However, late season nests produced fewer fledglings than early season nests when incubation was begun on the fourth or fifth egg ($\chi^2 > 13.0$, df = 3, $P < 0.01$ for both). This was a result of the greater decrease in viability as air temperature increased later in the nesting season (Veiga, 1992).

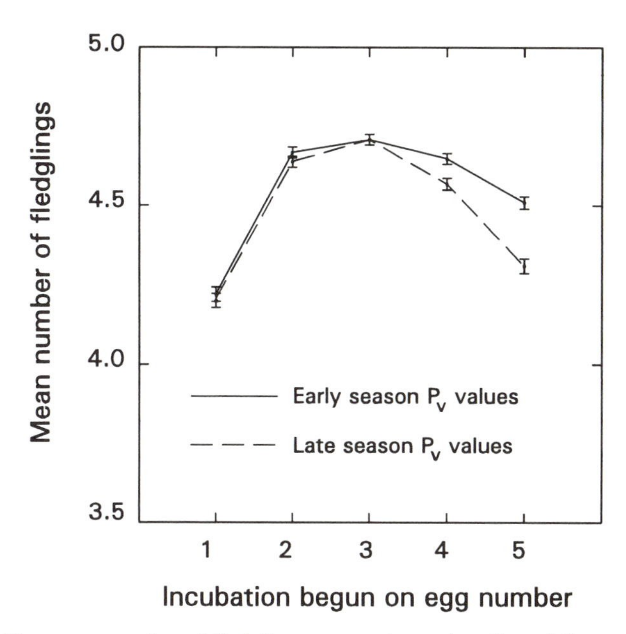

FIGURE 8. The mean number of fledglings per nest as a function of the egg on which incubation was begun, from a stochastic simulation of breeding in the House Sparrow. Simulations were run using viability probabilities (P_v) for early season and late season nests separately. All points represent the mean values of 1000 iterations, and bars represent standard errors of the means.

The relative importance of different factors in the simulation on reproductive success can be assessed by conducting a sensitivity analysis. The sensitivity of a parameter is measured as the proportional change in the expected yield of fledglings that results from changing the value of that parameter. We decreased each parameter by 10% for beginning on the first, third, and fifth egg (Table IV). For these simulations, we used an average of the early and late season viability values.

The impact of changing individual transition probabilities varied with the incubation strategy. Those parameters based on the period of egg laying before the onset of incubation, such as P_n and P_v values, had little effect on fledging success when incubation started on the first egg, and became progressively more important as the preincubation period lengthened. In contrast, P_{br} values became progressively less critical as incubation started later. In the House Sparrow, it appears that changes in nest failure rates (P_n and P_f) have little effect on overall success, perhaps because the original rates were very low, reflecting the species' preference for nesting in cavities.

TABLE IV
Sensitivity of the Expected Yield of Fledglings under Three Incubation Strategies to a 10% Decrease in Each Model Parameter[a]

	Incubation begun on:		
Parameter	Egg 1	Egg 3	Egg 5
P_n	0.0	>0.1	0.3
P_f	>0.1	>0.1	0.0
P_v			
E_1	10.3	6.1	2.3
E_2	0.0	2.1	2.3
E_3	0.0	2.1	1.9
E_4	0.0	0.0	1.7
E_5	0.0	0.0	1.1
P_{br}			
N_1	2.7	6.6	10.0
N_2	2.1	1.9	0.0
N_3	2.4	1.9	0.0
N_4	2.1	0.0	0.0
N_5	1.5	0.0	0.0

[a]Expressed as percentage change from the initial value for that strategy.

The effects of viability and brood reduction were of similar magnitude in this model (Table IV). This relatively even trade-off is the reason that the intermediate incubation strategy, beginning on the third egg, has the greatest reproductive success. With this strategy, the costs of delaying incubation, expressed as a decline in the viability of first-laid eggs, counterbalanced the costs of initiating incubation early, expressed as a reduced survivorship of later-hatched chicks (Fig. 6). The balanced nature of this trade-off in the House Sparrow has some empirical support. Compared to relatively synchronous broods, relatively asynchronous broods experienced greater hatching success and lower survivorship of later-hatched nestlings, but overall fledging success did not differ between the two groups (Veiga and Viñuela, 1993).

In the model, the probability of nestling N_i surviving the brood reduction period (P_{br}) was strictly a function of the difference in age between N_i and N_1. In other words, the death of a nestling within the brood was assumed to have no effect on the survival of its nestmates. This simplistic approach was used because no data on conditional survival probabilities were available for the House Sparrow. However, nestling survival may be affected by brood size as well as age differences within a brood (e.g., Ricklefs, 1965; Werschkul, 1979; Stouffer, 1989). In particular, if an older nestling dies, the smallest nestling may have a greatly increased chance of survival (but see Stanback, 1991). This "insurance" effect was added to the House Sparrow model by incorporating a facultative adjustment to the size rank of nestlings based on the fate of their older sibs. For example, in a completely asynchronous brood, if N_2 dies, then N_3, N_4, and N_5 are given the transition probabilities for N_2, N_3, and N_4, respectively. Simulations were run of this modified age and brood size dependent model using an average of early and late season viability values.

Results from the age and brood size dependent model showed two major changes from the age dependent model (Fig. 9). The mean number of fledglings produced was greater in the age and brood size dependent model, which included the insurance effect, than in the age de-

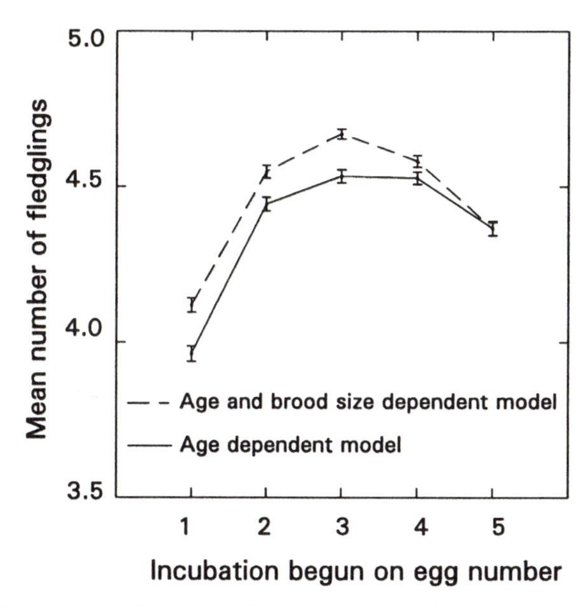

FIGURE 9. The mean number of fledglings per nest as a function of the egg on which incubation is begun, from a stochastic simulation of breeding in the House Sparrow using two different models of offspring survival during the brood reduction period. In the basic model, P_{br} was age-dependent. The second model was modified to include a facultative adjustment in the probability of surviving brood reduction based on the fate of older siblings, so that P_{br} was dependent on both age and brood size. See text for details of the models. Simulations used an average of early- and late-season viability values. All points represent the mean values of 1000 iterations, and bars represent standard errors of the means.

pendent model, for all incubation strategies except for starting on egg 5, when all nestlings had the same size rank. Because the probability of survival for younger nestlings increased if an older sib died, the inherent costs of brood reduction were reduced. When an average of early and late viability values were used, the age dependent model suggested similar fledging success when incubation was begun on egg 3 or 4 ($\chi^2 = 99.7$, $P = 0.53$; Fig. 9). Because the inherent costs of early incubation are mitigated in the age and brood size dependent model, beginning incubation on the third egg became a better strategy than beginning on the fourth egg ($\chi^2 = 202$, df $= 3$, $P < 0.001$). In general, when an insurance effect of clutch size can be demonstrated, early incubation becomes relatively more advantageous.

This approach of modeling multiple factors may provide a clear and consistent methodology for exploring the impact and interactions of different factors that potentially affect the onset of incubation and subsequent hatching patterns. It may help to identify the trade-offs that parents face in trying to maximize their reproductive success. To do so, however, will require that different factors or hypotheses (i.e., Section 4) can be measured in relation to their effects on egg or nestling survival. Not all hypotheses, however, are easily related or measured in those terms.

7. CONCLUSIONS

Asynchronous hatching of bird eggs has intrigued and puzzled ornithologists for decades, because of the *Paradox of Hatching Asynchrony:* parents primarily control the onset of development for eggs, yet appear to choose a pattern of hatching that frequently results in offspring mortality. The *Paradox of Hatching Asynchrony* has been the subject of considerable research, and 17 hypotheses have been proposed to account for this phenomenon (Table I). The *Brood Reduction Hypothesis* appeared to provide a solution. Most studies have tested this hypothesis or others that also seek an adaptive function for asynchronous hatching and the resulting nestling size hierarchy, i.e., those hypotheses for which the nestling period is the critical period of the nesting cycle (Table I). Such tests have provided little evidence that early incubation leading to asynchronous hatching confers advantages for offspring survival after hatching or reduces parental investments (Table II). These studies have ignored the physiological, environmental, and social constraints that directly affect the onset of incubation, and the possible adaptive significance of early incubation.

We feel it is important for studies of hatching asynchrony to refocus attention from the search for adaptive hatching patterns during the nestling period to the events surrounding the onset of incubation during egg-laying. Hatching patterns are primarily determined by the timing of the onset of incubation (Figs. 1 and 2). But the onset of incubation may reflect phylogenetic or physiological constraints. Asynchronous hatching appears to be a derived condition, and there is little variation in hatching patterns in many orders and families (Figs. 3 and 4). It is not clear if these similarities among species are due to similar selection regimes or because the traits are phylogenetically fixed. Hormonal controls of egg-laying and incubation behavior may circumscribe when incubation can begin. Likewise, fluctuations in the food available to laying females may also facultatively influence the initiation of incubation. These factors may constrain the range of possible incubation patterns and the resulting asynchrony of hatching patterns.

The onset of incubation involves a decision that results in trade-offs among promoting a parent's own survival, promoting survival of the embryos, and the costs and benefits inherent to a particular hatching pattern that affect nestling survival. There are a variety of benefits that birds may derive from the early onset of incubation, such as protecting eggs from adverse effects of the environment, predators, brood parasites, and reproductive interference. In many instances, it appears that asynchronous hatching may itself be the cost of initiating incubation early. This cost is manifested through increased offspring mortality due to age differences among offspring, from starvation, accidents, or passive competition among nestlings.

If birds derive benefits from early incubation itself, then the resulting nestling size hierarchy and the mortality of the smallest nestlings may simply be epiphenomena. One might then expect mechanisms to evolve to offset the competitive disadvantages experienced by smallest nestlings. Indeed, some species do exhibit such behaviors. In a number of species, one or both parents preferentially fed younger nestlings (Ferguson and Sealy, 1983; Stamps *et al.*, 1985; Gottlander, 1987; Sasvári, 1990). Maguari Stork (*Ciconia maguari*) parents fed their broods a variety of different-sized food items simultaneously; larger young tended to eat the largest items, allowing smaller young to feed on the smaller items (Thomas, 1984). No brood reduction occurs in this species despite an extreme disparity in nestling sizes. Finally, parents may even be more aggressive towards larger, more competitive chicks to allow smaller chicks more access to food (Leonard *et al.*, 1988). These observations are consistent with the premise that hatching asynchrony may be in part a consequence, and not the cause, of early incubation.

It has become increasingly obvious that no one hypothesis is sufficient to explain hatching patterns in all species. It is also likely that single species may be subject to multiple factors that influence the onset of incubation and the resulting hatching patterns. Variation in the onset of incubation among species is likely to be a function of the costs and benefits derived from these factors, and from their interactions. Thus, effects of the onset of incubation on survival of *both* eggs during the laying period (as a result of parental neglect) and chicks during the nestling period (as a result of size differences) must be considered, as shown in the House Sparrow examples (Figs. 8 and 9). Multiple factors should be tested, one at a time in a coordinated manner, to determine their effects on reproductive success as a function of different incubation strategies. Results should be integrated in models (e.g., Fig. 7) to evaluate the relative importance of each factor and the combinations that lead to greatest reproductive success. The effects on fitness of different factors affecting the onset of incubation may become evident only when considered jointly in this manner. If the interactions of different factors are simple enough, they can be evaluated analytically (e.g., Arnold *et al.*, 1987). More likely, interactions of different factors will be probablistic in nature, necessitating the use of stochastic models such as the one we presented.

Finally, important experimental results are most likely to be produced from studies of taxa where asynchrony is not merely a result of phylogenetic or physiological constraints. For example, it is not clear whether a small degree of asynchrony (24–48 hr), as is typical of passerines, represents an adaptation or is an incidental effect of the mechanics of egg-laying. Greater hatch spreads are more likely to indicate true adaptations. The majority of hatching asynchrony studies have been done with carnivorous or insectivorous species in temperate areas. Rapid advances are likely to occur by studying species with fundamentally different selective regimes. In taxonomic groups that exhibit variation in hatch spreads (e.g., the Rallidae), the variation in asynchrony among species may illustrate different factors affecting the onset of incubation in those species. Such taxa are especially amenable to comparative analyses. By rigorously assessing multiple factors in appropriate species and using modern comparative approaches, studies of hatching asynchrony will progress out of the "stagnant backwater of population biology" and into the mainstream.

8. SUMMARY

1. Birds are unique among animals in being able to influence the birthing intervals of their young through the timing of the onset incuba-

tion. However, many species hatch their young asynchronously, frequently resulting in reduced survivorship for later-hatched young. This is the *Paradox of Hatching Asynchrony.*

2. The *Brood Reduction Hypothesis* provided a resolution to the paradox by suggesting an adaptive function to the offspring mortality that results from asynchrony. Experimental tests have provided little support, and 16 alternative hypotheses have been proposed, but few have been tested. Most experimental tests have not measured important parameters such as parental effort and postfledging survival. Many have lacked adequate controls or sufficient statistical power.

3. We divide the hypotheses for hatching asynchrony into four categories based on the effects of intrinsic or extrinsic factors during a critical period of the nesting cycle which constrains reproductive success. Hatching asynchrony could be simply the consequence of the early onset of incubation during egg-laying, either as a result of physiological constraints on incubation or because parents derive fitness benefits from the protective function of early incubation. During the nestling period, hatching asynchrony could be adaptive if it allowed parents to eliminate one or more nestlings selectively, or increased parental efficiency. Alternatively, parents could manipulate the duration of the different periods of the nesting cycle to maximize benefits.

4. Because the onset of incubation generally determines hatching patterns, we encourage refocusing attention from the search for adaptive hatching patterns during the nestling period to the events surrounding the onset of incubation during egg-laying. Many factors can affect when incubation is begun, including physiology, and interactions with the environment, predators, competitors, and mates.

5. Patterns of the onset of incubation are difficult to determine and to quantify, in part because many birds begin incubating gradually, or at night. In some species, the onset of incubation varies with clutch size, but not in others.

6. The onset of incubation is the principle proximate control of hatching patterns, but other factors, such as egg size, embryonic vocalizations, and time of year may also affect hatching patterns.

7. Synchronous hatching is the primitive condition in birds, and is widespread in the lower, primarily precocial taxa. Most altricial species hatch their eggs asynchronously, although some exhibit synchrony as a secondarily derived trait. Hatching patterns show wide variation within some orders and families.

8. Patterns of the onset of incubation and hatching in a species may reflect the influence of multiple factors. The relative importance of those factors may depend on the trade-offs associated with the poten-

tial benefits of early incubation to the survival of eggs and the potential costs to the survivor of later-hatching young associated with nestling size hierarchies.

9. The relative effects of multiple factors can be examined by integrating the results of empirical tests of single factors through modeling.

10. We demonstrated the use of a stochastic model by using empirical data from the House Sparrow. Results revealed the trade-offs inherent in the onset of incubation from differences in egg viability and nestling survivorship. An intermediate onset of incubation produced the greatest fledging success.

11. Other factors may be integrated into such models if they can be measured in terms of their effects on fledging success. Different factors, represented by different hypotheses, vary in how readily they may be modeled.

ACKNOWLEDGMENTS. We thank D. J. Anderson, G. Bortolotti, R. Bowman, P. N. Hébert, B. J. Ploger, M. S. Stanback, P. C. Stouffer, and K. Wiebe for generously providing unpublished manuscripts and data, and T. Amundsen and T. Slagsvold for providing reprints. The outstanding holdings of the Ornithological Library of the Peabody Museum facilitated our literature research. Our parrotlet work was funded by the Smithsonian Institution's International Environmental Sciences Program in Venezuela, the National Geographic Society, the Frank M. Chapman fund of the American Museum of Natural History, the American Ornithologists' Union, Sigma Xi, NSF grant IBN 9407349, and Yale University. Reviews and comments by Jerram Brown, Doug Mock, Martin Morton, Bonnie Ploger, Dennis M. Power, Tore Slagsvold, and Pamela L. L. Stoleson improved this paper.

REFERENCES

Alatalo, R., Gottlander, K., and Lundberg, A., 1988, Conflict or cooperation between parents in feeding nestlings in the Pied Flycatcher *Ficedula hypoleuca*, *Ornis Scand.* **19**:31–34.

Alerstam, T., and Högstedt, G., 1981, Evolution of hole-nesting in birds, *Ornis Scand.* **12**:188–193.

Alexander, R. D., 1974, The evolution of social behavior, *Annu. Rev. Ecol. Syst.* **5**:325–384.

Alvarez del Toro, M., 1971, On the biology of the American Finfoot in southern Mexico, *Living Bird* **10**:79–88.

Amundsen, T., and Slagsvold, T., 1991a, Asynchronous hatching in the Pied Flycatcher: an experiment, *Ecology* **72**:797–804. (T1 & T2)

Amundsen, T., and Slagsvold, T., 1991b, Hatching asynchrony: facilitating adaptive or maladaptive brood reduction?, *Acta XX Cong. Inter. Ornith.* **3**:1707–1719. (T1 & T2)

Amundsen, T., and Stokland, J. N., 1988, Adaptive significance of asynchronous hatching in the shag: a test of the brood reduction hypothesis, *J. Anim. Ecol.* **57**:329–344. (T1 & T2)

Amundsen, T., and Stokland, J. N., 1990, Egg size and parental quality influence nestling growth in the Shag, *Auk* **107**:410–413.

Anderson, D. J., 1989, The role of hatching asynchrony in siblicidal brood reduction of two booby species, *Behav. Ecol. Sociobiol.* **25**:363–368. (T1 & T2)

Anderson, D. J., 1990, Evolution of obligate siblicide in boobies. 1. A test of the insurance-egg hypothesis, *Am. Nat.* **135**:334–350. (T1)

Anthony, A. J., and Sherry, B. Y., 1980, Openbill Storks breeding in the southwestern lowveld of Zimbabwe Rhodesia, *Ostrich* **58**:1–6.

Arnold, T. W., 1993, Factors affecting egg viability and incubation time in prairie dabbling ducks, *Can. J. Zool.* **71**:1146–1152.

Arnold, T. W., Rohwer, F. C., and Armstrong, T., 1987, Egg viability, nest predation, and the adaptive significance of clutch size in prairie ducks, *Am. Nat.* **130**:643–653.

Astheimer, L. B., 1991, Embryo metabolism and egg neglect in Cassin's Auklets, *Condor* **93**:486–495.

van Balen, J. H., and Cavé, A. J., 1970, Survival and weight loss of nestling Great Tits, *Parus major,* in relation to brood size and air temperature, *Neth. J. Zool.* **20**:464–474.

Ball, G. F., 1991, Endocrine mechanisms and the evolution of avian parental care, *Acta XX Cong. Inter. Ornith.*: 984–991.

Balph, C. P., 1975, Life style of *Coccyzus pumilus*, a tropical cuckoo, *Condor* **77**:60–72.

Balthazart, J., 1983, Hormonal correlates of behavior, in: *Avian Biology,* Vol. 7, (D. S. Farner, J. R. King, and K. C. Parkes, eds.), Academic Press, New York, pp. 221–365.

Bancroft, G. T., 1984, Growth and sexual dimorphism of the Boat-tailed Grackle, *Condor* **86**:423–432.

Bancroft, G. T., 1985, The influence of total nest failure and partial losses on the evolution of asynchronous hatching, *Am. Nat.* **126**:495–504. (T1)

Baur, B., and Baur, A., 1986, Proximate factors influencing egg cannibalism in the land snail *Arianata arbustorum* (Pulmonata, Helicidae), *Oecologia* **70**:283–287.

Bednarz, J. C., and Hayden, T. J., 1991, Skewed brood sex-ratio and sex-biased hatching sequence in Harris' Hawks, *Am. Nat.* **137**:116–132.

Beissinger, S. R., 1987, Anisogamy overcome: female strategies in Snail Kites, *Am. Nat.* **129**:486–500.

Beissinger, S. R., and Bucher, E. H., 1992, Sustainable harvesting of parrots for conservation, in: *New World Parrots in Crisis: Solutions from Conservation Biology,* (S. R. Beissinger and N. F. R. Snyder, eds.), Smithsonian Institution Press, Washington, D.C., pp. 73–115.

Beissinger, S. R., and Stoleson, S. H., 1991, Nestling mortality patterns in relation to brood size and hatching asynchrony in Green-rumped Parrotlets, *Acta XX Cong. Inter. Ornith.*:1727–1733. (T1 & T2)

Beissinger, S. R., and Waltman, J. R., 1991, Extraordinary clutch size and hatching asynchrony of a neotropical parrot, *Auk* **108**:863–871.

Bengtsson, H., and Rydén, O., 1981, Development of parent-young interaction in asynchronously hatched broods of altricial young, *Z. Tierpsychol.* **56**:255–272.

Bengtsson, H., and Rydén, O., 1983, Parental feeding rate in relation to begging behavior in asynchronously hatched broods of the Great Tit *Parus major, Behav. Ecol. Sociobiol.* **12**:243–251.

Bennett, A. F., and Dawson, W. R., 1979, Physiological responses of embryonic Heermann's Gulls to temperature, *Physiol. Zool.* **52**:413–421.

Bent, A. C., 1961, *Life Histories of North American Birds of Prey*, Dover Publications, New York.

Biebach, H., 1981, Energetic costs of incubation on different clutch sizes in Starlings (*Sturnus vulgaris*), *Ardea* **69**:141–142.

Boag, D., 1982, *The Kingfisher*, Blanford Press, Dorset, UK.

Boag, D. A., and Schroeder, M. A., 1992, Spruce Grouse, in: *The Birds of North America*, No. 5, (A. Poole, P. Stettenheim, and F. Gills, eds.), The Academy of Natural Sciences, Philadelphia; The American Ornithologists' Union, Washington, D.C.

Bollinger, P. B., Bollinger, E. K., and Malecki, R. A., 1990, Tests of three hypotheses of hatching asynchrony in the Common Tern, *Auk* **107**:696–706. (T1 & T2)

Bolton, M., 1991, Determinants of chick survival in the Lesser Black-backed Gull: relative contributions of egg size and nestling quality, *J. Anim. Ecol.* **60**:949–960.

Bortolotti, G. R., 1986, Influence of sibling competition on nestling sex ratios of sexually dimorphic birds, *Am. Nat.* **127**:495–507.

Bortolotti, G. R., and Wiebe, K. L., 1993, Incubation behaviour and hatching patterns in the American Kestrel *Falco sparverius*, *Ornis Scand.* **24**:41–47.

Bortolotti, G. R., Wiebe, K. L., and Iko, W. M., 1991, Cannibalism of nestling American Kestrels by their parents and siblings, *Can. J. Zool.* **69**:1447–1453.

Bowman, R., 1992, Asynchronous hatching and brood reduction in the White-crowned Pigeon (*Columba leucocephala*), Ph.D. Dissertation, University of South Florida, Tampa. (T1 & T2)

Bradley, J. S., Wooller, R. D., Skira, I. J., and Serventy, D. L., 1990, The influence of mate retention and divorce upon reproductive success in Short-tailed Shearwaters *Puffinus tenuirostris*, *J. Anim. Ecol.* **59**:487–496.

Braun, B. M., and Hunt, G. L., Jr., 1983, Brood reduction in Black-legged Kittiwakes, *Auk* **100**:469–476.

Brawn, J. D., and Balda, R. P., 1988, Population biology of cavity nesters in northern Arizona: do nest sites limit breeding densities?, *Condor* **90**:61–71.

Breitwisch, R., 1989, Mortality patterns, sex ratios, and parental investment in monogamous birds, in: *Current Ornithology*, Vol. 6, (D. M. Power, ed.), Plenum Press, New York, pp. 1–50.

Briskie, J. V., and Sealy, S. G., 1989, Nest-failure and the evolution of hatching asynchrony in the Least Flycatcher, *J. Anim. Ecol.* **58**:653–665. (T1)

Brooks, D. R., and McLennan, D. A., 1991, *Phylogeny, Ecology, and Behavior*, University of Chicago Press, Chicago.

Brown, J. L., 1987, *Helping and Communal Breeding in Birds*, Princeton University Press, Princeton.

Brown, L., and Amadon, D., 1968, *Eagles, Hawks, and Falcons of the World*, McGraw Hill, New York.

Bryant, D. M., 1978, Establishment of weight hierarchies in the broods of House Martins *Delichon urbica*, *Ibis* **120**:16–26.

Bryant, D. M., 1988, Energy expenditure and body mass changes as measures of reproductive costs in birds, *Funct. Ecol.* **2**:23–34.

Bryant, D. M., and Gardiner, A., 1979, Energetics of growth in House Martins (*Delichon urbica*), *J. Zool., London.* **189**:275–304.

Bryant, D. M., and Hails, C. J., 1983, Energetics and growth patterns of three tropical bird species, *Auk* **100**:425–439.

Bryant, D. M., and Tatner, P., 1990, Hatching asynchrony, sibling competition and siblicide in nestling birds: studies of swiftlets and bee-eaters, *Anim. Behav.* **39**:657–671. (T1 & T2)

Bryant, D. M., and Tatner, P., 1991, Intraspecific variation in avian energy expenditure: correlates and constraints, *Ibis* **133**:236–245.

Burley, N., 1986, Sex-ratio manipulation in color-banded populations of Zebra Finches, *Evolution* **40**:1191–1206.

Butler, R. W., 1992, Great Blue Heron, in: *The Birds of North America*, No. 25, (A. Poole, P. Stettenheim, and F. Gill, eds.), The Academy of Natural Sciences, Philadelphia; The American Ornithologists' Union, Washington, D.C.

Caine, L. A., and Marion, W. R., 1991, Artificial addition of snags and nestboxes to slash pine plantations, *J. Field. Ornithol.* **62**:97–106.

Calder, W. A., and Calder, L. L., 1992, Broad-tailed Hummingbird, in: *The Birds of North America*, No. 16, (A. Poole, P. Stettenheim, and F. Gill, eds.), The Academy of Natural Sciences, Philadelphia; The American Ornithologists' Union, Washington, D.C.

Candy, M., 1984, Habits and breeding biology of the Great Blue Turaco, *J. East Afr. Nat. Hist. Soc. Natl. Mus.* **37**:1–19 (No. 180).

Cannon, M. E., Carpenter, R. E., and Ackerman, R. A., 1986, Synchronous hatching and oxygen consumption of Darwin's Rhea eggs (*Pterocnemia pennata*), *Physiol. Zool.* **59**:95–108.

Carey, C., 1980, The ecology of avian incubation, *BioScience* **30**:819–824.

Cargill, S. M., and Cooke, F., 1981, Correlation of laying and hatching sequences in clutches of the Lesser Snow Goose *Anser caerulescens caerulescens*), *Can. J. Zool.* **59**:1201–1204.

Cash, K. J., and Evans, R. M., 1986, Brood reduction in the American White Pelican (*Pelecanus erythrorhynchos*), *Behav. Ecol. Sociobiol.* **18**:413–418.

Clark, A. B., and Wilson, D. S., 1981, Avian breeding adaptations: hatching asynchrony, brood reduction, and nest failure, *Q. Rev. Biol.* **56**:253–277.

Clark, A. B., and Wilson, D. S., 1985, The onset of incubation in birds, *Am. Nat.* **125**:603–611.

Clutton-Brock, T. H., 1986, Sex ratio variation in birds, *Ibis* **128**:317–329.

Clutton-Brock, T. H., 1991, *The Evolution of Parental Care*, Princeton University Press, Princeton.

Cohen, J., 1988, *Statistical Power Analysis for the Behavioral Sciences*, 2nd ed., Lawrence Erlbaum Associates, Hillsdale, New Jersey.

Copi, I. M., 1972, *Introduction to Logic*, 4th ed., Macmillan, New York.

Courtney, J., and Marchant, S., 1971, Breeding details of some common birds in south-eastern Australia, *Emu.* **71**:121–133.

Cramp, S., 1977, *Handbook of the Birds of Europe, the Middle East and North Africa. The Birds of the Western Palearctic. Vol. I. Ostrich to Ducks*, Oxford University Press, Oxford.

Cramp, S., 1980, *Handbook of the Birds of Europe, the Middle East and North Africa. The Birds of the Western Palearctic. Vol. II. Hawks to Bustards*, Oxford University Press, Oxford.

Cramp, S., 1983, *Handbook of the Birds of Europe, the Middle East and North Africa. The Birds of the Western Palearctic. Vol. III. Waders to Gulls*, Oxford University Press, Oxford.

Cramp, S., 1985, *Handbook of the Birds of Europe, the Middle East and North Africa. The Birds of the Western Palearctic. Vol. IV. Terns to Woodpeckers*, Oxford University Press, Oxford.

Cronmiller, J. R., and Thompson, C. F., 1981, Sex-ratio adjustment in malnourished Red-winged Blackbird broods, *J. Field Ornithol.* **52:**65–67.

Csada, R. D., and Brigham, R. M., 1992, Common Poorwill, in: *The Birds of North America*, No. 32. (A. Poole, P. Stettenheim, and F. Gill, eds.), The Academy of Natural Sciences, Philadelphia; The American Ornithologists' Union, Washington, D.C.

Davies, J. C., and Cooke, F., 1983, Intraclutch hatch synchronization in the Lesser Snow Goose, *Can. J. Zool.* **61:**1398–1401.

Deeming, D. C., and Ferguson, M. W. J., 1992, Physiological effects of incubation temperature on embryonic development in reptiles and birds, in: *Egg Incubation: Its Effects on Embryonic Development in Birds and Reptiles*, (D. C. Deeming and M. W. J. Ferguson, eds.), Cambridge University Press, Cambridge, pp. 147–173.

De Steven, D., 1980, Clutch size, breeding success, and parental survival in the Tree Swallow (*Iridoprocne bicolor*), *Evolution* 34:278–291.

Dorward, D. F., 1962, Comparative biology of the White Booby and the Brown Booby *Sula* spp. at Ascension, *Ibis* **103b:**174–220.

Douthwaite, R. J., 1978, Breeding biology of the Pied Kingfisher *Ceryle rudis* on Lake Victoria, *J. East Afr. Nat. Hist. Soc. Natl. Mus.* 31 (No. 166).

Drent, R. H., 1970, Functional aspects of incubation in the Herring Gull, *Behav.* (*Suppl.*) **17:**1–125.

Drent, R. H., 1973, The natural history of incubation, in: *Breeding Biology of Birds*, (D. S. Farner, ed.), National Academy of Science, Washington D.C., pp. 262–322.

Drent, R. H., 1975, Incubation, in: *Avian Biology*, Vol. 5, (D. S. Farner and J. R. King, eds.), Academic Press, New York, pp. 333–420.

Drummond, H., 1987, A review of parent-offspring conflict and brood reduction in the Pelecaniformes, *Col. Waterbirds* **10:**1–15.

Drummond, H., and García Chavelas, C., 1989, Food shortage influences sibling aggression in the Blue-footed Booby, *Anim. Behav.* **37:**806–819.

Drummond, H., Osorna, J. L., Torres, R., García Chavelas, C., and Merchant Larios, H., 1991, Sexual size dimorphism and sibling competition: implications for avian sex ratios, *Am. Nat.* **138:**623–641.

Dunlop, E., 1910, On incubation, *British Birds* **4:**137–145.

Dunn, E. H., 1975, The timing of endothermy in the development of altricial birds, *Condor* **77:**288–293.

Edwards, T. C., and Collopy, M. W., 1983, Obligate and facultative brood reduction in eagles: an examination of factors that influence fratricide, *Auk* **100:**630–635.

Eickwort, K. R., 1973, Cannibalism and kin selection in *Labidomera clivicollis* (Coleoptera: Chrysomelidae), *Am. Nat.* **101:**452–453.

Emlen, S. T., and Oring, L. W., 1977, Ecology, sexual selection, and the evolution of mating systems, *Science* **197:**215–223.

Emlen, S. T., Wrege, P. H., Demong, N. J., and Hegner, R. E., 1991, Flexible growth rates in nestling White-fronted Bee-eaters: a possible adaptation to short-term food shortage, *Condor* **93:**591–597.

Emslie, S. D., Sydeman, W. J., and Pyle, P., 1992, The importance of mate retention and experience on breeding success in Cassin's Auklet (*Ptychoramphus aleuticus*), *Behav. Ecol.* **3:**189–195.

Enemar, A., and Arheimer, O., 1989, Developmental asynchrony and onset of incubation among passerine birds in a mountain birch forest of Swedish Lapland, *Ornis Fenn.* **66:**32–40.

Epple, A., and Stetson, M. H., 1980, *Avian Endocrinology*, Academic Press, New York.

Evans, R. M., 1988, Embryonic vocalizations as care-soliciting signals, with particular reference to the American White Pelican, *Proc. Int. Orn. Cong.* **19:**1467–1475.

Evans, R. M., 1990a, Vocal regulation of temperature by avian embryos: a laboratory study with pipped eggs of the American White Pelican, *Anim. Behav.* **40**:969–979.
Evans, R. M., 1990b, Terminal egg neglect in the American White Pelican, *Wilson Bull.* **102**:684–692.
Evans, R. M., 1990c, Effects of low incubation temperatures during the pipped egg stage on hatchability and hatching times in domestic chickens and Ring-billed Gulls, *Can. J. Zool.* **68**:836–840.
Ewert, M. A., 1992, Cold torpor, diapause, delayed hatching and aestivation in reptiles and birds, in: *Egg Incubation: Its Effects on Embryonic Development in Birds and Reptiles,* (D. C. Deeming and M. W. J. Ferguson, eds.), Cambridge University Press, Cambridge, pp. 173–191.
Ferguson, R. S., and Sealy, S. G., 1983, Breeding ecology of the Horned Grebe, *Podiceps auritus,* in southwest Manitoba, *Can. Field. Nat.* **97**:401–408.
Fisher, R. A., 1930, *The Genetical Theory of Natural Selection,* Oxford Press, London.
Forbes, L. S., 1990, Insurance offspring and the evolution of avian clutch size, *J. Theor. Biol.* **147**:345–359.
Forbes, L. S., 1991, Insurance offspring and brood reduction in a variable environment: the costs and benefits of pessimism, *Oikos* **62**:325–332.
Forbes, L. S., and Ydenberg, R. D., 1992, Sibling rivalry in a variable environment, *Theor. Pop. Biol.* **41**:135–160.
Forbes, M. L. R., and Ankney, C. D., 1987, Hatching asynchrony and food allocation within broods of Pied-billed Grebes, *Podilymbus podiceps, Can. J. Zool.* **65**:2872–2877.
Forshaw, J. M., 1989, *Parrots of the World,* Landsdowne Editions, Willoughby, Australia.
Freese, C. H., 1975, Notes on nesting in the Double-striped Thick-knee (*Burhinus bistriatus*) in Costa Rica, *Condor* **77**:353–354.
Fujioka, M., 1985, Food delivery and sibling competition in experimentally even-aged broods of the Cattle Egret, *Behav. Ecol. Sociobiol.* **17**:67–74. (T1 & T2)
Gargett, V., 1982, Synchronous hatching and the Cain and Abel struggle in the Black Eagle, *Ostrich* **53**:147–150. (T1 & T2)
Gaston, A. J., and Jones, I. L., 1989, The relative importance of stress and programmed anorexia in determining mass loss by incubating Ancient Murrelets, *Auk* **106**:653–658.
Gaston, A. J., and Powell, D. W., 1989, Natural incubation, egg neglect, and hatchability in the Ancient Murrelet, *Auk* **106**:433–438.
Gibb, J., 1950, The breeding biology of the Great and Blue Titmice, *Ibis* **92**:507–539.
Gibbons, D. W., 1987, Hatching asynchrony reduces parental investment in the Jackdaw, *J. Anim. Ecol.* **56**:403–414. (T1 & T2)
Gibson, F., 1971, The breeding biology of the American Avocet (*Recurvirostra americana*) in central Oregon, *Condor* **73**:444–454.
Gilbert, A. B., 1971, Transport of the egg through the oviduct and oviposition, in: *Physiology and Biochemistry of the Domestic Fowl,* Vol. 3, (D. J. Bell and B. M. Freeman, eds.), Academic Press, New York, pp. 1345–1352.
Godfray, H. C. J., and Harper, A. B., 1990, The evolution of brood reduction by siblicide in birds, *J. Theor. Biol.* **145**:163–175.
Godfary, H. C. J., Partridge, L., and Harvey, P. H., 1991, Clutch size, *Annu. Rev. Ecol. Syst.* **22**:409–429.
Götmark, F., 1992, The effects of investigator disturbance on nesting birds, in *Current Ornithology,* Vol. 9, (D. M. Power, ed.), Plenum Press, New York, pp. 63–104.
Goldsmith, A. R., 1983, Prolactin in avian reproductive cycles, in: *Hormones and Behavior in Higher Vertebrates,* (J. Balthazart, E. Pröve, and R. Gilles, eds.), Springer-Verlag, Berlin, pp. 375–387.

Goldsmith, A. R., and Williams, D. M., 1980, Incubation in Mallards (*Anas platyrhynchos*): changes in plasma levels of prolactin and luteinizing hormone, *J. Endocrinol.* **86**:371–379.
Gottlander, K., 1987, Parental feeding behavior and sibling competition in the Pied Flycatcher *Ficedula hypoleuca*, *Ornis Scand.* **18**:269–276.
Gowaty, P. A., 1991, Facultative manipulation of sex ratios in birds: rare or rarely observed?, in *Current Ornithology*, Vol. 8, (D. M. Power, ed.), Plenum Press, New York, pp. 141–171.
Gowaty, P. A., and Lennartz, M. R., 1985, Sex ratios of nestling and fledgling Red-cockaded Woodpeckers (*Picoides borealis*) favor males, *Am. Nat.* **126**:347–353.
Grant, G. S., 1982, Avian incubation: egg temperature, nest humidity, and behavioral thermoregulation in a hot environment, *Ornith. Monogr.* 30.
Gratto-Trevor, C. L., 1992, Semipalmated Sandpiper, in: *The Birds of North America*, No. 6, (A. Poole, P. Stettenheim, and F. Gill, eds.), The Academy of Natural Sciences, Philadelphia; The American Ornithologists' Union, Washington, D.C.
Graves, J., 1991, Comments on the sample sizes used to test the effect of experimental brood enlargement on adult survival, *Auk* **108**:967–969.
Graves, J., Whiten, A., and Henzi, P., 1984, Why does the Herring Gull lay three eggs?, *Anim. Behav.* **32**:798–805.
Greig-Smith, P., 1985, Weight differences, brood reduction, and sibling competiton among nestling Stonechats *Saxicola torquata* (Aves: Turdidae), *J. Zool., London.* **205**:453–465.
Groves, S., 1984, Chick growth, sibling rivalry, and chick production in American Black Oystercatchers, *Auk* **101**:525–531.
Haftorn, S., 1981, Incubation during the egg-laying period in relation to clutch-size and other aspects of reproduction in the Great Tit *Parus major*, *Ornis Scand.* **12**:169–185.
Haftorn, S., 1986, Clutch size, intraclutch egg size variation, and breeding strategy in the Goldcrest *Regulus regulus*, *J. Ornithol.* **127**:291–301.
Hahn, D. C., 1981, Asynchronous hatching in the Laughing Gull: cutting losses and reducing rivalry, *Anim. Behav.* **29**:421–427. (T1 & T2)
Haig, D., 1990, Brood reduction and optimal parental investment when offspring differ in quality, *Am. Nat.* **136**:550–566.
Haig, D., 1992, Brood reduction in gymnosperms, in: *Cannibalism: Ecology and Evolution Among Diverse Taxa*, (M. A. Elgar and B. J. Crespi, eds.), Oxford University Press, Oxford, pp. 63–84.
Haig, S. M., 1992, Piping Plover, in: *The Birds of North America*, No. 2, (A. Poole, P. Stettenheim, and F. Gill, eds.), The Academy of Natural Sciences, Philadelphia; The American Ornithologists' Union, Washington, D.C.
Hall, M. R., and Goldsmith, A. R., 1983, Factors affecting prolactin secretion during breeding and incubation in the domestic duck (*Anas platyrhynchus*), *Gen. Comp. Endocr.* **49**:270–276.
Hamilton, W. D., 1964, The genetical evolution of social behaviour, *J. Theor. Biol.* **7**:1–52.
Harper, R. G., Juliano, S. A., and Thompson, C. F., 1992, Hatching asynchrony in the House Wren, *Troglodytes aedon*: a test of the brood-reduction hypothesis, *Behav. Ecol.* **3**:76–83. (T1 & T2)
Harvey, P. H., and Pagel, M. D., 1991, *The Comparative Method in Evolutionary Biology*, Oxford University Press, New York.
Haydock, J., and Ligon, J. D., 1986, Brood reduction in the Chihuahuan Raven: an experimental study, *Ecology* **67**:1194–1205. (T1 & T2)
Hébert, P. N., 1993a, An experimental study of brood reduction and hatching asynchrony in Yellow Warblers, *Condor* **95**:362–371. (T1 & T2)

Hébert, P. N., 1993b, Hatching asynchrony in the Yellow Warbler: a test of the sexual conflict hypothesis, *Am. Nat.* **142:**881–892. (T1)
Hébert, P. N., and Barclay, R. M. R., 1986, Asynchronous and synchronous hatching: effect on early growth and survivorship of Herring Gull, *Larus argentatus*, chicks, *Can. J. Zool.* **64:**2357–2362. (T1 & T2)
Hébert, P. N., and Sealy, S. G., 1992, Onset of incubation in Yellow Warblers: a test of the hormonal hypothesis, *Auk* **109:**249–255. (T1)
Hébert, P. N., and Sealy, S. G., 1993, Hatching asynchrony in Yellow Warblers: a test of the nest-failure hypothesis, *Ornis Scand.* **24:**10–14. (T1)
Hewston, N., 1984, Breeding of the White-cheeked Touraco, *Tauraco leucotis leucotis*, *Avicul. Mag.* **90:**209–215.
Hilborn, R., and Stearns, S. C., 1982, On inference in ecology and evolutionary biology: the problem of multiple causes, *Acta Biotheor.* **31:**145–164.
Horsfall, J. A., 1991, Brood reduction and brood division in Coots, *Anim. Behav.* **32:**216–225.
Howe, H. F., 1976, Egg size, hatching asynchrony, sex, and brood reduction in the Common Grackle, *Ecology* **57:**1195–1207. (T1 & T2)
Howe, H. F., 1978, Initial investment, clutch size, and brood reduction in the Common Grackle *Quiscalus quiscula*, *Ecology* **59:**1109–1122.
Husby, M., 1986, On the adaptive value of brood reduction in birds: experiments with the magpie *Pica pica*, *J. Anim. Ecol.* **55:**75–83.
Hussell, D. J. T., 1972, Factors affecting clutch size in Arctic passerines, *Ecol. Monogr.* **42:**317–364.
Hussell, D. J. T., 1985, On the adaptive basis for hatching asynchrony: brood reduction, nest failure and asynchronous hatching in Snow Buntings, *Ornis Scand.* **16:**205–212. (T1)
Hussell, D. J. T., 1991, Regulation of food provisioning in broods of altricial birds, *Acta XX Cong. Inter. Ornith.*:946–960.
Hussell, D. J. T., and Page, G. W., 1976, Observations on the breeding biology of Black-bellied Plovers on Devon Island, Northwest Territories, Canada, *Wilson Bull.* **88:**632–653.
Hussell, D. J. T., and Quinney, T. E., 1987, Food abundance and clutch size of Tree Swallows *Tachycineta bicolor*, *Ibis* **129:**243–258.
Ingram, C., 1959, Cannibalism in the breeding biology of certain birds of prey, *Auk* **76:**218–226.
Jackson, H. D., 1985, Aspects of the breeding biology of the Fierynecked Nightjar, *Ostrich* **56:**263–276.
Jackson, W. M., 1993, Causes of conspecific nest parasitism in the northern masked weaver, *Behav. Ecol. Sociobiol.* **32:**119–126. (T1 & T2)
Jehl, J. R., Jr., 1973, Breeding biology and systematic relationships of the Stilt Sandpiper, *Wilson Bull.* **85:**115–147.
Jehl, J. R., Jr., and Mahoney, S. A., 1987, The roles of thermal environment and predation in habitat choice in the California Gull, *Condor* **89:**850–862.
Jover, L., Ruiz, X., and González-Martín, M., 1993, Significance of intraclutch egg size variation in the Purple Heron, *Ornis Scand.* **24:**127–134.
Kahl, M. P., 1966, A contribution to the ecology and reproductive biology of the Marabou Stork (*Leptoptilus crumeniferos*) in East Africa, *J. Zool., London.* **148:**289–311.
Kaufmann, G. W., 1989, Breeding ecology of the Sora, *Porzana carolina*, and the Virginia Rail, *Rallus limicola*, *Can. Field Nat.* **103:**270–282.
Kavanau, J. L., 1987, *Lovebirds, Cockatiels, Budgerigars: Behavior and Evolution*, Science Software Systems, Inc., Los Angeles.

Kemp, A. C., 1978, A review of the hornbills: biology and radiation, *Living Bird* **17**:105–136.

Kendra, P. E., Roth, R. R., and Tallamy, D. W., 1988, Conspecific brood parasitism in the House Sparrow, *Wilson Bull.* **100**:80–90.

Kennamer, R. A., Harvey, W. F., IV, and Hepp, G. R., 1990, Embryonic development and nest attentiveness of Wood Ducks during egg laying, *Condor* **92**:587–592.

Kepler, A. K., 1977, Publications of the Nuttall Ornithological Club, No. 18, *Comparative Study of Todies (Todidae), With Emphasis on the Puerto Rican Tody, Todus mexicanus*, Nuttall Ornithological Club, Cambridge, Mass.

Klomp, H., 1970, The determination of clutch size in birds: a review, *Ardea* **58**:1–124.

Koenig, W. D., 1982, Ecological and social factors affecting hatchability of eggs, *Auk* **99**:526–536.

Konarzewski, M., 1993, The evolution of clutch size and hatching asynchrony in altricial birds: the effect of environmental variability, egg failure and predation, *Oikos* **67**:97–106.

Kuitunen, M., and Aleknonis, A., 1992, Nest predation and breeding success in Common Treecreepers nesting in boxes and natural cavities, *Ornis Fenn.* **69**:7–12.

Lack, D., 1947, The significance of clutch size, *Ibis* **89**:302–352.

Lack, D., 1954, *The Natural Regulation of Animal Numbers*, Oxford University Press, London.

Lack, D., 1966, *Population Studies of Birds*, Clarendon Press, Oxford.

Lack, D., 1968, *Ecological Adaptations for Breeding in Birds*, Methuen and Co., London.

Lack, D., 1973, *Swifts in a tower*, Chapman and Hall, London.

Lamey, T. C., 1990, Hatch asynchrony and brood reduction in penguins, in: *Penguin Biology*, (L. S. Davis and J. T. Darby, eds.), Academic Press, San Diego, pp. 399–416.

Lamey, T. C., and Mock, D. W., 1991, Nonaggressive brood reduction in birds, *Acta XX Cong. Inter. Ornith.*:1741–1751.

Land, D., Marion, W. R., and O'Meara, T. E., 1989, Snag availability and cavity nesting birds in slash pine plantations, *J. Wild. Man.* **53**:1165–1171.

Leonard, M. L., Horn, A. G., and Eden, S. F., 1988, Parent-offspring aggression in moorhens, *Behav. Ecol. Sociobiol.* **23**:265–270.

Lessells, C. M., and Avery, M. I., 1989, Hatching asynchrony in European Bee-eaters *Merops apiaster*, *J. Anim. Ecol.* **58**:815–835.

Lifjeld, J. T., and Slagsvold, T., 1986, The function of courtship feeding during incubation in the Pied Flycatcher *Ficedula hypoleuca*, *Anim. Behav.* **34**:1441–1453.

Lifson, N., and McClintock, R., 1966, Theory of use of the turnover rates of body water for measuring energy and material balance, *J. Theor. Biol.* **12**:46–74.

Ligon, J. D., 1968, Miscellaneous Publications of the Museum of Zoology, No. 136, *The Biology of the Elf Owl, Micrathene whitneyi*, The University of Michigan, Ann Arbor.

Ligon, J. D., and Ligon, S. H., 1990, Female-biased sex-ratios at hatching in the Green Woodhoopoe, *Auk* **107**:765–771.

Lloyd, D. G., 1980, Sexual strategies in plants. I. An hypothesis of sexual adjustment of maternal investment during one reproductive session, *New Phytol.* **86**:69–79.

Lloyd, D. G., 1987, Selection of offspring size at independence and other size-versus-number strategies, *Am. Nat.* **129**:800–817.

Lombardo, M. P., Power, H. W., Stouffer, P. C., Romagnano, L. C., and Hoffenberg, A. S., 1989, Egg removal and intraspecific brood parasitism in the European Starling (*Sturnus vulgaris*), *Behav. Ecol. Sociobiol.* **24**:214:223.

Lowther, P. E., and Cink, C. L., 1992, House Sparrow, in: *The Birds of North America*, No. 12, (A. Poole, P. Stettenheim, and F. Gill, eds.), The Academy of Natural Sciences, Philadelphia; The American Ornithologists' Union, Washington, D.C.

Lyon, B. E., and Montgomerie, R. D., 1985, Incubation feeding in Snow Buntings: female manipulation or indirect male parental care?, *Behav. Ecol. Sociobiol.* **17**:279–284.

Mader, W. J., 1982, Ecology and breeding habits of the Savanna Hawk in the llanos of Venezuela, *Condor* **84**:261–271.

Magrath, R. D., 1988, Hatching asynchrony in altricial birds: nest failure and adult survival, *Am. Nat.* **131**:893–900.

Magrath, R. D., 1989, Hatching asynchrony and reproductive success in the Blackbird, *Nature* **339**:536–538. (T1 & T2)

Magrath, R. D., 1990, Hatching asynchrony in altricial birds, *Biol. Rev.* **65**:587–622.

Magrath, R. D., 1992, Roles of egg mass and incubation pattern in establishment of hatching hierarchies in the Blackbird (*Turdus merulus*), *Auk* **109**:474–487.

Manly, B. F. J., 1991, *Randomization and Monte Carlo Methods in Biology*, Chapman and Hall, London.

Marion, W. R., and Fleetwood, R. J., 1978, Nesting ecology of the Plain Chachalaca in south Texas, *Wilson Bull.* **90**:386–395.

Martin, T. E., 1992, Interaction of nest predation and food limitations in reproductive strategies, *Current Ornithology*, Vol. 9, (D. M. Power, ed.), Plenum Press, New York, pp. 163–197.

Martin, T. E., 1993, Evolutionary determinants of clutch size in cavity-nesting birds: nest predation or limited breeding opportunities?, *Am. Nat.* **142**:937–946.

McClure, P. A., 1981, Sex-biased litter reduction in food-restricted wood rats (*Neotoma floridana*), *Science* **211**:1058–1060.

McNab, B. K., 1989, Laboratory and field studies of the energy expenditure of endotherms: a comparison, *TREE* **4**:111–112.

McRae, S. B., Weatherhead, P. J., and Montgomerie, R., 1993, American Robin nestlings compete by jockeying for position, *Behav. Ecol. Sociobiol.* **33**:101–106.

Mead, P. S., and Morton, M. L., 1985, Hatching asynchrony in the Mountain White-crowned Sparrow (*Zonotrichia leucophrys oriantha*): a selected or incidental trait?, *Auk* **102**:781–792.

Meanley, B., 1992, King Rail, in: *The Birds of North America*, No. 3, (A. Poole, P. Stettenheim, and F. Gill, eds.), The Academy of Natural Sciences, Philadelphia; The American Ornithologists' Union, Washington, D.C.

Meathrel, C. E., and Ryder, J. P., 1987, Intraclutch variation in the size, mass and composition of Ring-billed Gull eggs, *Condor* **89**:364–368.

Meijer, T., 1990, Incubation development and clutch size in the Starling, *Ornis Scand.* **21**:163–168.

Meyburg, B. U., 1978, Sibling aggression and cross-fostering eagles, in: *Endangered Birds: Symposium on Management Techniques for Preserving Endangered Birds, University of Wisconsin-Madison 1977*, (S. A. Temple, ed.), University of Wisconsin Press, Madison, pp. 195–200.

Milne, I. S., and Calow, P., 1990, Costs and benefits of brooding in Glossiphoniid leeches with special reference to hypoxia as a selection pressure, *J. Anim. Ecol.* **59**:41–56.

Mock, D. W., 1984a, Infanticide, siblicide, and avian nestling mortality, in: *Infanticide: Comparative and Evolutionary Perspectives*, (G. Hausfater and S. B. Hrdy, eds.), Aldine, New York, pp. 3–30.

Mock, D. W., 1984b, Siblicidal aggression and resource monopolization in birds, *Science* **225**:731–733.

Mock, D. W., 1985a, An introduction to the neglected mating system, *Ornith. Monogr.* **37**:1–10.

Mock, D. W., 1985b, Siblicidal brood reduction: The prey-size hypothesis, *Am. Nat.* **125**:327–343.

Mock, D. W., 1987, Siblicide, parent-offspring conflict, and unequal parental investment by egrets and herons, *Behav. Ecol. Sociobiol.* **20:**247–256.

Mock, D. W., and Parker, G. A., 1986, Advantages and disadvantages of egret and heron brood reduction, *Evolution* **40:**459–470.

Mock, D. W., and Ploger, B. J., 1987, Parental manipulations of optimal hatch asynchrony in cattle egrets: an experimental study, *Anim. Behav.* **35:**150–160. (T1 & T2)

Mock, D. W., and Schwagmeyer, P. L., 1990, The peak load reduction hypothesis for avian hatching asynchrony, *Evol. Ecol.* **4:**249–260.

Mock, D. W., Drummond, H., and Stinson, C. H., 1990, Avian siblicide, *Am. Sci.* **78:**438–449.

Møller, A. P., 1989, Parasites, predators and nest boxes: facts and artefacts in nest box studies of birds?, *Oikos* **56:**421–423.

Moreno, J., 1989a, Energetic constraints on uniparental incubation in the Wheatear *Oenanthe oenanthe* (L.), *Ardea* **77:**107–115.

Moreno, J., 1989b, Strategies of mass change in breeding birds, *Biol. J. Linn. Soc.* **37:**297–310.

Moreno, J., and Carlson, A., 1989, Clutch size and the costs of incubation in the Pied Flycatcher *Ficedula hypoleuca, Ornis Scand.* **20:**123–128.

Morton, M. L., and Pereyra, M. E., 1985, The regulation of egg temperatures and attentiveness patterns in the Dusky Flycatcher (*Empidonax oberholseri*), *Auk* **102:**25–37.

Mueller, A. J., 1992, Inca Dove, in: *The Birds of North America*, No. 28, (A. Poole, P. Stettenheim, and F. Gill, eds.), The Academy of Natural Sciences, Philadelphia; The American Ornithologists' Union, Washington, D.C.

Murphy, M. T., 1986, Body size and condition, timing of breeding, and aspects of egg production in Eastern Kingbirds, *Auk* **103:**465–476.

Murray, B. G., 1985, Evolution of clutch size in tropical species of birds, *Ornith. Monogr.* **36:**505–519.

Murray, B. G., In press, Effect of selection for successful reproduction on hatching synchrony and asynchrony, *Auk.*

Murton, R. K., and Westwood, N. J., 1977, *Avian Breeding Cycles*, Clarendon Press, Oxford.

Myers, J. H., 1978, Sex ratio adjustment under food stress: maximization of quality or numbers of offspring, *Am. Nat.* **112:**381–388.

Myers, S. A., Millam, J. R., and El Halawani, M. E., 1989, Plasma LH and prolactin levels during the reproductive cycle of the Cockatiel (*Nymphicus hollandicus*), *Gen. Comp. Endocr.* **73:**85–91.

Navarro, J. L., and Bucher, E. H., 1990, Growth of Monk Parakeets, *Wilson Bull.* **102:**520–524.

Nelson, J. B., 1964, Factors influencing clutch-size and chick growth in the North Atlantic Gannet *Sula bassana, Ibis* **106:**63–77.

Newton, I., and Marquiss, M., 1984, Seasonal trend in the breeding performance of sparrowhawks, *J. Anim. Ecol.* **53:**809–829.

Nice, M. M., 1962, Development of behavior in precocial birds, *Trans. Linn. Soc. N.Y.* **8:**1–211.

Nilsson, J.-Å., 1993a, Bisexual incubation facilitates hatching asynchrony, *Am. Nat.* **142:**712–717.

Nilsson, J.-Å., 1993b, Energetic constraints on hatching asynchrony, *Am. Nat.* **141:**158–166. (T1)

Nilsson, J.-Å., and Smith, H. G., 1988, Incubation feeding as a male tactic for early hatching, *Anim. Behav.* **36:**641–647.

Nilsson, S., 1986, Evolution of hole-nesting in birds: on balancing selection pressures, *Auk* **103:**432–435.
Nisbet, I. C. T., and Cohen, M. E., 1975, Asynchronous hatching in Common and Roseate Terns, *Sterna hirundo* and *Sterna dougallii*, *Ibis* **117:**374–379.
Norberg, R. Å., 1981, Temporary weight decrease in breeding birds may result in more fledged young, *Am. Nat.* **118:**838–850.
Neuchterlein, G. L., 1981, Asynchronous hatching and sibling competition in Western Grebes, *Can. J. Zool.* **59:**994–998.
Neuchterlein, G. L., and Johnson, A., 1981, The downy young of the Hooded Grebe, *Living Bird* **19:**69–71.
O'Connor, R. J., 1978, Brood reduction in birds: selection for fratricide, infanticide and suicide?, *Anim. Behav.* **26:**79–96.
O'Connor, R. J., 1984, *The Growth and Development of Birds*, John Wiley and Sons, New York.
Olver, M. D., 1984, Breeding biology of the Reed Cormorant, *Ostrich* **55:**133–140.
Orejuela, J. E., 1977, Comparative biology of Turquoise-browed and Blue-crowned Motmots in the Yucatán Peninsula, México, *Living Bird* **16:**193–208.
Oring, L., 1982, Avian mating systems, in: *Avian Biology*, Vol. 6, (D. S. Farner, J. R. King, and K. C. Parkes, eds.), Academic Press, New York, pp. 1–92.
Osborne, D. R., and Bourne, G. R., 1977, Breeding behavior and food habits of the Wattled Jacana, *Condor* **79:**98–105.
Palmer, R. S., 1962, *Handbook of North American Birds*, Vol. 1, Yale University Press, New Haven.
Parmalee, D. F., 1970, Breeding behavior of the Sanderling in the Canadian high arctic, *Living Bird* **9:**97–146.
Parry, V., 1973, The auxiliary social system and its effects on territory and breeding in Kookaburras, *Emu.* **73:**81–100.
Parsons, J., 1975, Asynchronous hatching and chick mortality in the Herring Gull *Larus argentatus*, *Ibis* **117:**517–520.
Parsons, J., 1976, Factors determining the number and size of eggs laid by the Herring Gull, *Condor* **78:**481–492.
Perrins, C. M., and McCleery, R. H., 1989, Laying dates and clutch size in the Great Tit, *Wilson Bull.* **101:**236–253.
Peterman, R. M., 1990, Statistical power analysis can improve fisheries research and management, *Can. J. Fish. Aquat. Sci.* **47:**2–15.
Pettingill, O. S., Jr., 1985, *Ornithology in Laboratory and Field*, 5th ed., Academic Press, Orlando.
Pierotti, R., and Bellrose, C. A., 1986, Proximate and ultimate causation of egg size and the "third-chick disadvantage" in the Western Gull, *Auk* **103:**401–407.
Pietiäinen, H., Saurola, P., and Väisänen, R. A., 1986, Parental investment in clutch size and egg size in the Ural Owl *Strix uralensis*, *Ornis Scand.* **17:**309–325.
Pijanowski, B. C., 1992, A review of Lack's brood reduction hypothesis, *Am. Nat.* **139:**1270–1292.
Ploger, B. J., 1992, Proximate and ultimate causes of brood reduction in Brown Pelicans (*Pelecanus occidentalis*), Ph.D. Dissertation, University of Florida, Gainsville. (T1)
Poole, A., 1985, Courtship feeding and Osprey reproduction, *Auk* **102:**479–492.
Power, H. W., Litovich, E., and Lombardo, M. P., 1981, Male Starlings delay incubation to avoid being cuckolded, *Auk* **98:**386–389.
Redondo, T., and Castro, F., 1992, Signalling of nutritional need by Magpie nestlings, *Ethology* **92:**193–204.

Ricklefs, R. E., 1965, Brood reduction in the Curve-billed Thrasher, *Condor* **67**:505–510.

Ricklefs, R. E., 1969, An analysis of nesting mortality in birds, *Smithsonian Cont. Zool.* **9**:1–48.

Ricklefs, R. E., 1976, Growth rates of birds in the humid new world tropics, *Ibis* **118**:179–207.

Ricklefs, R. E., 1983, Avian postnatal development, in: *Avian Biology*, Vol. 7, (D. S. Farner, J. R. King, and K. C. Parkes, eds), Academic Press, New York, pp. 1–83.

Ricklefs, R. E., 1987, Characterizing the development of homeothermy by rate of body cooling, *Funct. Ecol.* **1**:151–157.

Ricklefs, R. E., 1992, Embryonic development period and the prevalence of avian blood parasites, *Proc. Natl. Acad. Sci. USA* **89**:4722–4725.

Ricklefs, R. E., 1993, Sibling competition, hatching asynchrony, incubation period, and lifespan in altricial birds, in: *Current Ornithology*, Vol. 11, (D. M. Power, ed.), Plenum Press, New York, pp. 199–276.

Ricklefs, R. E., and Hussell, D. J. T., 1984, Changes in adult mass associated with the nesting cycle in the European Starling, *Ornis Scand.* **15**:155–161.

Ridley, M., 1978, Paternal care, *Anim. Behav.* **26**:904–932.

Robertson, R. J., and Rendell, W. B., 1990, A comparison of the breeding ecology of a secondary cavity nesting bird, the Tree Swallow (*Tachycineta bicolor*), in nest boxes and natural cavities, *Can. J. Zool.* **68**:1046–1052.

Roby, D. D., and Ricklefs, R. E., 1984, Observations on the cooling tolerance of embryos of the Diving Petrel *Pelecanoides georgicus*, *Auk* **101**:160–161.

Rol'nik, V. V., 1970, *Bird Embryology*, Israeli Program for Scientific Translations, Jerusalem.

Romagnano, L., Hoffenberg, A. S., and Power, H. W., 1990, Intraspecific brood parasitism in the European Starling, *Wilson Bull.* **102**:279–291.

Romanoff, A. L., and Romanoff, A. J., 1972, *Pathogenesis of the Avian Embryo*, J. Wiley and Sons, New York.

Roskaft, E., and Slagsvold, T., 1985, Differential mortality of male and female offspring in experimentally manipulated broods of the Rook, *J. Anim. Ecol.* **54**:261–266.

Rowan, M. K., 1967, A study of the Colies of southern Africa, *Ostrich* **38**:63–115.

Rowan, M. K., 1983, *The Doves, Parrots, Louries, and Cuckoos of Southern Africa*, Croom Helm, Beckenham, Kent, UK.

Rydén, O., and Bengtsson, H., 1980, Differential begging and locomotory behaviour by early and late hatched nestlings affecting the distribution of food in asynchronously hatched broods of altricial young, *Z. Tierpsychol.* **53**:209–224.

Salfert, I. G., and Moodie, G. E. E., 1984, Filial egg-cannibalism in the brook stickleback, *Culaea inconstans* (Kirtland), *Behav.* **93**:82–100.

Sasvári, L., 1990, Feeding response of mated and widowed bird parents to fledglings: an experimental study, *Ornis Scand.* **21**:287–292.

Saunders, D. A., 1986, Breeding season, nesting success, and nestling growth in Carnaby's Cockatoo, *Calyptorhynchus funereus latirostris*, over sixteen years at Coomalla Creek, and a method for assessing the viability of populations in other areas, *Aust. Wildl. Res.* **13**:261–273.

Schwagmeyer, P. L., Mock, D. W., Lamey, T. C., Lamey, C. S., and Beecher, M. D., 1991, Effects of sibling contact on hatch timing in an asynchronously hatching bird, *Anim. Behav.* **41**:887–894.

Scott, P. E., and Martin, R. F., 1986, Clutch size and fledging success in the Turquoise-browed Motmot, *Auk* **103**:8–13.

Seddon, P. J., and van Heezik, Y. M., 1991a, Effects of hatching order, sibling asymmetries, and nest site on survival analysis of Jackass Penguin chicks, *Auk* **108**:548–555.

Seddon, P. J., and van Heezik, Y. M., 1991b, Hatching asynchrony and brood reduction in the Jackass Penguin: an experimental study, *Anim. Behav.* **42**:347–356. (T1 & T2)
Sedgwick, J. A., and Knopf, F. L., 1990, Habitat relationships and nest site characteristics of cavity-nesting birds in cottonwood floodplains, *J. Wild. Man.* **54**:112–124.
Shaw, P., 1985, Brood reduction in the Blue-eyed Shag *Phalacrocorax atriceps*, *Ibis* **127**:476–494. (T1 & T2)
Sibley, C. G., and Ahlquist, J. E., 1990, *Phylogeny and Classification of Birds: a Study in Molecular Evolution*, Yale University Press, New Haven.
Sibley, C. G., and Monroe, B. L., Jr., 1990, *Distribution and Taxonomy of Birds of the World*, Yale University Press, New Haven.
Silver, R., Andrews, H., and Ball, G. F., 1985, Parental care in an ecological perspective: a quantitative analysis of avian subfamilies, *Amer. Zool.* **25**:823–840.
Skagen, S. K., 1987, Hatching asynchrony in American Goldfinches: an experimental study, *Ecology* **68**:1747–1759. (T1 & T2)
Skagen, S. K., 1988, Asynchronous hatching and food limitation: a test of Lack's Hypothesis, *Auk* **105**:78–88. (T1 & T2)
Skutch, A. F., 1957, The incubation patterns of birds, *Ibis* **99**:69–93.
Skutch, A. F., 1964, Life history of the Scaly-breasted Hummingbird, *Condor* **66**:186–198.
Skutch, A. F., 1969, Pacific Coast Avifauna, No. 35, *Life Histories of Central American Birds*, Vol. 2, Cooper Ornithological Society, Berkeley.
Skutch, A. F., 1972, Publications of the Nuttall Ornithological Club, No. 10, *Studies of Tropial American Birds*, Nuttall Ornithological Club, Cambridge, Mass.
Skutch, A. F., 1983, *Birds of Tropical America*, University of Texas Press, Austin.
Skutch, A. F., 1985, Clutch size, nesting success, and predation on nests of Neotropical birds, reviewed, *Ornith. Monogr.* **36**:575–594.
Slagsvold, T., 1982, Clutch size, nest size, and hatching asynchrony in birds: experiments with the Fieldfare (*Turdus pilaris*), *Ecology* **63**:1389–1399. (T1 & T2)
Slagsvold, T., 1985, Asynchronous hatching in passerine birds: influence of hatching failure and brood reduction, *Ornis Scand.* **16**:81–87.
Slagsvold, T., 1986a, Asynchronous versus synchronous hatching in birds: experiments with the Pied Flycatcher, *J. Anim. Ecol.* **55**:1155–1134. (T1 & T2)
Slagsvold, T., 1986b, Hatching asynchrony: interspecific comparisons of altricial birds, *Am. Nat.* **128**:120–125.
Slagsvold, T., 1990, Fisher's sex ratio theory may explain hatching patterns in birds, *Evolution* **44**:1009–1017.
Slagsvold, T., and Amundsen, T., 1992, Do Great Tits adjust hatching spread, egg size and offspring sex ratio to changes in clutch size?, *J. Anim. Ecol.* **61**:249–258.
Slagsvold, T., and Lifjeld, J. T., 1989a, Constraints on hatching asynchrony and egg size in Pied Flycatchers, *J. Anim. Ecol.* **58**:837–845.
Slagsvold, T., and Lifjeld, J. T., 1989b, Hatching asynchrony in birds: the hypothesis of sexual conflict over parental investment, *Am. Nat.* **134**:239–253. (T1)
Slagsvold, T., Sandvik, J., Rofstad, G., Lorentsen, O., and Husby, M., 1984, On the adaptive value of intraclutch egg-size variation in birds, *Auk* **101**:685–697.
Slagsvold, T., Roskaft, E., and Engen, S., 1986, Sex ratio, differential cost of rearing young, and differential mortality between the sexes during the period of parental care: Fisher's theory applied to birds, *Ornis Scand.* **17**:117–125.
Slagsvold, T., Husby, M., and Sandvik, J., 1992, Growth and sex ratio of nestlings in two species of crow: how important is hatching asynchrony?, *Oecologia* **90**:43–49.
Slagsvold, T., Amundsen, T., and Dale, S., 1994, Selection by sexual conflict for evenly spaced offspring in blue tits, *Nature* **370**:136–138. (T1 & T2)

Smith, G. T., and Saunders, D. A., 1986, Clutch size and productivity in three sympatric species of Cockatoo (Psittaciformes) in the southwest of Australia, *Aust. Wildl. Res.* **13**:275–285.

Smith, H. G., and Montgomerie, R., 1991, Nestling American Robins compete with siblings by begging, *Behav. Ecol. Sociobiol.* **29**:307–312.

Smith, K. G., 1988, Clutch-size dependent asynchronous hatching and brood reduction in *Junco hyemalis*, *Auk* **105**:200–203.

Snow, D. W., 1961, The natural history of the Oilbird, *Steatornis caripensis*, in Trinidad, W.I. Part 1. General behavior and breeding habits, *Zool.* **46**:27–48.

Snyder, N. F. R., Wiley, J. W., and Kepler, C. B., 1987, *The Parrots of Luquillo: Natural History and Conservation of the Puerto Rican Parrot*, Western Foundation of Vertebrate Zoology, Los Angeles.

Speakman, J. R., and Racey, P. A., 1988, The doubly-labelled water technique for measurement of energy expenditure in free-living animals, *Sci. Prog., Oxf.* **72**:227–237.

Spellerberg, I. F., 1971, Breeding behavior of the McCormick Skua, *Ardea* **59**:189–230.

Stamps, J., Clark, A., Arrowood, P., and Kus, B., 1985, Parent-offspring conflict in Budgerigars, *Behav.* **94**:1–40.

Stamps, J., Clark, A., Kus, B., and Arrowood, P., 1987, The effects of parent and offspring gender on food allocation in Budgerigars, *Behav.* **101**:177–199.

Stanback, M. T., 1991, Causes and consequences of nestling size variation in the cooperatively breeding Acorn Woodpecker *Melanerpes formicivorus*, Ph.D. Dissertation, University of California, Berkeley.

Stanback, M. T., and Koenig, W. D., 1992, Cannibalism in birds, in: *Cannibalism: Ecology and Evolution Among Diverse Taxa*, (M. A. Elgar and B. J. Crespi, eds.), Oxford University Press, Oxford, pp. 276–298.

Stearns, S. C., 1980, A new view of life-history evolution, *Oikos* **35**:266–281.

Steidl, R. J., and Griffin, C. R., 1991, Growth and brood reduction of mid-Atlantic coast Ospreys, *Auk* **108**:363–370.

Stinson, C. H., 1979, On the selective advantage of fratricide in raptors, *Evolution* **33**:1219–1225.

Stokland, J. N., and Amundsen, T., 1988, Initial size hierarchy in broods of the Shag: relative significance of egg size and hatching asynchrony, *Auk* **105**:308–315.

Stouffer, P. C., 1989, Hatching asynchrony in the European Starling, Ph.D. Dissertation, University of Pennsylvania, Philadelphia.

Stouffer, P. C., and Power, H. W., 1990, Density effects on asynchronous hatching and brood reduction in European Starlings, *Auk* **107**:359–366.

Stouffer, P. C., and Power, H. W., 1991, An experimental test of the brood-reduction hypothesis in European Starlings, *Auk* **108**:519–531. (T1 & T2)

Strahl, S. D., 1988, The social organization and behavior of the Hoatzin *Opisthocomus hoazin* in central Venezuela, *Ibis* **130**:483–502.

Strehl, C., 1978, Asynchrony of hatching in Red-winged Blackbirds and survival of late and early hatching birds, *Wilson Bull.* **90**:653–655.

Sturkie, P. D., and Mueller, W. J., 1976, Reproduction in the female and egg production, in: *Avian Physiology*, 3rd ed., (P. D. Sturkie, ed.), Springer-Verlag, New York, pp. 302–330.

Stutchbury, B. J., and Robertson, R. J., 1988, Within-season and age-related patterns of reproductive performance in female Tree Swallows (*Tachycineta bicolor*), *Can. J. Zool.* **66**:827–834.

Sullivan, J. P., 1988, Effects of provisioning rates and number fledged on nestling aggression in Great Blue Herons, *Col. Waterbirds* **11**:220–226.

Swanberg, P. O., 1950, On the concept of "incubation period," *Vår Fågalvärd* **9**:63–77.

Sydeman, W. J., and Emslie, S. D., 1992, Effects of parental age on hatching asynchrony, egg size, and third-chick disadvantage in Western Gulls, *Auk* **109**:242–248.
Tacha, T. C., Nesbitt, S. A., and Vohs, P. A., 1992, Sandhill Crane, in: *The Birds of North America*, No. 31, (A. Poole, P. Stettenheim, and F. Gill, eds.), The Academy of Natural Sciences, Philadelphia; The American Ornithologists' Union, Washington, D.C.
Tallamy, D. W., 1984, Insect parental care, *BioScience* **34**:20–24.
Taplin, A., and Beurteaux, Y., 1992, Aspects of the breeding biology of the Pheasant Coucal, *Emu*. **92**:141–146.
Taylor, B. L., and Gerrodette, T., 1993, The uses of statistical power in conservation biology: the Vaquita and the Northern Spotted Owl, *Conserv. Bio.* **7**:489–500.
Teather, K. L., and Weatherhead, P. J., 1988, Sex-specific energy requirements of Great-tailed Grackle (*Quiscalus mexicanus*) nestlings, *J. Anim. Ecol.* **57**:659–668.
Teather, K. L., and Weatherhead, P. J., 1989, Sex-specific mortality in nestling Great-tailed Grackles, *Ecology* **70**:1485–1493.
Thomas, B. T., 1984, Maguari Stork nesting: juvenile growth and behavior, *Auk* **101**:812–823.
Thomas, B. T., and Strahl, S. D., 1990, Nesting behavior of Sunbitterns (*Eurypyga helias*) in Venezuela, *Condor* **92**:576–581.
Thompson, S. C., and Raveling, D. G., 1987, Incubation behavior of Emperor Geese compared with other geese: interactions of predation, body size, and energetics, *Auk* **104**:707–716.
Trivers, R. L., 1972, Parental investment and sexual selection, in: *Sexual Selection and the Descent of Man*, (B. Campbell, ed.), Aldine, Chicago, pp. 136–179.
Trivers, R. L., 1974, Parent-offspring conflict, *Amer. Zool.* **14**:249–265.
Trivers, R. L., and Willard, D. E., 1973, Natural selection of parental ability to vary the sex ratio of offspring, *Science* **179**:90–92.
van Heezik, Y. M., and Davis, L., 1990, Effects of food variability on growth rates, fledgling sizes and reproductive success in the Yellow-eyed Penguin *Megadyptes antipodes*, *Ibis* **132**:354–365.
van Tienhoven, A., 1983, *Reproductive Physiology of Vertebrates*, Cornell University Press, Ithaca.
Veiga, J. P., 1990, A comparative study of reproductive adaptations in House and Tree Sparrows, *Auk* **107**:45–59.
Veiga, J. P., 1992, Hatching asynchrony in the House Sparrow: a test of the egg-viability hypothesis, *Am. Nat.* **139**:669–675. (T1)
Veiga, J. P., and Viñuela, J., 1993, Hatching asynchrony and hatching success in the House Sparrow: evidence for the egg viability hypothesis, *Ornis Scand.* **24**:237–242.
Vince, M. A., 1964, Social facilitation of hatching in the Bobwhite Quail, *Anim. Behav.* **12**:531–534.
Vince, M. A., 1968, Retardation as a factor in the synchronization of hatching, *Anim. Behav.* **16**:332–335.
Viñuela, J., 1991, Ecología reproductiva del Milano Negro *Milvus migrans* en el Parque Nacional Doñana, PhD. Dissertation, Universidad Complutense de Madrid, Spain.
Viñuela, J., and Bustamante, J., 1993, Effect of growth and hatching asynchrony on the fledging age of Black and Red Kites, *Auk* **109**:748–757.
Vleck, C. M., 1981, Energetic costs of incubation in the Zebra Finch, *Condor* **83**:229–237.
Voous, K. H., 1975, *Owls of the Northern Hemisphere*, William Collins Sons & Co., Ltd., London.
Walsberg, G. E., and King, J. R., 1978, The energetic consequences of incubation for two passerine species, *Auk* **95**:644–655.

Walsberg, G. E., and Voss-Roberts, K. A., 1983, Incubation in desert-nesting doves: mechanisms for egg-cooling, *Physiol. Zool.* **56:**88–93.

Waltman, J. R., and Beissinger, S. R., 1992, Breeding behavior of the Green-rumped Parrotlet, *Wilson Bull.* **104:**65–84.

Ward, D., 1990, Incubation temperatures and behavior of Crowned, Black-winged, and Lesser Black-winged Plovers, *Auk* **107:**10–17.

Weatherhead, P. J., and Teather, K. L., 1991, Are skewed fledgling sex ratios in sexually dimorphic birds adaptive?, *Am. Nat.* **138:**1159–1172.

Weathers, W. W., 1979, Climatic adaptation in avian standard metabolic rate, *Oecologia* **42:**81–89.

Weathers, W. W., and Sullivan, K. A., 1989, Nest attentiveness and egg temperature in the Yellow-eyed Junco, *Condor* **91:**628–633.

Webb, D. R., 1987, Thermal tolerance of avian embryos: a review, *Condor* **89:**874–898.

Werschkul, D. F., 1979, Nestling mortality and the adaptive significance of early locomotion in the Little Blue Heron, *Auk* **96:**116–130. (T1 & T2)

Westerterp, K. R., and Bryant, D. M., 1984, Energetics of free existence in swallows and martins (Hirundinidae) during breeding: a comparative study using doubly-labelled water, *Oecologia* **62:**376–381.

White, F. N., and Kinney, J. L., 1974, Avian incubation, *Science* **186:**107–115.

Wiebe, K. L., and Bortolotti, G. R., 1994, Food supply and hatching spans of birds: energy constraints or facultative manipulation?, *Ecology* **75:**813–823.

Wiebe, K. L., and Bortolotti, G. R., 1994b, Energetic efficiency of reproduction: the benefits of asynchronous hatching for American Kestrels, *J. Anim. Ecol.* **63:**551–560. (T1 & T2)

Wiley, R. H., and Wiley, M. S., 1980, Spacing and timing in the nesting ecology of a tropical blackbird: comparison of populations in different environments, *Ecol. Monogr.* **50:**153–178.

Williams, J. A., 1980, Aspects of the breeding biology of the Subantarctic Skua, *Ostrich* **58:**160–167.

Williams, J. B., 1991, On the importance of energy considerations to small birds with gynelateral intermittent incubation, *Acta XX Congr. Inter. Ornith.*:1964–1975.

Wilson, R. T., Wilson, M. P., and Durkin, J. W., 1986, Breeding biology of the Barn Owl *Tyto alba* in central Mali, *Ibis* **128:**81–90.

Wilson, R. T., Wilson, M. P., and Durkin, J. W., 1987, Aspects of the reproductive ecology of the Hamerkop *Scopus umbretta* in central Mali, *Ibis* **129:**382–388.

Woodard, A. E., and Mather, F. B., 1964, The timing of ovulation, movement of the ovum through the oviduct, pigmentation and shell deposition in Japanese Quail (*Coturnix coturnix japonica*), *Poultry Sci.* **43:**1427–1432.

Wrege, P. H., and Emlen, S. T., 1991, Breeding seasonality and reproductive success of White-fronted Bee-eaters in Kenya, *Auk* **108:**673–687.

Yom-Tov, Y., Ar, A., and Mendelssohn, H., 1978, Incubation behavior of the Dead Sea Sparrow, *Condor* **80:**340–343.

Zach, R., 1982, Hatching asynchrony, egg size, growth, and fledging in Tree Swallows, *Auk* **99:**695–700.

Zerba, E., and Morton, M. L., 1983, The rhythm of incubation from egg laying to hatching in Mountain White-crowned Sparrows, *Ornis Scand.* **14:**188–197.

Zwickel, F. C., 1992, Blue Grouse, in: *The Birds of North America*, No. 15, (A. Poole, P. Stettenheim, and F. Gill, eds.), The Academy of Natural Sciences, Philadelphia; The American Ornithologists' Union, Washington, D.C.

INDEX